Geometry and Arithmetic

Around Euler Partial Differential Equations

Mathematics and Its Applications (*East European Series*)

Managing Editor:

M. HAZEWINKEL
Centre for Mathematics and Computer Science, Amsterdam, The Netherlands

Editorial Board:

A. BIALYNICKI-BIRULA, *Institute of Mathematics Warsaw University, Poland*
H. KURKE, *Humboldt University Berlin, D.D.R.*
J. KURZWEIL, *Mathematics Institute, Academy of Sciences, Prague, Czechoslovakia*
L. LEINDLER, *Bolyai Institute, Szeged, Hungary*
L. LOVÁSZ, *Bolyai Institute, Szeged, Hungary*
D. S. MITRINOVIĆ, *University of Belgrade, Yugoslavia*
S. ROLEWICZ, *Polish Academy of Sciences, Warsaw, Poland*
BL. H. SENDOV, *Bulgarian Academy of Sciences, Sofia, Bulgaria*
I. T. TODOROV, *Bulgarian Academy of Sciences, Sofia, Bulgaria*
H. TRIEBEL, *University of Jena, D.D.R.*

Rolf-Peter Holzapfel

Institute of Mathematics of Academy of Science of the G.D.R., Berlin

Geometry and Arithmetic

Around Euler Partial Differential Equations

D. Reidel Publishing Company

A MEMBER OF THE KLUWER ACADEMIC PUBLISHERS GROUP

Dordrecht / Boston / Lancaster

Library of Congress Cataloging in Publication Data

CIP

Holzapfel, Rolf-Peter, 1942 –
Geometry and arithmetic around Euler partial differential equations.

(Mathematics and its applications. East European series; v.)
Bibliography: p.
Includes index.
1. Differential equations, Partial. 2. Functions, Hypergeometric. 3. Geometry, Algebraic. I. Title.
QA377.H576 1985 515.3'53 84-22330
ISBN 90-277-1827-X

Distributors for the Socialist Countries
VEB Deutscher Verlag der Wissenschaften, Berlin

Distributors for the U.S.A. and Canada
Kluwer Academic Publishers,
190 Old Derby Street, Hingham, MA 02043, U.S.A.

Distributors for all remaining countries
Kluwer Academic Publishers Group,
P.O. Box 322, 3300 AH Dordrecht, Holland

Printed in the German Democratic Republic

Contents

Editor's Preface

Approach your problems from the right end and begin with the answers. Then one day, perhaps you will find the final question.

'The Hermit Clad in Crane Feathers' in R. van Gulik's The Chinese Maze Murders.

It isn't that they can't see the solution. It is that they can't see the problem.

G. K. Chesterton. The Scandal of Father Brown 'The Point of a Pin'.

Growing specialization and diversification have brought a host of monographs and textbooks on increasingly specialized topics. However, the "tree" of knowledge of mathematics and related fields does not grow only by putting forth new branches. It also happens, quite often in fact, that branches which were thought to be completely disparate are suddenly seen to be related.

Further, the kind and level of sophistication of mathematics applied in various sciences has changed drastically in recent years: measure theory is used (non-trivially) in regional and theoretical economics; algebraic geometry interacts with physics; the Minkowsky lemma, coding theory and the structure of water meet one another in packing and covering theory; quantum fields, crystal defects and mathematical programming profit from homotopy theory; Lie algebras are relevant to filtering; and prediction and electrical engineering can use Stein spaces. And in addition to this there are such new emerging subdisciplines as "completely integrable systems", "chaos, synergetics and largescale order", which are almost impossible to fit into the existing classification schemes. They draw upon widely different sections of mathematics.

This program, Mathematics and Its Applications, is devoted to such (new) interrelations as exempla gratia:

— a central concept which plays an important role in several different mathematical and/or scientific specialized areas;
— new applications of the results and ideas from one area of scientific endeavor into another;
— influences which the results, problems and concepts of one field of enquiry have and have had on the development of another.

The Mathematics and Its Applications programme tries to make available a careful selection of books which fit the philosophy outlined above. With such books, which are stimulating rather than definitive, intriguing rather than encyclopaedic, we hope to contribute something towards better communication among the practitioners in diversified fields.

The hypergeometric function and its associated second-order differential equation are prime examples of objects which do not want to fit neatly into any particular recognized mathematical specialism. Besides, with complex function theory, they are heavily involved in special functions, representation theory, automorphic form theory, elliptic curves and integrals and more algebraic geometry, and number theory.

One way or another, this book is concerned with two-dimensional versions of all this, with at least as rich a phenomenology. Things centre around the study of a beautiful concrete hypergeometric object of complex dimension 2 and the (complete solution of the) associated (systems of) Euler partial differential equations.

Both results and methods involve several parts of mathematics, notably algebraic geometry (theory of surfaces, cohomology and vanishing theorems, Gauss-Manin connection), authomorphic form theory, monodromy groups of differential equations, equivalent K-theory and number theory (L-series). A book thus which treats of interactions between various parts of mathematics, be it mostly "pure" parts, though of course the Euler differential equation itself governs the propagation of air waves and, hence is, important in acoustics.

In one way this book is quite unique. To a great extent it is not a book on a particular topic (as virtually all books are); instead it is a book on one very rich specific example. And as the author indicates: it is often easier to develop a general theory than to understand completely some given concrete mathematical object. Yet, in the final analysis, such completely understood concrete objects are the wellsprings of future mathematics.

The unreasonable effectiveness of mathematics in science ...

Eugene Wigner

Well, if you knows of a better 'ole, go to it.

Bruce Bairnsfather

What is now proved was once only imagined.

William Blake

As long as algebra and geometry proceeded along separate paths, their advance was slow and their applications limited.
But when these sciences joined company they drew from each other fresh vitality and thenceforward marched on at a rapid pace towards perfection.

Joseph Louis Lagrange

Amsterdam, October 1984

Michiel Hazewinkel

Preface

It is a great pleasure for me to write this monograph during the Euler year 1983. Leonhard Euler (1707—1783) was able to a high degree to read off differential equations directly from our natural physical environment. This monograph deals with differential equations of type

$$0 = \frac{\partial^2}{\partial t_1\, \partial t_2} F + \frac{1}{t_2 - t_1} \left(a \frac{\partial}{\partial t_1} - b \frac{\partial}{\partial t_2}\right) F, \qquad a, b \text{ constant},$$

which are called Euler partial differential equations. This name was already given by Darboux (1842—1917). One hundred years ago, Emil Picard (1856—1941) discovered a lot of them together with some solutions among certain algebraic families of complex curves (Riemann surfaces), depending on two parameters. The solutions are expressed as integrals $\int \frac{\mathrm{d}x}{\sqrt[n]{Q(x)}}$ on the corresponding Riemann surfaces of equations of the type $Y^n = Q(X, t_1, t_2)$, with $Q(X) = Q(X, t_1, t_2)$ polynomials in X depending on two complex parameters t_1, t_2. In special cases Picard saw also a connection between the monodromy of these (multivalued) functions, (arithmetic) lattices acting on the two-dimensional complex unit ball and automorphic (periodic) functions on the ball. But he had to leave open a lot of mysterious problems.

Recent developments in mathematics, especially in algebraic geometry, throw a new light on this classical theme.

In Sections 0.1 and 0.2 of the introduction we give an outline of the historical development beginning with dimension 1 and then passing to dimension 2. We insert quotations of Gauss, Riemann and Klein because their deep mathematical ideas did not lose in vision and power to these days.

In Section 0.3 of the introduction we present the main results of this monograph about general systems of partial differential equations of Euler-Picard type in arbitrary dimension, their solutions of type $\int \frac{\mathrm{d}x}{\sqrt[n]{Q(x)}}$ (Chapter II), and their connections with automorphic forms with respect to Eisenstein lattices in the two-dimensional complex unit ball in the special case of "Picard curves"

$$Y^3 = p_4(X) \qquad (p_n(X) \text{ denotes a polynomial in } X \text{ of degree } n)$$

(Chapter I). In the "methods of proofs" in part 0.3 the modern mathematical techniques are introduced, which shall intervene.

The monograph is not self-contained. Its main purpose is the presentation of new concrete results. But it can be understood as "almost self-contained" in the following sense: We develop the necessary techniques as far as we need them for a natural understanding of the results and their proofs. Adequate precise references are given for results needed from the other mathematical literature, and the reader will have no difficulties to find them if desired. In spite of having to deal with many different branches of mathematics I took trouble to preserve the algebraic geometric flavour of the monograph. On the other hand I hope that the book is readable also for mathematicians not exclusively concerned with the field of algebraic geometry. But the reader should have at his disposal some fundamental knowledges from algebraic geometry. This he can acquire from one of the famous foundation books of Šafarevič, Griffith/Harris and Hartshorne, respectively. At the end of the book we present a list of references, which will allow the reader to take part in actual current research in this field. The results presented here will serve as a good start for an undertaking.

A central theme of Chapter I is the classification of complex algebraic surfaces. During the last years I worked on classification of arithmetic ball quotient surfaces. The results in Chapter I are concrete applications of this work to the very important case of Picard modular surfaces of Eisenstein lattices.

The monograph should be also considered as a starting point for a deeper arithmetic study of special (cycloelleptic) curve families and automorphic forms of ball lattices. After the proof of the Mordell conjecture by Faltings'questions about effectivity concerning this theme become important. On the other hand we do not neglect the great influence which the earlier intensive arithmetic study of the elliptic curve family has had in the recent progress. In Chapter II the reader will learn that from our point of view the generalized Picard curves $Y^{n-1} = X^n + Z^n$ belong to much better cyclotomic families than the Fermat curves $Y^n = X^n + Z^n$. The families $\{Y^{n-1} = p_n(X)\}$ seem to be the best generalizations of the elliptic curve family. So the detailed study of the Picard curve family $\{Y^3 = p_4(X)\}$ in Chapter I is also justified from the arithmetic point of view.

Berlin, Dec. 1983 **Rolf-Peter Holzapfel**

Chapter 0

Herr von Ribbeck auf Ribbeck im Havelland,
Ein Birnbaum in seinem Garten stand, ...

Introduction

0.1. Historical Development in Dimension 1

The theory of *hypergeometric functions* has a long history. The first occurrence of *hypergeometric series* goes back to the 17th century (Wallis). Euler was the first to consider the series as a function of one variable x:

0.1. $$F(a, b; c; x) = 1 + \frac{a \cdot b}{1 \cdot c} x + \frac{a(a+1) \cdot b(b+1)}{1 \cdot 2 \cdot c(c+1)} x^2 + \cdots.$$

F is a solution of the *hypergeometric differential equation*

0.2. $$x(1-x)\, y'' + \big(c - (a+b+1)\, x\big)\, y' - aby = 0.$$

This was found by Euler [15]. Euler also found that up to a constant factor F can be expressed as a *hypergeometric integral* ([16], ch. X)

0.3. $$\int_0^1 u^{b-1}(1-u)^{c-b-1}\,(1-xu)^{-a}\,du.$$

The next developments are closely connected with the name of Gauss. In his paper [21] he considered the hypergeometric function as a function of a complex variable. Here one can find the first investigations of circles of convergence. His new modern point of view is outlined in a letter to Schumacher (17. September 1808, [22]):

* „Mir ist bei der Integralrechnung immer das weit weniger interessant gewesen, wo es nur auf Substituiren, Transformiren etc. kurz auf einen gewissen geschickt zu hand-

---

* To me, there is little interest in that aspect of integral calculus where we use substitutions, transformations, etc. — merely clever mechanical tricks — in order to reduce integrals to algebraic, logarithmic or trigonometric forms, as compared with the deeper study of those transcendental functions which cannot be so reduced. We are as familiar with circular and logarithmic functions as with one times one, but the magnificent goldmine which contains the secrets of higher functions is still almost completely unknown territory. I have, formerly, done a lot of work in this area and intend to devote a substantial treatise to it, of which I have given a glimpse in my Disquiss. Arithmeticae p. 593, Art. 335. One cannot help but be astounded at the great richness of the new and extremely interesting results and relations which these functions exhibit (the functions associated with rectification of the ellipse and hyperbola being included among them).

habenden Mechanismus ankommt, um Integrale auf algebraische oder Logarithmische oder Kreisfunctionen zu reduciren, als die genauere tiefere Betrachtung solcher Transcendenten Functionen, die sich auf jene nicht zurückführen lassen. Mit Kreisfunctionen und Logarithmischen wissen wir jetzt umzugehen, wie mit dem 1 mal 1, aber die herrliche Goldgrube, die das Innere der höheren Functionen enthält, ist noch fast ganz Terra Incognita. Ich habe darüber ehemals sehr viel gearbeitet und werde dereinst ein eigenes großes Werk darüber geben, wovon ich bereits in meinen Disquiss. arithm. p. 593, Art. 335, einen Wink gegeben habe. Man geräth in Erstaunen über den überschwenglichen Reichthum an neuen höchst interessanten Wahrheiten und Relationen, die dergleichen Functionen darbieten (wohin u. a. auch diejenigen gehörigen, mit denen die Rectification der Ellipse und Hyperbel zusammenhängt)."

Gauss could not carry out the programme mentioned in the letter. For this project the work of many outstanding mathematicians was needed: Abel, Cauchy, Dedekind, Fuchs, Jacobi, Klein, Kummer, Poincaré, Riemann, Schwarz, Weierstrass and others. The theory of hypergeometric functions was a central theme in the creation of the fundaments of function theory of one complex variable. The celebrated lectures of F. Klein in Göttingen 1893/94 can be regarded as a summary of the work on Gauss' programme concerning hypergeometric functions, at least for the 19th century. Looking back at the fruitful developments Klein corroborates Gauss' point of view ([39], p. 24):

* „In der Behandlung unserer Differentialgleichung (und der Differentialgleichungen überhaupt) kann man zweierlei wesentlich entgegengesetzte Betrachtungsweisen unterscheiden, nämlich einen elementaren, antiquierten Standpunkt, ..., und den modernen funktionentheoretischen Standpunkt.

Der elementare Standpunkt läßt sich etwa folgendermaßen charakterisieren: ... Man sucht ... (die Lösungen) ... durch ‚elementare Funktionen' auszudrücken; ... Auf diesem Standpunkt ist die Auflösung unsrer (und überhaupt irgendeiner) Differentialgleichung lediglich Sache der Routine und der Übung in der Anwendung gewisser Kunstgriffe.

---

* In the treatment of our differential equation (and of differential equations generally) one can distinguish two essentially opposite points of view; there is the elementary, old-fashioned standpoint ... and the modern, function-theoretic one.

The elementary standpoint can be characterised as follows: ... we seek to express ... (solutions) ... in terms of 'elementary functions', ... from this point of view, the solution of our differential equation (and any differential equation) is a routine exercise in the application of known results.

The modern theory takes quite a different approach. We ask for the following:

1. Proof of the 'existence' of functions ... which satisfy our differential equation.
2. Investigation of the properties of these functions. We look upon every differential equation as defining a specific (possibly new) class of transcendental functions.

In the study of properties of functions arising from this definition, these must appear as a corollary statement concerning when our functions reduce to simpler functions, particularly to elementary functions. The principal interest lies, however, not in the special cases when we obtain these lower functions, but precisely in the general case when we obtain new functions.

Ganz anders ist die Betrachtungsweise der modernen Theorie. Man verlangt da folgendes:

1. Beweis für die ‚Existenz' von Funktionen ..., welche unserer Differentialgleichung genügen.
2. Erforschung der Eigenschaften dieser Funktionen. Man sieht eben jede Differentialgleichung als Definition einer bestimmten (eventuell neuen) Gattung transzendenter Funktionen an.

Bei der von dieser Definition ausgehenden Untersuchung der Eigenschaften der Funktionen muß sich als Nebenresultat ergeben, wann sich unsere Funktionen, insbesondere auf elementare Funktionen, reduzieren.
Das Hauptinteresse ruht aber hier nicht auf den Spezialfällen der niederen Funktionen, sondern gerade auf dem allgemeinen Fall, den neuen Funktionen."

Starting from ideas of Riemann H. A. Schwarz published in 1873 his remarkable article [61]. In this paper one finds a rigorous geometric approach to the problems of hypergeometric functions. He considered quotients $\eta = y_1/y_2$ of two independent solutions of a hypergeometric differential equation. Omitting the singular points of the hypergeometric differential equation one obtains a punctured Riemann sphere $\mathbb{P}^1 \setminus \{0, 1, \infty\}$. If C consists of two simple paths joining 0, 1 and 1, ∞, respectively, then $\mathbb{P}^1 \setminus C$ is simply connected, η maps $\mathbb{P}^1 \setminus C$ bijectively onto a triangle in $\mathbb{P}^1$, and η becomes multivalued by analytic extension crossing the omitted paths. In many cases the image points of the multivalued function η essentially fill an open disc in the η-plane. More precisely, the disc is covered by infinitely many image triangles. For example one finds for $(a, b, c) = \left(\frac{1}{2}, \frac{1}{2}, 1\right)$ and $(a, b, c) = \left(\frac{1}{12}, \frac{1}{12}, \frac{2}{3}\right)$ in (0.1), (0.2), (0.3) the beautiful Figures 0.A and 0.B.

At this place we quote again F. Klein from his lectures ([39], p. 286):

* „Jetzt wollen wir unsere Dreiecksfiguren nach einer anderen Seite zur Geltung bringen, indem wir uns fragen: Wann ist die (im allgemeinen mehrdeutige) Funktion $\eta(x)$ eindeutig umkehrbar, d. h. wann ist x eine eindeutige Funktion von η?

---

* Now we want to apply our triangle-figures in another direction, by asking: When is the (generally multi-valued) function $\eta(x)$ uniquely invertible, that is, when is x a single-valued function of η?

This is a special case of a quite general problem in function theory, arising from the desire to avoid operations with multi-valued functions as far as possible. In general, if one has a functional dependence $y(x)$ in which many, perhaps infinitely many, values of y correspond to a given value of x, and conversely many values of x correspond to a given value of y, then one can ask whether one cannot express both y and x as single-valued functions of a third variable, t. A case in point is the example of elliptic functions, in which we can express, not merely the algebraic functions on a Riemannian surface, but also the integrals of the first and second kinds, as single-valued functions of a third ('uniformizing') variable, namely the integral of the first kind. We can say, therefore: It is a general problem of analysis, in the discussion of analytical relationships, always to find a 'uniformizing parameter'.

0.A.

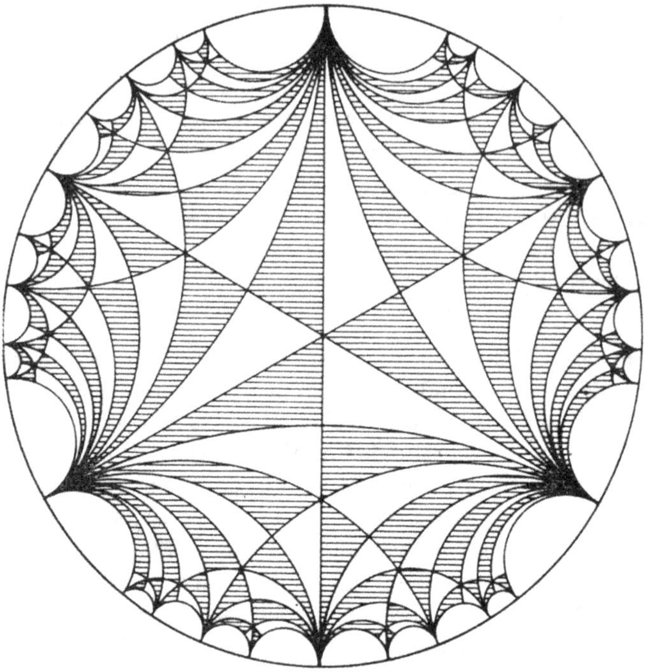

$\mathbb{S}l_2(\mathbb{Z})$

$$(a, b, c) = \left(\frac{1}{12}, \frac{1}{12}, \frac{2}{3}\right)$$

0.B.

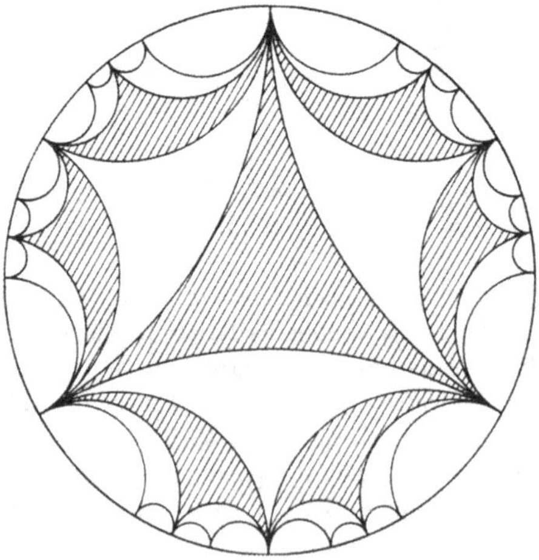

congruence subgroup
$\mathbb{S}l_2(\mathbb{Z})\,(2)$

$$(a, b, c) = \left(\frac{1}{2}, \frac{1}{2}, 1\right)$$

Es ist dies ein spezieller Fall einer ganz allgemeinen Fragestellung der Funktionentheorie, welche darauf hinausgeht, das Rechnen mit mehrdeutigen Funktionen möglichst zu vermeiden. Hat man überhaupt eine funktionale Abhängigkeit $y(x)$, bei der einem Werte von x mehrere, vielleicht unendlich viele, Werte von y entsprechen und umgekehrt einem Werte von y mehrere Werte von x, dann kann man allemal fragen, ob man nicht sowohl y wie x als eindeutige Funktionen einer dritten Veränderlichen t darstellen kann. Man denke nur an das Beispiel der elliptischen Funktionen, wo nicht nur die algebraischen Funktionen einer Riemannschen Fläche, sondern auch die Integrale erster und zweiter Gattung als eindeutige Funktionen einer dritten (‚uniformisierenden') Veränderlichen, nämlich des Integrals erster Gattung, dargestellt werden können. Wir sagen also:

Es ist eine allgemeine Aufgabe der Analysis, bei der Diskussion analytischer Beziehungen jedesmal eine ‚uniformisierende Hilfsveränderliche' aufzufinden."

From now we consider only the cases where η is really a uniformizing variable of x. Then the triangle is the fundamental domain of a *lattice subgroup* Λ of the group $\mathrm{Aut}_{\mathrm{hol}}\ \mathfrak{D}$ of biholomorphic automorphisms of our disc $\mathfrak{D}$. We depict by 0.a

0.a

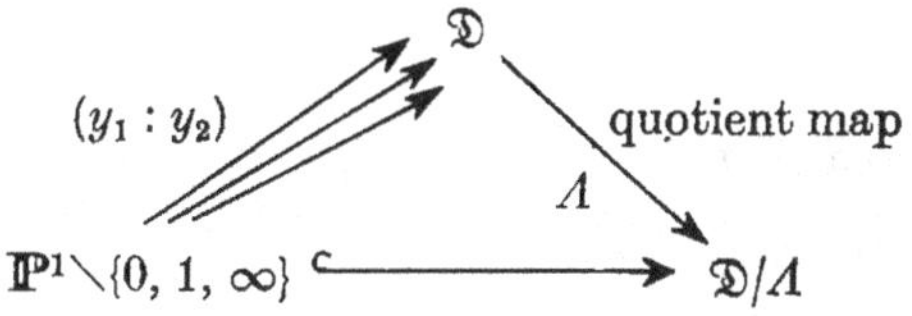

the multivalued map $(y_1 : y_2)$ into $\mathfrak{D} \subset \mathbb{P}^1$ and the inverse quotient map $\mathfrak{D} \to \mathfrak{D}/\Lambda$. The lattice Λ is called the *monodromy group* of the corresponding differential equation. Its conjugacy class is uniquely determined by the differential equation alone. Now the problem of inversion of the function η can be formulated in the following manner: Describe the image map $\mathfrak{D} \to \mathfrak{D}/\Lambda$ in 0.a by means of holomorphic functions on $\mathfrak{D}$. A suitable quotient of such functions could be a Λ-periodic (meromorphic) function, which is the inverse of η. The Λ-periodic (meromorphic) functions on $\mathfrak{D}$ are called *Λ-automorphic functions*. Entire numerators and denominators can be found among the Λ-automorphic forms. The $\mathbb{C}$-vector space $[\Lambda, m]$ of *Λ-automorphic forms* of weight m consists of all holomorphic functions $f\colon \mathfrak{D} \to \mathbb{C}$ satisfying

$$f(\lambda(z)) = j_\lambda^m(z) \cdot f(z), \qquad z \in \mathfrak{D}, \qquad \lambda \in \Lambda, \qquad j_\lambda = \frac{\mathrm{d}\lambda}{\mathrm{d}z},$$

together with a meromorphy condition at the cusps.

Central examples in the theory of hyperelliptic differential equations are the cases $(a, b, c) = \left(\frac{1}{12}, \frac{1}{12}, \frac{2}{3}\right)$ and $(a, b, c) = \left(\frac{1}{2}, \frac{1}{2}, 1\right)$ (see 0.A, 0.B). The disc $\mathfrak{D}$ is biholomorphically equivalent to the Siegel upper half plane $\mathfrak{H} = \{z \in \mathbb{C};\ \mathrm{Im}\, z > 0\}$. The monodromy group in case 0.A is the arithmetic group $\mathrm{Sl}_2(\mathbb{Z})$ acting on $\mathfrak{H}$ by fractional linear transformations. In case 0.B the monodromy group is the congruence subgroup $\mathrm{Sl}_2(\mathbb{Z})(2)$ defined by the exact sequence

$$1 \to \mathrm{Sl}_2(\mathbb{Z})(2) \to \mathrm{Sl}_2(\mathbb{Z}) \to \mathrm{Sl}_2(\mathbb{Z}/2\mathbb{Z}) \to 1 .$$

The rings of automorphic forms are

$$\bigoplus_{m=0}^{\infty} [\mathrm{Sl}_2(\mathbb{Z}), m] = \mathbb{C}[g_2, g_3], \qquad \bigoplus_{m=0}^{\infty} [\mathrm{Sl}_2(\mathbb{Z})(2), m] = \mathbb{C}[\varepsilon_1, \varepsilon_2],$$

where g_i is of weight i, and $\varepsilon_1, \varepsilon_2$ are of weight 1. For a suitable choice of y_1, y_2 one has the inversion diagrams

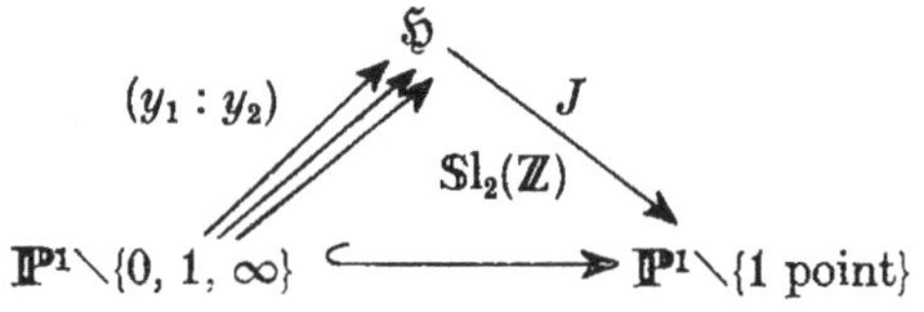

$$J := \frac{g_2^3}{g_2^3 - 27 g_3^2}$$

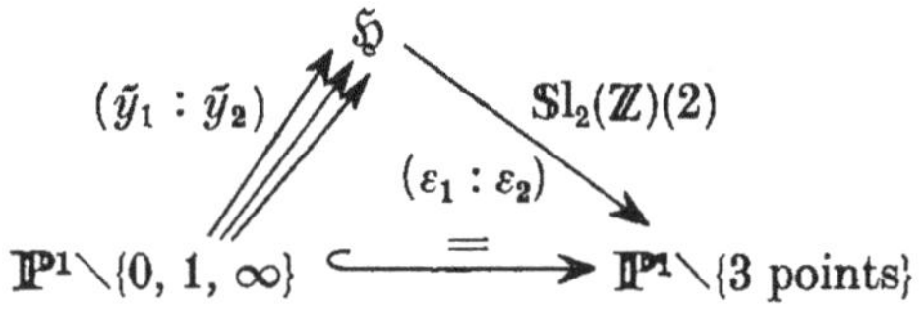

in the cases 0.A and 0.B. Here g_2 and g_3 are the second and third elementary symmetric functions of three suitable functions $\varepsilon_1, \varepsilon_2, \varepsilon_3 \in [\mathrm{Sl}_2(\mathbb{Z})(2), 1]$ with $\varepsilon_1 + \varepsilon_2 + \varepsilon_3 = 0$ and the symmetric group $S_3 = \mathrm{Sl}_2(\mathbb{Z}/2\mathbb{Z})$ acts on $\{\varepsilon_1, \varepsilon_2, \varepsilon_3\}$ by permutations (for details see 6.2).

There is a deep connection with number theory. Namely g_2, g_3 are up to a constant factor the *Eisenstein series* E_2, E_3 (see e.g. [62], ch. VII):

$$\left.\begin{aligned} E_2 &= 1 + 240 \sum_{n=1}^{\infty} \sigma_3(n)\, q^n, \\ E_3 &= 1 - 504 \sum_{n=1}^{\infty} \sigma_5(n)\, q^n, \end{aligned}\right\} \qquad q = e^{2\pi i z}, \qquad \sigma_k(n) := \sum_{d \mid n} d^k .$$

Another interesting aspect is the algebraic geometric point of view, namely the classification up to isomorphism of elliptic curves. One has the following commutative diagram (see e.g. [62]):

$$\begin{array}{ccc} \tau & \longmapsto & C(\tau):\ Y^2 = X^3 + g_2(\tau)\, X + g_3(\tau) \\ \mathfrak{H} & \longrightarrow & \{\text{elliptic curves}\} \\ {\scriptstyle \mathrm{Sl}_2(\mathbb{Z})}\big\downarrow {\scriptstyle J} & & \big\downarrow \\ \mathbb{P}^1 \setminus \{\infty\} & \longleftrightarrow & \{\text{elliptic curves}\}/\mathrm{Iso} \end{array}$$

The fundamental modular function J (Hauptmodul) was first constructed by Dedekind in 1877 and independently by Klein one year later (see [10], [38]). Every other $\mathrm{Sl}_2(\mathbb{Z})$-automorphic function (*modular function*) is a rational function of J.

Because of the importance of the modular case 0.A we quote for $(a, b, c) = \left(\frac{1}{2}, \frac{1}{2}, 1\right)$ all the relevant hypergeometric expressions:

hypergeometric function (series):

$$F\left(\frac{1}{2}, \frac{1}{2}; 1; x\right) = 1 + \sum_{n=1}^{\infty} \left(\frac{1 \cdot 3 \cdot 5 \cdots (2n-1)}{2^n \cdot n!}\right)^2 \cdot x^n,$$

hypergeometric differential equation:

$$y'' + \frac{1-2x}{x(1-x)}\, y' - \frac{1}{4x(1-x)}\, y = 0,$$

hypergeometric integrals:

$$\int_g^h \frac{du}{\sqrt{u(u-1)(xu-1)}}, \qquad g, h \in \{0, 1, \infty\}$$

on the elliptic curve $C(x): y^2 = u(u-1)(xu-1)$.

0.2. Historical Development in Dimension 2

As we have seen in 0.1 the 19th century was an era of high mathematical creativity in my country. The reasons may be described as follows:

1. The work on concrete mathematical objects;
2. these objects posed challenges and ask for new mathematical ideas and theories;
3. the objects interrelated different branches of mathematics.

What happened in higher dimensional periodic function theory? Already the first steps made it clear that one has to overcome a lot of difficulties. The work of many specialists was (and is) necessary for advancing into new fields. The speed of the development of general theories seems to be greater than the development of understanding concrete mathematical objects. The aim of Chapter I is to illustrate the application of algebraic geometry to a concrete hypergeometric object of complex dimension 2, which seems to play the same role as the modular case in dimension 1.

This object is closely connected with a partial differential equation of second order. In connection with some questions of acoustics Euler already considered the differential equation

$$(\mathrm{E}_a) \qquad \frac{\partial^2 z}{\partial x\,\partial y} - \frac{a}{x-y}\cdot\frac{\partial z}{\partial x} + \frac{a}{x-y}\cdot\frac{\partial z}{\partial y} = 0$$

([16], III, ch. 3, 4, 5). Darboux called it the *Euler partial differential equation* (see [9]). Much later, Riemann again considered air waves and the Euler partial differential equation. In the introduction of his paper „Ueber die Fortpflanzung ebener Luftwellen von endlicher Schwingungsweite" [60] we find the following words:

* „... wenn auch zur Erklärung der bisjetzt experimentell festgestellten Erscheinungen die bisherige Behandlung vollkommen ausreicht, so könnten doch, bei den großen Fortschritten, welche in neuester Zeit durch Helmholtz auch in der experimentellen Behandlung akustischer Fragen gemacht worden sind, die Resultate dieser genaueren Rechnung in nicht allzu ferner Zeit vielleicht der experimentellen Forschung einige Anhaltspunkte gewähren; und dies mag, abgesehen von dem theoretischen Interesse, welche die Behandlung nichtlinearer partieller Differentialgleichungen hat, die Mittheilung derselben rechtfertigen."

This paper of Riemann has two interesting mathematical aspects.

1. Riemann constructs two auxiliary functions r, s starting from the density and the velocity of a gas particle. Then there follows a remarkable inversion step. He changes the roles of the pairs (r, s) and (x, t), where x, t are the variables of place and time. That means that x, t are locally understood as functions of r and s. Integrating a complete differential one finds a function of r, s satisfying the Euler partial differential equation.

At this point I pose the following purely mathematical question: Is there a description of solutions of special Euler partial differential equations by means of "periodic functions" and an inversion procedure?

The main result of Chapter I is a positive answer for the Euler partial differential equation $(\mathrm{E}) = (\mathrm{E}_{1/3})$.

2. Some reduction steps in Riemann's paper lead to hypergeometric functions. For direct purely mathematical considerations concerning this fact we refer the reader to Darboux's book [9], ch. III, IV.

On the other hand Appell (1880) considered hypergeometric functions (series) in two variables ("Appell series", [1]). In 1881 Picard (see [54]) found a hyper-

---

* Although the mathematical treatment up to this point has proved to be sufficient to explain the experimental phenomena observed so far, it may well be justified — when one looks at the great advances made recently by Helmholtz in the experimental treatment of acoustic problems, to hope that the results of this more precise study may, in the not too distant future, provide some starting points for this new experimental research. This hope, apart from the intrinsic theoretical interest of treating non-linear partial differential equations, may justify the present communication.

geometric integral expression for the Appell series:

$$\int_0^1 u^\alpha(1-u)^\beta\,(1-xu)^\gamma\,(1-yu)^\delta\,du\,.$$

There exists a monograph of Appell about hypergeometric functions of several variables with applications to mathematical physics (see [2]).

Starting in the first volume of Acta mathematica (1882) E. Picard published a series of articles in this journal. He showed that his hypergeometric integrals are particular solutions of partial differential equations of order 2 with algebraic coefficients. Stimulated by the paper of H. A. Schwarz mentioned in 0.1 he called the functions "hyperfuchsiennes". The following special case occupies a central place (see [55]):

0.2.1. $$I(g,h;t_1,t_2)=\int_g^h\frac{dx}{\sqrt[3]{x(x-1)\,(x-t_1)\,(x-t_2)}}\,,\qquad g,h\in\{0,1,t_1,t_2,\infty\}\,.$$

These functions are particular solutions of the partial Euler differential equation $(\mathrm{E})=(\mathrm{E}_{1/3})$. Three of them form a fundamental system of solutions of a system of three partial differential equations containing (E). They map a dense subset of $\mathbb{P}^2$ (locally injectively) into the two-dimensional complex unit ball

$$\mathfrak{B}=\{(z_1,z_2)\in\mathbb{C}^2;\;|z_1|^2+|z_2|^2<1\}\subset\mathbb{P}^2\,.$$

The monodromy group acting on $\mathfrak{B}$ is generated by five elements of $\mathbb{U}\big((2,1);\mathfrak{O}_K\big)$, where $K=\mathbb{Q}\big(\sqrt{-3}\big)$ is the field of *Eisenstein numbers* and $\mathfrak{O}_K$ is its ring of integers. For more details together with a description of recent results of Mostow/Deligne we refer the reader to § 6 and § 7 of Chapter I of the present book.

Picard called the five generators "fundamental substitutions". This is one of the origins of the notion of "fundamental group". Indeed, the five elements generate the fundamental group of the space $\mathscr{P}$ of nonsingular points of the differential equations:

$$\mathscr{P}=\{(t_1,t_2)\in\mathbb{C}^2;\;t_i\neq 0,1,\;t_1\neq t_2\}\,.$$

At the same time $\mathscr{P}$ is the space of "distinguished" smooth Picard curves

$$C(t_1,t_2)\colon\;Y^3=X(X-1)\,(X-t_1)\,(X-t_2)\,.$$

Picard was aware of the central role of his model for further investigations. In [55] he remarks:

* « Nos fonctions de u et de v (ball coordinates) jouent dans la théorie des ces fonctions abéliennes le même rôle que la fonction modulaire dans la théorie des fonctions elliptiques. »

---

* Our functions of u and v play in the theory of these abelian functions the same role as the modular function in the theory of elliptic functions.

Twenty years later R. Alezais, stimulated by Picard, published a more detailed treatment about this object. In 1931 Picard referred in his book [58] (note IV, 3) again to this model hoping that the situation would be clarified in the future:*

* « Nous avons cherché, M. Alezais et moi, à préparer le terrain pour une telle recherche dans un cas très particulier, en faisant une étude arithmétique préliminaire. »

It should be remarked that there are a lot of other hypergeometric integrals of several variables where Picard found a similar state of affairs (see [57]). Recently Mostow proved that in some cases the corresponding monodromy groups are lattices in the automorphism group of the ball.

An algebraic geometric approach to Picard curves can be found in a paper of Shimura (see [63]). Using Jacobians he showed that the moduli field is the function field of the surface $\mathfrak{B}/\mathbb{U}\big((2,1), \mathfrak{O}_K\big)$. Shimura proved that this field is rational.

0.3. Results and Methods

In the first chapter we will prove the following main results:

1. *The monodromy group of the hypergeometric differential equation system considered by Picard is the principal congruence subgroup* $\tilde{\Gamma}' = \tilde{\Gamma}(1-\omega)$ *of the full Eisenstein lattice of the ball* $\tilde{\Gamma} = \mathbb{U}\big((2,1), \mathfrak{O}_{(\mathbb{Q}\sqrt{-3})}\big)$ *with respect to the principal ideal* $\mathfrak{O}_{\mathbb{Q}(\sqrt{-3})} \cdot (1-\omega)$, $\omega = e^{2\pi i/3}$ (Theorem 6.3.12(i)).

2. *The rings of automorphic forms of* $\tilde{\Gamma}'$ *and* $\tilde{\Gamma}$ (of Nebentypus χ) *are polynomial rings* (Propositions 4.4.3 and 4.4.9):

$$\bigoplus_{m=0}^{\infty} [\tilde{\Gamma}', m]_\chi = \mathbb{C}[\xi_1, \xi_2, \xi_3], \qquad \bigoplus_{m=0}^{\infty} [\tilde{\Gamma}, m]_\chi = \mathbb{C}[G_2, G_3, G_4],$$

with ξ_1, ξ_2, ξ_3 *of weight* 1, G_k *of weight* k, $k = 2, 3, 4$.

A key result is the following one (Theorem 4.3.9):

$$\bigoplus_{m=0}^{\infty} [\Gamma', m] = \mathbb{C}[\xi_1, \xi_2, \xi_3, \zeta], \qquad \Gamma' = \mathbb{SU}\big((2,1), \mathfrak{O}_{\mathbb{Q}(\sqrt{-3})}\big)\,(1-\omega),$$

with ξ_k as above for $k = 1, 2, 3$, ζ of weight 2,

$$\zeta^3 = (\xi_3^2 - \xi_2^2)\,(\xi_3^2 - \xi_1^2)\,(\xi_2^2 - \xi_1^2).$$

3. $\mathfrak{B}/\tilde{\Gamma}' \cong \mathbb{P}^2 \setminus \{4 \text{ points in general position}\}$ (Remark 3.3.6).

4. *If* F_1, F_2, F_3 *are three fundamental solutions of Picard's system of differential equations, then there is a basis* ξ_1', ξ_2', ξ_3' *of the space of automorphic forms of weight* 1 *with respect to a group isomorphic (similar in* $\mathbb{Gl}_3(\mathbb{C})$*) to* $\tilde{\Gamma}'$, *such that* $(\xi_1' : \xi_2' : \xi_3')$

---

* We have sought, Mr. Alezais and I, to prepare the ground for such investigations in a very special case, by doing some preliminary arithmetic studies.

is the inverse of the multivalued map $(F_1 : F_2 : F_3)$:

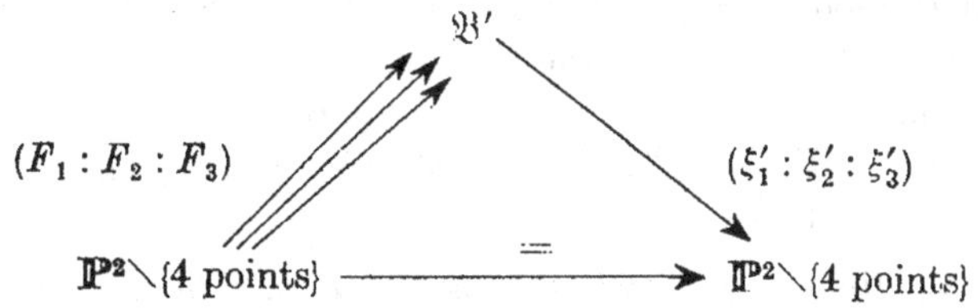

where $\mathfrak{B}'$ is projectively equivalent to $\mathfrak{B}$ in $\mathbb{P}^2$ (Theorem 6.3.12(ii)).

5. Let $\widehat{\mathfrak{B}/\bar{\Gamma}} = \mathfrak{B}^*/\bar{\Gamma}$ be the Baily-Borel compactification, $\mathfrak{B}^* = \mathfrak{B} \cup \partial_{\bar{\Gamma}}\mathfrak{B}$, $\partial_{\bar{\Gamma}}\mathfrak{B} \subset \partial\mathfrak{B}$ the set of $\bar{\Gamma}$-cusps. Then we have the following commutative diagram:

$$P \longmapsto C(P) : Y^3 = X^4 + G_2(P) X^2 + G_3(P) X + G_4(P)$$

$$\mathfrak{B}^* \longrightarrow \{Picard\ curves\}/proj.\ equivalence$$

quotient map $\quad \bar{\Gamma}$

$$\mathfrak{B}^*/\bar{\Gamma} = \widehat{\mathfrak{B}/\bar{\Gamma}}$$

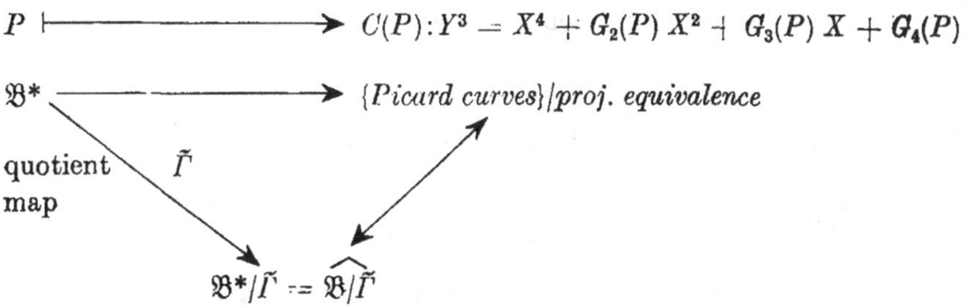

($Y^3 = X^4$ is excluded from the set of Picard curves. On $\mathfrak{B} \setminus \bar{\Gamma} \cdot \mathfrak{D} \subset \mathfrak{B}^*$, $\mathfrak{D} = \{(z, 0); |z| < 1\}$ a subdisc of $\mathfrak{B}$, projective equivalence can be replaced by isomorphism.) *The image of* $\mathfrak{B} \setminus \bar{\Gamma} \cdot \mathfrak{D}$ *is the set of isomorphism classes of all smooth Picard curves* (Theorem 6.1.3).

6. *The image of the* $\bar{\Gamma}$-*fixed point set of B is the set of classes of Picard curves with larger automorphism groups* (larger than $\mathbb{Z}/3\mathbb{Z}$) (Proposition 6.1.4, Table 6.1.α).

7. The Lagrange resolvent map, which makes correspond to each polynomial of degree 4 a polynomial of degree 3, admits by a change of fibres procedure over the "moduli space" of Picard curves (change Picard curves to elliptic curves) a geometric interpretation as a morphism $\mathfrak{B}/\bar{\Gamma} \dashrightarrow \mathfrak{H}/\mathrm{Sl}_2(\mathbb{Z})$. There is a lift along the automorphic form (quotient) morphisms $\mathfrak{B} \to \mathfrak{B}/\bar{\Gamma}$, $\mathfrak{H} \to \mathfrak{H}/\mathrm{Sl}_2(\mathbb{Z}(2))$ (Diagram 6.2.a).

This lifting procedure should be studied in more detail. It seems to be of number theoretic interest. Geometrically it gives a connection between the arithmetic classification of Picard curves and the arithmetic classification of elliptic curves.

The main tools of Chapter I are methods from the theory of algebraic surfaces. By a general theorem of Baily-Borel (see [4]) one knows that $\widehat{\mathfrak{B}/\Gamma} = \mathrm{Proj}\left(\bigoplus_{m=0}^{\infty}[\Gamma, m]\right)$ for an arithmetic lattice group Γ acting on $\mathfrak{B}$. If Γ is neat (this implies especially that Γ acts without fixed points), then it follows from the proportionality theory of Hirzebruch-Mumford (see [50]) that $[\Gamma, m] \cong H^0\big(\overline{\mathfrak{B}/\Gamma}, \big(\Omega^2_{\overline{\mathfrak{B}/\Gamma}}(\log \mathfrak{T})\big)^m\big)$, where $\overline{\mathfrak{B}/\Gamma}$ is the minimal smooth compactification of $\mathfrak{B}/\Gamma$, $\mathfrak{T} = \overline{\mathfrak{B}/\Gamma} \setminus \mathfrak{B}/\Gamma$ the compactification divisor and $\Omega^2_{\overline{\mathfrak{B}/\Gamma}}(\log \mathfrak{T})$ the so-called logarithmic canonical sheaf. For arbitrary arithmetic lattices Γ one knows that there exists a neat normal subgroup of finite index Γ' (Borel, [5]). Now the rough programme is clear. It consists of the following steps:

(i)   Find a concrete (almost) neat normal subgroup Γ' of $\tilde{\Gamma}$ of small index.

(ii) Classify the surface $\widetilde{\mathfrak{B}/\Gamma'}$ (minimal resolution of singularities of $\widehat{\mathfrak{B}/\Gamma'}$).

(iii) Find a logarithmic canonical divisor D on $\widetilde{\mathfrak{B}/\Gamma'}$, that means

$$\mathfrak{O}^2_{\widetilde{\mathfrak{B}/\Gamma'}}(D) \cong \Omega^2_{\widetilde{\mathfrak{B}/\Gamma'}}(\log \mathfrak{T}).$$

(iv) Study the logarithmic pluricanonical maps

$$\Phi_{mD}: \widetilde{\mathfrak{B}/\Gamma'} \dashrightarrow \mathbb{P}^{N(m)}.$$

(v)   Find the ring structure of the logarithmic pluricanonical ring

$$\bigoplus_{m=0}^{\infty} H^0\big(\widetilde{\mathfrak{B}/\Gamma'}, \mathfrak{O}_{\widetilde{\mathfrak{B}/\Gamma'}}(mD)\big).$$

(vi) Apply invariant theory of the finite group $\tilde{\Gamma}/\Gamma'$ to the logarithmic canonical ring.

Carrying out this programme one gets results 3. and 2. The reader who is familar with problems of this kind observes immediately that there are some difficulties, which have to be overcome. We have no troubles with (i). Γ' will be the principal congruence subgroup $\Gamma(1 - \omega)$ of $\Gamma = \mathbf{SU}\big((2, 1), \mathfrak{O}_{\mathbb{Q}(\sqrt{-3})}\big)$. (ii) is managed by means of a study of the Chern numbers of the surface and a classifying constellation of curves. Here one needs the proportionality formulas for Chern numbers of (not necessarily neat) ball lattices (see [32]). They were found by a combination of Mumford's proportionality theorem (for neat auxiliary groups), the equivariant K-theory of Atiyah-Singer and Riemann-Roch-Hirzebruch-Grothendieck theory. A proportionality formula for the selfintersection number of arithmetic curves (coming from Γ-rational subdiscs of B) can be found in my paper [33]. The proportionality formulas involve the volumes of the fundamental domains. By number theoretic methods I found an L-series expression for the volumes for all Picard modular groups $\mathbf{SU}\big((2, 1), \mathfrak{O}_{\mathbb{Q}(\sqrt{-d})}\big)$ in [30].

The classifying curve is the "fundamental curve" $\mathfrak{F}$ in $\overline{\mathfrak{B}/\Gamma'}$ in Picture 1.6.A. It consists of the resolution of singularities curve of $\widehat{\mathfrak{B}/\Gamma'}$ and some arithmetic curves coming from $\tilde{\Gamma}$-reflection discs. After doing the fine classification of $\widetilde{\mathfrak{B}/\Gamma'}$ we find logarithmic canonical divisors with support on $\mathfrak{F}$. So we get (iii).

For (iv) and (v) the dimension formula in Proposition 4.2.1 for $[\Gamma', m] \cong H^0\big(\widetilde{\mathfrak{B}/\Gamma'}, \big(\Omega^2_{\widetilde{\mathfrak{B}/\Gamma'}}(\log \mathfrak{T})\big)^m\big)$ is very important. The proof consists of a combination of Mumford's proportionality theory and a vanishing theorem due also to Mumford. Using only the dimension formula it is easy to see that the ring of Γ'-automorphic forms has four generators satisfying exactly one algebraic relation. In order to find this relation we use result 3., which plays a very central role for all further results. $\widehat{\mathfrak{B}/\Gamma'}$ is a three-sheeted cyclic Galois cover of $\mathbb{P}^2 = \widehat{\mathfrak{B}/\tilde{\Gamma}'}$ branched along a very simple divisor on $\mathbb{P}^2$. By means of the techniques of cyclic covers we will find the exact relation between the generating Γ'-automorphic forms. In this way we realize steps (iv) and (v). Knowing that $\tilde{\Gamma}/\Gamma' \cong (\mathbb{Z}/3\mathbb{Z}) \times S_4$, $\tilde{\Gamma}/\tilde{\Gamma}' \cong \mathbb{Z}/3\mathbb{Z}$, it is quick easy to deal with last step (vi) in the above list.

Let us now say something about the main results 1. and 4. By the work of Picard, Mostow and Deligne we know that the quotient map $\mathfrak{B} \to \mathfrak{B}/\Sigma$ is the inversion of the multivalued (almost everywhere locally biholomorphic) map $(F_1 : F_2 : F_3) : \mathbb{P}^2 \setminus \{4 \text{ points}\} \Rrightarrow \mathfrak{B}$ for suitable linear combinations F_1, F_2, F_3 of integrals $\displaystyle\int_g^h \frac{dx}{\sqrt[3]{x(x-1)(x-t_1)(x-t_2)}}$ where $\Sigma = \langle S_1, \ldots, S_5 \rangle$ is the monodromy group generated by Picard's five "fundamental substitutions". Fortunately $S_1, \ldots, S_5$ are in $\tilde{\Gamma}'$. Comparing the surfaces $\widehat{\mathfrak{B}/\Sigma}$ and $\widehat{\mathfrak{B}/\tilde{\Gamma}'}$ it turns out that they have to be equal. Moreover, $\mathfrak{B} \to \mathfrak{B}/\tilde{\Gamma}'$ can be realized by means of a basis ξ_1, ξ_2, ξ_3 of $[\Gamma', 1]$. In this way the results 1. and 4. are obtained.

After a suitable affine transformation $X \mapsto aX + b$ each polynomial of degree 4

$$p_4(X) = a_4X^4 + a_3X^3 + a_2X^2 + a_1X + a_0, \qquad a_4 \neq 0 \quad \left(p_4(X) \neq a_4X^4\right)$$

can be described by means of $\tilde{\Gamma}'$-automorphic forms $\xi_1, \xi_2, \xi_3, \xi_4$, $\sum_{i=1}^{4} \xi_i = 0$, of weight 1:

$$p_4(X) \underset{\text{affine equiv.}}{=\!=\!=} \prod_{i=1}^{4} \left(X - \xi_i(P)\right), \qquad P \in \mathfrak{B}^*.$$

This is a consequence of 3. and the description of the quotient map $\mathfrak{B}^* \to \widehat{\mathfrak{B}/\tilde{\Gamma}'}$ by means of $\tilde{\Gamma}'$-automorphic forms of weight 1. This factorization result extends to Picard curves $Y^3 - p_4(X) = 0$. In § 5 we prove that $\mathbb{P}^2/S_4$, $S_4 = \operatorname{Aut}(\mathbb{P}^2, \{P_1, \ldots, P_4\})$, yields essentially the classification up to isomorphism of Picard curves. The correspondence is given by

$$\mathbb{P}^3 \supset \mathbb{P}^2 = \left\{(x_1 : x_2 : x_3 : x_4) \in \mathbb{P}^3;\ \sum_{i=1}^{4} x_i = 0\right\} \to \{\text{Picard polynomials}\},$$

$$(x_1 : x_2 : x_3 : x_4) \qquad \mapsto Y^3 - \prod_{i=1}^{4} (X - x_i).$$

The main tool in § 5 is the Hesse functor acting on plane projective curves. Substituting automorphic forms for x_1, x_2, x_3, x_4 then yields us result 5.

The fact that there is an analogy of the situation for polynomials of degree 3 and that of modular forms gave the idea for deducing result 7. together with beautiful geometric interpretations.

6. is the result of a careful study of special Picard curves and the $\tilde{\Gamma}$-fixed point set on $\mathfrak{B}^*$.

The Γ- and $\tilde{\Gamma}$-fixed point sets were first found by Feustel in 1976 (see [17]). He used hermitean lattices in his classification method. Unfortunately this result is not published. So I present another proof. It consists of a combination of Feustel's method applied to the much more simple (almost neat) subgroup Γ' (see I.1.4) and geometric considerations on the elliptic ruled surface $\widetilde{\mathfrak{B}/\Gamma'}$ and on $\widehat{\mathfrak{B}/\tilde{\Gamma}'} = \mathbb{P}^2$.

We present the structure of the rings of automorphic forms and surface classifications not only for $\Gamma', \tilde{\Gamma}', \tilde{\Gamma}$ but also for all normal subgroups of $\tilde{\Gamma}$ containing Γ'. This seems to be necessary for a deeper number theoretic exploitation of result 7. in the near future.

The main result of Chapter II is the complete solution of systems of partial differential equations which are the natural generalizations of the Picard-Fuchs equations. We obtain the best results for the following systems of r^2 differential equations:

0.3.1. $$\frac{\partial^2}{\partial t_i\,\partial t_j} F + \frac{l}{n(t_j - t_i)} \left(\frac{\partial}{\partial t_i} - \frac{\partial}{\partial t_j}\right) F = 0, \qquad 1 \leqq i, j \leqq r, \quad i \neq j,$$

0.3.2. $$\frac{\partial^2}{\partial t_k^2} F + \frac{l}{n(t_k - 1)\, t_k} \left\{ \sum_{\substack{i=1\\ i\neq k}}^{r} \frac{t_i^2 - t_i}{t_i - t_k} \frac{\partial}{\partial t_i} - \left[\sum_{\substack{i=1\\ i\neq k}}^{r} \frac{(t_k - 1)\, t_i}{t_i - t_k} - (r+2)(t_k - 1) - t_k - 1 \right] \frac{\partial}{\partial t_k} + \frac{l(r+2)}{n} - 1 \right\} F = 0, \qquad k = 1, \ldots, r,$$

where $F = F(t_1, \ldots, t_r)$ denotes a holomorphic function (solution) and l, n are natural numbers such that $0 < l < n$, $(r + 2, n) = 1$. For historical reasons we call the equations of mixed type (0.3.1) Euler partial differential equations. The other ones, i.e. those in (0.3.2) are called Picard equations. The whole system (0.3.1), (0.3.2) is called an Euler-Picard system of partial differential equations. The main results about these systems are the following ones:

1. *The integral functions* $\int_{\alpha_t} \mathrm{d}x/y^l$ *depending on* $t = (t_1, \ldots, t_r)$ *and a family of cycles* α_t *on the compact Riemann surfaces of planar equations of the type*

0.3.3. $$Y^n = (X - 1)\, X(X - t_1)\, (X - t_2) \cdot \ldots \cdot (X - t_r),$$

are solutions of the Euler-Picard system.

2. *Each solution of the Euler-Picard system is a $\mathbb{C}$-linear combination of $r + 1$ linearly independent solutions of integral type described in* 1. (*change families of cycles*).

3. *The Euler-Picard equations* (0.3.1), (0.3.2) *are equations of smallest degree with respect to the partial derivatives* $\dfrac{\partial}{\partial t_i}$, *which have all the functions of integral type* $\int_{\alpha_t} dx/y^l$ *as solutions. They are the only "simple" quadratic equations*

$$\left\{ \frac{\partial^2}{\partial t_i \, \partial t_j} + \text{linear differential operator} \right\} F = 0$$

with this property.

4. *Each differential operator with holomorphic coefficients, which annihilates the functions* $\int_{\alpha_t} dx/y^l$, *lies in the sheaf of ideals generated by the Euler-Picard operators.*

For the proofs we have to go through several stages. Mainly we use the methods of modern algebraic geometry. The equations (0.3.3) describe an algebraic family $\mathfrak{Y}/S$ of affine curves over the manifold S of the parameters $t_1, \ldots, t_r$. We call it the general affine cycloelliptic curve family of type (n, r). The relative de Rham cohomology group $H^1_{\mathrm{DR}}(\mathfrak{Y}/S)$ of this family is a module over the ring

$$\mathbb{C}[S, \mathrm{D}] = \mathbb{C}[S] \left[\frac{\partial}{\partial t_1}, \ldots, \frac{\partial}{\partial t_r} \right]$$

of differential operators on S. Moreover, $H^1_{\mathrm{DR}}(\mathfrak{Y}/S)$ has a Galois module structure coming from the cyclic Galois coverings $\tilde{\mathfrak{Y}}_t \to \mathbb{P}^1$, $(x, y) \mapsto x$, where $\tilde{\mathfrak{Y}}_t$ denotes the smooth compactification of the curve $\mathfrak{Y}_t$ over $t \in S$. The module $H^1_{\mathrm{DR}}(\mathfrak{Y}/S)$ has an isotypical decomposition with respect to this Galois group action. $H^1_{\mathrm{DR}}(\mathfrak{Y}/S)_l$ denotes the $\mathbb{C}[S, \mathrm{D}]$-submodule on which a fixed generator of the Galois group acts by multiplication with $e^{2\pi i l/n}$. In 1.4 it is shown that the element dx/y^l of $H^1_{\mathrm{DR}}(\mathfrak{Y}/S)_l$ is annihilated by all the Euler-Picard operators in (0.3.1), (0.3.2) (Proposition 1.4.1). For the proof to this result we generalize classical ideas going back to Picard in the two-parameter case. We translated clever complicated integral substitutions from the classical to the modern language of relative de Rham cohomology. In particular it turns out that our road leads us through differential operators of order 3.

In the second part of 1.4 it will be proved that $H^1_{\mathrm{DR}}(\mathfrak{Y}/S)_l$ is generated by dx/y^l as a $\mathbb{C}[S, \mathrm{D}]$-module. Moreover, it is a free $\mathbb{C}[S]$-module of rank $r + 1$ (Theorem 1.4.23). Here I borrowed heavily from Katz' article [36], where one can find structure results for one-parameter families

$$Y^n = (X - 1)^{b_{-1}} \cdot X^{b_0} \cdot (X - t_1)^{b_1}.$$

In § 2 we switch from the affine curve families to the corresponding cycloelliptic smooth projective curve families $\tilde{\mathfrak{Y}}/S$. There we investigate the hypercohomology sheaves $\underline{\mathcal{H}}^1_{\mathrm{DR}}(\tilde{\mathfrak{Y}}/S)$. The sheaves $\mathcal{D}$ of differential operators over S operate on these sheaves. The $\mathscr{D}$-module sheaf $\mathscr{H}^1_{\mathrm{DR}}(\mathfrak{Y}/S)$ is known as the Gauss-Manin connection of the curve family. The global construction of the Gauss-Manin connection goes essentially back to Katz (see e.g. [37]). We use some general structure theorems (base change, locally free) due to Deligne. It will be shown that $\mathscr{H}^1_{\mathrm{DR}}(\mathfrak{Y}/S)_l = \mathscr{D} \cdot \overline{\mathrm{d}x/y^l}$ is a $\mathscr{D}$-direct summand of $\mathscr{H}^1_{\mathrm{DR}}(\mathfrak{Y}/S)$ (Theorem 2.4.3(vi)). Passing to the analytic category the use of de Rham duality for curve families will lead us to result 2.

Our curve families (0.3.3) with the condition $(r + 2, n) = 1$ are called primitive. We define also primitive differential forms $x^k\mathrm{d}x/y^l$, primitive $\mathbb{C}[S, \mathrm{D}]$-modules, primitive connections and primitive systems of partial differential equations. On a primitive general cycloelliptic curve family (0.3.3) all relative differential forms $\mathrm{d}x/y^l$, $0 < l < n$, are primitive and of the second kind on the fibres $\mathfrak{Y}_t$. We show that primitive relative differential forms lead to primitive modules, primitive connections and primitive differential equation systems of Euler-Picard type. The primitivity of the system implies its integrability on the one hand and leads to the structure results 3. and 4. on the other hand.

We investigate the structure of the Gauss-Manin connection not only for the general families (0.3.3) but more generally for the cycloelliptic curve families of equation type

$$Y^n = (X - 1)^{b_{-1}} X^{b_0}(X - t_1)^{b_1} \cdot \ldots \cdot (X - t_r)^{b_r}, \qquad b_i \text{ natural numbers.}$$

So Chapter II contains also the results of Katz concerning one-parameter families with their link to Picard-Fuchs equations. On the other hand we also investigate in this way the Euler equations of the more general type

$$\frac{\partial^2}{\partial t_i\, \partial t_j} F + \frac{l}{n(t_j - t_i)} \left(b_j \frac{\partial}{\partial t_i} - b_i \frac{\partial}{\partial t_j}\right) F = 0\,.$$

... So spendet Segen noch immer die Hand
Des von Ribbeck auf Ribbeck im Havelland.

Theodor Fontane

Chapter I

The Picard Curve Family and Eisenstein Lattices of the Complex Unit Ball

§ 1. PRELIMINARY RESULTS AND NOTATIONS

1.1. *Eisenstein Lattices in the Ball*

Let

$$\mathfrak{B} = \{(z_1, z_2) \in \mathbb{C}^2; |z_1|^2 + |z_2|^2 < 1\}$$

be the *two-dimensional complex unit ball*. If V is a vector space over the field of complex numbers $\mathbb{C}$, we denote the corresponding projective space $(V \setminus \{0\})/\mathbb{C}^*$, $\mathbb{C}^* = \mathbb{C} \setminus \{0\}$, by $\mathbb{P}V$ and the projection of a subset $M \subset V$ by $\mathbb{P}M$. Let Φ be the hermitian form represented by the matrix $\begin{pmatrix} 1 & 0 & 0 \\ 0 & 1 & 0 \\ 0 & 0 & -1 \end{pmatrix}$ with respect to canonical coordinates in $\mathbb{C}^3$ and

$$\tilde{\mathfrak{B}} = \left\{ \mathfrak{z} = \begin{pmatrix} z_1 \\ z_2 \\ z_3 \end{pmatrix} \in \mathbb{C}^3; \Phi(\mathfrak{z}, \mathfrak{z}) < 0 \right\}.$$

Then the *unitary group* $\mathbb{U}\big((2, 1), \mathbb{C}\big) = \mathbb{U}(\Phi, \mathbb{C})$ acts on $\tilde{\mathfrak{B}}$ and (transitively) on $\mathfrak{B} = \mathbb{P}\tilde{\mathfrak{B}}$. For each squarefree natural number d we define the (*special*) *Picard modular group* of the imaginary quadratic field $K = \mathbb{Q}\big(\sqrt{-d}\big)$ as

$$\Gamma^{(d)} = \mathbb{SU}\big((2, 1), \mathfrak{O}_K\big), \qquad \mathfrak{O}_K = \text{ring of integers in } K.$$

This is a lattice subgroup of $\mathbb{SU}\big((2, 1), \mathbb{C}\big)$ acting properly discontineously on $\mathfrak{B}$. For an ideal $\mathfrak{a} \subseteq \mathfrak{O}_K$ the *congruence subgroup* $\Gamma^{(d)}(\mathfrak{a})$ is defined by the exact sequence

$$1 \to \Gamma^{(d)}(\mathfrak{a}) \to \Gamma^{(d)} \to \mathbb{SU}\big((2, 1), \mathfrak{O}_K/\mathfrak{a}\big).$$

For $\mathfrak{a} = \alpha\mathfrak{O}_K$ we write simply $\Gamma^{(d)}(\alpha)$ instead of $\Gamma^{(d)}(\alpha\mathfrak{O}_K)$. For the special case $K = \mathbb{Q}(\omega)$, $\omega = e^{2\pi i/3}$, $\mathfrak{O} = \mathfrak{O}_K = \mathbb{Z} + \mathbb{Z}\omega$, $\mathfrak{a} = \mathfrak{O}(1 - \omega)$ we fix in this paper the following notations:

$\Gamma = \Gamma^{(3)}$	$= \mathbb{SU}\big((2, 1), \mathfrak{O}\big)$	the *special Eisenstein lattice*,
$\tilde{\Gamma}$	$= \mathbb{U}\big((2, 1), \mathfrak{O}\big)$	the *full Eisenstein lattice*,
Γ'	$= \Gamma(1 - \omega)$,	$\tilde{\Gamma}' = \tilde{\Gamma}(1 - \omega)$.

Generally we call a discrete subgroup of $\mathbb{U}((2,1),\mathbb{C})$ an *Eisenstein lattice* if it is commensurate with Γ; that means that the intersection with Γ is a subgroup of finite index in both of the intersecting groups. Sometimes it is convenient to identify without change of notations our Eisenstein lattices $\Gamma, \ldots$ with groups $\mathbb{P}\Gamma$ which act effectively on $\mathfrak{B} = \mathbb{P}\tilde{\mathfrak{B}}$. In each case the meaning will be clear.

1.2. Cusps and Cusp Bundles

The *parabolic group* $P_\varkappa$ of a boundary point $\varkappa \in \partial\mathfrak{B}$ of the ball $\mathfrak{B}$ is the isotropy group

$$P_\varkappa := \{g \in \mathbb{SU}((2,1),\mathbb{C});\ g\varkappa = \varkappa\}.$$

Its unipotent radical is denoted by $U_\varkappa$. For a lattice Ω of $\mathbb{U}((2,1),\mathbb{C})$ we call $\varkappa \in \partial\mathfrak{B}$ an *Ω-cusp* if and only if

$$\Omega_{\varkappa,u} =: \Omega \cap U_\varkappa \text{ is a lattice in } U_\varkappa.$$

This means that $U_\varkappa/\Omega_{\varkappa,u}$ is compact. $\partial_\Omega\mathfrak{B}$ denotes the set of all Ω-cusps. At least for arithmetic lattices Ω one knows that $\partial_{\mathfrak{B}}\Omega/\Omega$ is a finite set. The set

$$\widehat{\mathfrak{B}/\Omega} = (\mathfrak{B} \cup \partial_\Omega\mathfrak{B})/\Omega,$$

endowed with a natural complex space structure, is a complex compact normal algebraic surface. It is called the *Baily-Borel compactification* of $\mathfrak{B}/\Omega$. The points of $\partial_\Omega\mathfrak{B}/\Omega$ are the *cusp singularities* of the surface. In order to study the complex structure around a cusp singularity $\hat{\varkappa} = \Omega\varkappa/\Omega$ it is useful to introduce the *cusp bundles* $\mathbb{F}_\varkappa$. To do this we turn from $\mathfrak{B}$ to the *Siegel domain*

$$\mathfrak{B} = \{(z, u) \in \mathbb{C}^2;\ 2\,\mathrm{Im}\, z - |u|^2 > 0\}$$

by means of the projective automorphism of $\mathbb{P}^2 = \mathbb{P}\mathbb{C}^3$

$$\mathbb{P}A: \mathfrak{B} \xrightarrow{\sim} \mathfrak{B}, \qquad A = \frac{1}{\sqrt{2}}\begin{pmatrix} 1 & 0 & 1 \\ 0 & -\sqrt{2}\,\mathrm{i} & 0 \\ \mathrm{i} & 0 & -\mathrm{i} \end{pmatrix} \in \mathbb{G}\mathrm{l}_3(\mathbb{C}).$$

Because the action of $\mathbb{SU}((2,1),\mathbb{C})$ on $\partial\mathfrak{B}$ is transitive we can assume that $\varkappa$ is transformed to the special boundary point

$$\infty := \mathbb{P}\begin{pmatrix} 1 \\ 0 \\ 0 \end{pmatrix} \in \partial\mathfrak{B}.$$

So we can assume that $\varkappa = \infty$ and that Ω acts on $\mathfrak{B}$.

$$1 \subset [U_\varkappa, U_\varkappa] \subset U_\varkappa \subset P_\varkappa = \left\{\alpha = \begin{pmatrix} * & \bar{a} & \frac{\mathrm{i}}{2}|a|^2 + r \\ 0 & * & a \\ 0 & 0 & * \end{pmatrix} \in \mathrm{Aut}_{\mathrm{hol}}\,\mathfrak{B}\right\}$$

is a chain of normal subgroups with natural isomorphisms

$$U_\varkappa/[U_\varkappa, U_\varkappa] \xrightarrow{\sim} \mathbb{C}, \quad [U_\varkappa, U_\varkappa] \xrightarrow{\sim} \mathbb{R},$$
$$U_\varkappa \ni \alpha \longmapsto \omega, \qquad \alpha \longmapsto r.$$

Taking intersections with Ω we get a chain of normal subgroups

$$1 \subset \Delta_\varkappa \subset \Omega_{\varkappa,u} \subset \Omega_\varkappa.$$

$\Lambda_\varkappa = \Omega_{\varkappa,u}/\Delta_\varkappa$ is a lattice in $\mathbb{C}$, $\Delta_\varkappa$ a lattice in $\mathbb{R}$ and $G_\varkappa = \Omega_\varkappa/\Omega_{\varkappa,u}$ is a finite cyclic group of order ≤ 6 $(\neq 5)$. By stepwise factorization and extension of group actions we obtain the local compactification diagram 1.2.a (see [29] for details):

1.2.a.

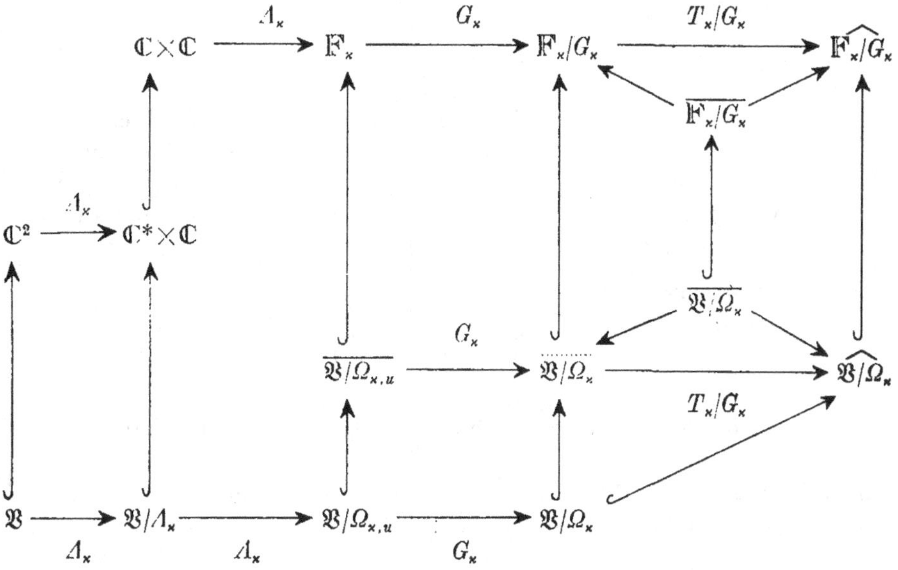

The change from $\ddot{X}$ to $\overline{X}$ is the minimal resolution of singularities, $\ddot{X} = \mathbb{F}_\varkappa/G_\varkappa$ or $\mathfrak{B}/\Omega_\varkappa$. Contracting $T_\varkappa/G_\varkappa$, where $T_\varkappa$ is the *elliptic (compactification) curve* $(0 \times \mathbb{C})/\Lambda_\varkappa$, we obtain $\hat{X}$ from $\ddot{X}$. The morphism $\overline{X} \to \hat{X}$ is called the *filled resolution of cusp singularity*. The image $\hat{\varkappa}$ of $T_\varkappa/G_\varkappa$ is the cusp singularity. $\mathbb{F}_\varkappa = \mathbb{F}_\varkappa(\Omega)$ is called the *cusp bundle of $\varkappa$ with respect to Ω*. It is a line bundle over the embedded elliptic curve $T_\varkappa$.

The compactifications $\mathbb{F}_\varkappa/G_\varkappa$, $\overline{\mathbb{F}_\varkappa/G_\varkappa}$, $\widehat{\mathbb{F}_\varkappa/G_\varkappa}$ define *global compactifications* $\mathfrak{B}/\Omega$, $\overline{\mathfrak{B}/\Omega}$, $\widehat{\mathfrak{B}/\Omega}$ and the smooth model $\widetilde{\mathfrak{B}/\Omega}$. The interrelations are illustrated by Diagram 1.2.b. The dotted arrows represent isomorphisms of germs of surfaces around curves and points, respectively. The solid arrows represent global morphisms (if one forgets the second member of each pair).

We return to the special case of Eisenstein lattices. There is then exactly one cusp singularity on $\widehat{\mathfrak{B}/\Gamma}$. This is a special case of Feustel's theorem asserting that

1.2.b.

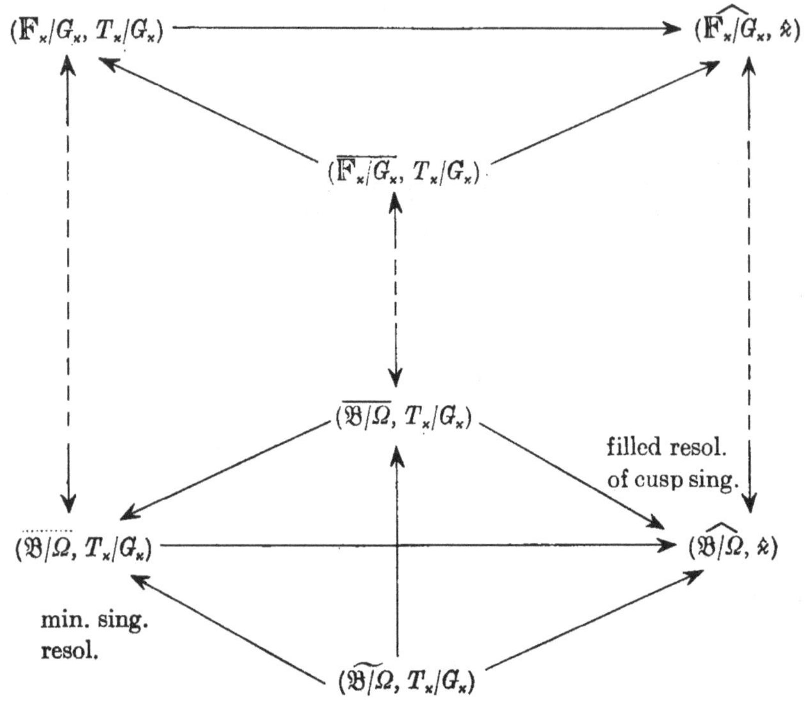

the Picard modular surfaces $\widehat{\mathfrak{B}/\Gamma^{(d)}}$ have exactly $h(d)$ cusp singularities, where $h(d)$ is the class number of $K = \mathbb{Q}(\sqrt{-d})$ (see [18] or [66]). Using $\varkappa = (0,1) = \mathbb{P}\begin{pmatrix}1\\0\\1\end{pmatrix}$ and the transformation $\mathbb{P}A: \mathfrak{B} \xrightarrow{\sim} \mathfrak{B}$ above ($\varkappa \mapsto \infty$) one finds that

$$\Gamma_{\varkappa,u} = \left\{ [\alpha, g] = \begin{pmatrix} 1 - \frac{1}{2}\left(|\alpha|^2 - g\sqrt{-3}\right) & \alpha & \frac{1}{2}\left(|\alpha|^2 - g\sqrt{-3}\right) \\ -\bar{\alpha} & 1 & \bar{\alpha} \\ \frac{1}{2}\left(-|\alpha|^2 + g\sqrt{-3}\right) & \alpha & 1 + \frac{1}{2}\left(|\alpha|^2 - g\sqrt{-3}\right) \end{pmatrix}; \right.$$

$$\left. \alpha \in \mathfrak{O},\ g \in \mathbb{Z},\ |\alpha|^2 \equiv g \bmod 2, \right\},$$

$$\Gamma'_{\varkappa,u} = \Gamma'_\varkappa = \{[\alpha, g] \in \Gamma_{\varkappa,u};\ 1 - \omega \mid \alpha\}.$$

With the notations $\Lambda'_\varkappa = \Lambda_\varkappa(\Gamma'_\varkappa)$, $\Lambda'_\varkappa = \Delta_\varkappa(\Gamma'_\varkappa)$ we have

1.2.1. $\Lambda'_\varkappa = \sqrt{-2}\,\mathfrak{O}(1-\omega)$, $\quad \Lambda_\varkappa = \sqrt{-2}\,\mathfrak{O}$, $\quad \Delta_\varkappa = \Delta'_\varkappa = 2\sqrt{3}\,\mathbb{Z}$.

Furthermore we have a commutative diagram

1.2.c.

$$\begin{array}{ccccc}
 & \mathbb{Z}/3\mathbb{Z} & & \mathbb{Z}/2\mathbb{Z} & \\
\mathbb{F}_\varkappa' & \xrightarrow[\text{étale}]{} & \mathbb{F}_\varkappa & \longrightarrow & \mathbb{F}_\varkappa/G_\varkappa \\
\uparrow & & \uparrow & & \| \\
\cup & & \cup & & \mathbb{F}_\varkappa'/G_\varkappa' \\
T_\varkappa' & \longrightarrow & T_\varkappa & &
\end{array}$$

where $\mathbb{F}'_\varkappa = \mathbb{F}_\varkappa(\Gamma')$, $G_\varkappa \simeq \mathbb{Z}/2\mathbb{Z}$, $G'_\varkappa = \Gamma_\varkappa/\Gamma'_\varkappa \simeq S_3$ (the symmetric group). Let $|\Lambda_\varkappa|$, $|\Delta_\varkappa|$ denote the volumes of the fundamental domains of the lattices $\Lambda_\varkappa$, $\Delta_\varkappa$ in $\mathbb{C}$ and $\mathbb{R}$, respectively. By the selfintersection number formula for $T_\varkappa$ in $\mathbb{F}_\varkappa$ (see [28] or [29]) we find

1.2.2. $$(T_\varkappa^2) = -\frac{2\,|\Delta_\varkappa|}{|\Lambda_\varkappa|} = -1 \quad \text{on} \quad \mathbb{F}_\varkappa,$$

$$(T_\varkappa'^2) = -\frac{2\,|\Delta'_\varkappa|}{|\Lambda'_\varkappa|} = -3 \quad \text{on} \quad \mathbb{F}'_\varkappa,$$

where $T_\varkappa \simeq \mathbb{C}/\Lambda_\varkappa \simeq \mathbb{C}/\mathfrak{O} \simeq T'_\varkappa$. Therefore $T_\varkappa$ and $T'_\varkappa$ are elliptic curves with complex multiplication with respect to the field $K = \mathbb{Q}(\omega)$. There is only one isogeny class of elliptic curves with this property and their equation type is (see [64], IV, § 4.5):

1.2.3. $T_\varkappa, T'_\varkappa$: $Y^2 = X^3 - 1$.

1.3. S_4-Action

1.3.1. The sequence

$$1 \to \Gamma' \to \Gamma \to \mathbf{SU}\big((2,1), \mathfrak{O}/(1-\omega)\mathfrak{O}\big) \to 1,$$
$$\wr\|$$
$$\mathbf{SO}\big((2,1), \mathbf{F}_3\big) \simeq S_4$$

where $\mathbf{F}_3$ is the Galois field $\mathbf{F}_3 = \mathbb{Z}/3\mathbb{Z}$, is exact.

Proof. Throughout this book we fix the following notations:

$$\varkappa_1 = (1,0) = \mathbb{P}\begin{pmatrix}1\\0\\1\end{pmatrix}, \qquad \varkappa_2 = (0,-1) = \mathbb{P}\begin{pmatrix}0\\-1\\1\end{pmatrix},$$

$$\varkappa_3 = (-1,0) = \mathbb{P}\begin{pmatrix}-1\\0\\1\end{pmatrix}, \qquad \varkappa_4 = (0,1) = \mathbb{P}\begin{pmatrix}0\\1\\1\end{pmatrix}.$$

Γ/Γ' acts transitively on the images $\bar{\varkappa}_1, \bar{\varkappa}_2, \bar{\varkappa}_3, \bar{\varkappa}_4$ of the four cusp vectors $\begin{pmatrix}1\\0\\1\end{pmatrix}$, $\begin{pmatrix}0\\-1\\1\end{pmatrix}$, $\begin{pmatrix}-1\\0\\1\end{pmatrix}$, $\begin{pmatrix}0\\1\\1\end{pmatrix}$ in $\mathfrak{O}^3/(1-\omega)\,\mathfrak{O}^3 = \mathbf{F}_3^3$ because Γ acts transitively on $\partial_\Gamma \mathfrak{B}$. The ineffective kernel is trivial. The group $\mathbf{SO}((2,1), \mathbf{F}_3)$ has 24 elements. Therefore the sequence in 1.3.1 is exact, and we have a faithful representation of Γ/Γ' as the symmetric group S_4 acting on $\{\varkappa_1, \varkappa_2, \varkappa_3, \varkappa_4\}$ by permutations.

As notation for the elements of S_4 we use their cycle decompositions. S_4 consists of five conjugacy classes (see [25]). They are denoted by C_i, $i = 1, \ldots, 5$:

1.3.2. C_1: 1,

C_2: (12) (34), (13) (24), (14) (23),

C_3: $\begin{cases}(123), (124), (134), (234),\\ (132), (142), (143), (243),\end{cases}$

C_4: $\begin{cases}(1234), (1243), (1324),\\ (1432), (1342), (1423),\end{cases}$

C_5: (12), (13), (14), (23), (24), (34).

For later use we fix at this time the notations for the normal subgroups of S_4 arranged in the chain

$$1 \subset K_4 \subset A_4 \subset S_4 .$$

$K_4 = C_1 \cup C_2$ is the Klein group of four elements, $A_4 = C_1 \cup C_2 \cup C_3^+ \cup C_4^+$ is the alternating group, where

$$C_i^+ = \{g \in C_i;\ \operatorname{sgn} g = +1\}.$$

In terms of group generators we note that

$$K_4 = \langle (12)\,(34), (13)\,(24)\rangle, \qquad A_4 = \langle K_4, (123)\rangle, \qquad S_4 = \langle A_4, (12)\rangle .$$

For $g \in S_4 \cong \Gamma/\Gamma'$ we denote by $g_\mathfrak{B} \in \Gamma$ a representative of g.

1.3.3. Each of the elements of C_2 and C_5 has a reflection $g_\mathfrak{B}$ as a representative; this means that $\mathrm{Fix}_\mathfrak{B}(g_\mathfrak{B})$ is a disc in $\mathfrak{B}$.

Proof. Set

$$(13)_\mathfrak{B} = \begin{pmatrix}-1 & 0 & 0\\ 0 & 1 & 0\\ 0 & 0 & -1\end{pmatrix}, \qquad (12)\,(34)_\mathfrak{B} = \begin{pmatrix}0 & 1 & 0\\ 1 & 0 & 0\\ 0 & 0 & -1\end{pmatrix}.$$

Since $(13) \in C_5$, $(12)\,(34) \in C_2$ and $\Gamma \to S_4$ is surjective, we get the result by conjugation.

For C_5 we define special representatives in the following unique manner: Let $\mathfrak{D}_{ij} = \mathfrak{D}_{ji}$ be the disc "going through" $\varkappa_i$ and $\varkappa_j$, $i \neq j$, more exactly $\mathfrak{D}_{ij}$ has $\varkappa_i$, $\varkappa_j$ as boundary points (see Figure 1.3.A). The $\mathfrak{D}_{ij}$'s are *Γ-reflection discs*:

$$\mathfrak{D}_{ij} = \mathrm{Fix}_{\mathfrak{B}}(kl)_{\mathfrak{B}}, \qquad i, j \neq k, l.$$

1.3.A.

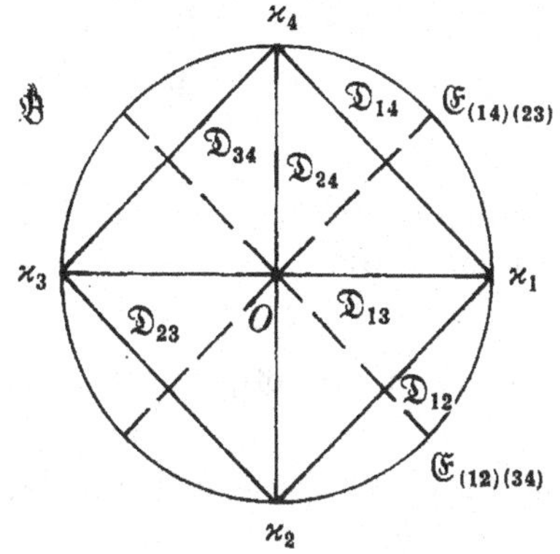

For example, using orthogonality with respect to $\mathbb{U}\big((2, 1), \mathbb{C}\big)$, one finds by the condition that $\varkappa_3$, $\varkappa_4$ have to be fixed

$$(1, 2)_{\mathfrak{B}} = A \begin{pmatrix} -1 & 0 & 0 \\ 0 & -1 & 0 \\ 0 & 0 & 1 \end{pmatrix} A^{-1} = \begin{pmatrix} 1 & -2 & 2 \\ -2 & 1 & -2 \\ -2 & 2 & -3 \end{pmatrix},$$

$$A = \begin{pmatrix} -1 & 0 & -1 \\ 0 & 1 & 1 \\ 1 & 1 & 1 \end{pmatrix}, \quad \begin{pmatrix} -1 \\ 1 \\ 1 \end{pmatrix} \perp \varkappa_3 = \begin{pmatrix} -1 \\ 0 \\ 1 \end{pmatrix}, \quad \varkappa_4 = \begin{pmatrix} 0 \\ 1 \\ 1 \end{pmatrix}.$$

Since the six elements (kl) are different in $S_4 = \Gamma/\Gamma''$, the discs $\mathfrak{D}_{ij}$ are not Γ'-equivalent.

Similarly we have three Γ''-non-equivalent discs

$$\mathfrak{E}_{(ij)(kl)} = \mathrm{Fix}(ij)\,(kl)_{\mathfrak{B}}.$$

Two of them have been drawn in Figure 1.3.A.

Let H be one of the four normal subgroups of S_4. We denote by $\Gamma' \cdot H$ the preimage of H with respect to the projection $\Gamma' \to S_4$. It is a normal subgroup of Γ. Let X be one of the surfaces $\overset{\cdots\cdots}{\mathfrak{B}/\Gamma' \cdot H}$, $\overline{\mathfrak{B}/\Gamma' \cdot H}$, $\widetilde{\mathfrak{B}/\Gamma' \cdot H}$, $\widehat{\mathfrak{B}/\Gamma' \cdot H}$. Each Γ-arithmetic disc $\mathfrak{D}$ defines after projection and topological closure an algebraic curve D_X on X (see [33]). If it is clear on which surface X we work, then we will usually omit the index X. For example we have nine curves

$$D_{ij}, E_{(ij)(kl)} \quad \text{on} \quad \widetilde{\mathfrak{B}/\Gamma''}.$$

We finish this elementary section with the remark that S_4 acts on each X with ineffective kernel H, that means that S_4/H acts effectively on X.

1.4. Γ'-Fixed Point Formation

1.4.1. DEFINITION. Let G be a group acting (effectively) on a topological space S. Then a system $\{C_i\}_{i\in I}$ of connected subspaces C_i of S is called a *G-fixed point formation on S* if and only if the following conditions are satisfied:

(i) C_i is a connected component of $\operatorname{Fix} g_i$ for a suitable $g_i \in G \setminus \{\mathrm{id}\}$.

(ii) If C is a connected component of $\operatorname{Fix} g$, $g \in G \setminus \{\mathrm{id}\}$, then C is G-equivalent to one and only one C_i, $i \in I$.

1.4.2. DEFINITION. A discrete subgroup Ω of $\mathrm{Aut}_{\mathrm{hol}}\,\mathfrak{B}$ is called *almost neat* if and only if the following conditions are satisfied:

(i) $\Omega_{\varkappa,u} = \Omega_\varkappa$ for each Ω-cusp $\varkappa$;

(ii) An Ω-fixed point formation on $\mathfrak{B}$ consists of a set of points (isolated elliptic points).

In [29] (I, Prop. 10.1) I proved that $\Gamma^{(d)}(\mathfrak{a})$ is almost neat if and only if $\mathfrak{a} \nmid 2$ (see also [31], Lemma 4.3 and Remark 4.4). Therefore we have

1.4.3. *Γ' is almost neat.*

We determine a Γ'-fixed point formation on $\mathfrak{B}$. The point $O = (0,0)$ is an *elliptic point* because $\begin{pmatrix}0\\0\\1\end{pmatrix}$ is an eigenvector of

$$\gamma = \begin{pmatrix}1 & 0 & 0\\ 0 & \omega & 0\\ 0 & 0 & \omega^2\end{pmatrix} \in \Gamma'.$$

$$\Gamma'_O = \langle\gamma\rangle = \mathbb{Z}/3\mathbb{Z}, \qquad \Gamma_O \simeq \mathbb{U}(2, \mathfrak{O}).$$

1.4.4. $\Gamma \cdot O/\Gamma'$ consists of three elements.

It suffices to check that we have a commutative exact diagram

$$\begin{array}{ccccccccc} 1 & \to & \Gamma' & \to & \Gamma & \to & S_4 & \to & 1\\ & & \uparrow & & \uparrow & & \uparrow & & \\ 1 & \to & \Gamma'_O & \to & \Gamma_O & \to & D_4 & \to & 1 \end{array}$$

where D_4 is the dihedral group

$$D_4 = \langle (13)_{\mathfrak{B}}, (24)_{\mathfrak{B}}, (14)\,(23)_{\mathfrak{B}}\rangle = \langle (13)_{\mathfrak{B}}, (1234)_{\mathfrak{B}}\rangle.$$

1.4.5. *$\Gamma \cdot O$ is the set of Γ'-elliptic points.*

Proof. It suffices to prove that for each elliptic $\gamma \in \Gamma'$ there is an element $\varrho \in \Gamma$ such that $\varrho\gamma\varrho^{-1} \in \Gamma_O$. The element γ has eigenvalues $1, \omega, \omega^2$. We choose primitive eigenvectors $\mathfrak{a}, \mathfrak{b}, \mathfrak{c} \in \mathfrak{O}^3$. Without loss of generality we can assume that

$$\gamma\mathfrak{a} = \omega\mathfrak{a}, \qquad \gamma\mathfrak{b} = \omega^2\mathfrak{b}, \qquad \gamma\mathfrak{c} = \mathfrak{c},$$

$$\mathfrak{a} \perp \mathfrak{b} \perp \mathfrak{c} \perp \mathfrak{a}, \qquad \mathfrak{a}^2, \mathfrak{b}^2 > 0, \qquad \mathfrak{c}^2 < 0 \quad \text{(with respect to } \Phi\text{)}.$$

We use the following two elementary lemmas:

1.4.6. LEMMA ([30], VI, Prop. 2.17). *Let γ be an automorphism of finite order of the unimodular hermitean lattice $(\mathfrak{O}_K^n, \Psi)$ with eigenvalues $\zeta_1, \ldots, \zeta_n$, $\zeta_n \neq \zeta_i$ for $i < n$. For an eigenvector $\mathfrak{z} \in \mathfrak{O}_{K(\zeta_n)}^n$ of γ one then has $\mathfrak{z}^2 \mid \prod_{i=2}^{n} (\zeta_n - \zeta_i)$.*

1.4.7. LEMMA ([19], I, Lemma 1.2). *It holds that*

$$(\mathfrak{c}^\perp)^\#/\mathfrak{c}^\perp \cong (\mathfrak{O}\mathfrak{c})^\#/\mathfrak{O}\mathfrak{c},$$

where $\mathfrak{c}^\perp = \{\mathfrak{x} \in \mathfrak{O}^3; \langle \mathfrak{x}, \mathfrak{c}\rangle = 0\}$ *and where for a sublattice M of $\mathfrak{O}^3$ we set*

$$M^\# = \{\mathfrak{x} \in M \otimes K; \langle \mathfrak{x}, \mathfrak{m}\rangle \in \mathfrak{O}_K \text{ for all } \mathfrak{m} \in M\}.$$

CASE 1: $\mathfrak{c}^2 = -1$.

Then $(\mathfrak{c}^\perp)^\#/\mathfrak{c}^\perp \simeq 0$ by Lemma 1.4.7. This means that $\mathfrak{c}^\perp$ is unimodular. Now we apply Lemma 1.4.6 to $\mathfrak{c}^\perp$ and $\gamma|_{\mathfrak{c}^\perp}$. We find $\mathfrak{a}^2, \mathfrak{b}^2 \mid \omega - \omega^2$, hence $\mathfrak{a}^2 = \mathfrak{b}^2 = 1$. Therefore

$$\varrho\colon \mathfrak{a} \mapsto (-\omega)^k \begin{pmatrix}1\\0\\0\end{pmatrix}, \ \mathfrak{b} \mapsto \begin{pmatrix}0\\1\\0\end{pmatrix}, \ \mathfrak{c} \mapsto \begin{pmatrix}0\\0\\1\end{pmatrix},$$

for a suitable $k \in \mathbb{Z}$, defines an element $\varrho \in \Gamma$ and $\varrho\gamma\varrho^{-1} \in \Gamma_0$.

CASE 2: $\mathfrak{c}^2 \neq -1$.

We will see that this case is excluded. From Lemma 1.4.6, applied to $\mathfrak{O}^3$, it follows that

$$\mathfrak{a}^2, \mathfrak{b}^2 \mid 3, \qquad \mathfrak{c}^3 = -3.$$

If $\mathfrak{a}^2 = 1$, then $\mathfrak{a}^\perp$ is unimodular by Lemma 1.4.7 (with $\mathfrak{a}$ instead of $\mathfrak{c}$), and Lemma 1.4.6 applied to $\mathfrak{a}^\perp$ and to $\gamma|_{\mathfrak{a}^\perp}$ yields $\mathfrak{c}^2 \mid 1 - \omega$, which is a contradiction. Therefore

$$\mathfrak{a}^2 = \mathfrak{b}^2 = 3, \qquad \mathfrak{c}^2 = -3$$

and

$$(\mathfrak{O}\mathfrak{a} \oplus \mathfrak{O}\mathfrak{b})^\# = (\mathfrak{O}\mathfrak{a})^\# \oplus (\mathfrak{O}\mathfrak{b})^\# = \frac{1}{3}\,\mathfrak{O}\mathfrak{a} \oplus \frac{1}{3}\,\mathfrak{O}\mathfrak{b}.$$

On the other hand $(\mathfrak{c}^\perp)^\#/\mathfrak{c}^\perp = \mathfrak{O}/3\mathfrak{O}$ by 1.4.7. So we have a chain of proper inclusions

$$\mathfrak{O}\mathfrak{a} \oplus \mathfrak{O}\mathfrak{b} < \mathfrak{c}^\perp < (\mathfrak{c}^\perp)^\# < (\mathfrak{O}\mathfrak{a} + \mathfrak{O}\mathfrak{b})^\#.$$

Let $\mathfrak{a}', \mathfrak{b}'$ be an $\mathfrak{O}$-base of $\mathfrak{c}^\perp$. It follows that

$$\det M = 3 \quad \text{for} \quad M = \begin{pmatrix} a & \beta \\ \bar{\beta} & b \end{pmatrix} = \begin{pmatrix} \mathfrak{a}'^2 & \langle \mathfrak{a}', \mathfrak{b}' \rangle \\ \langle \mathfrak{b}', \mathfrak{a}' \rangle & \mathfrak{b}'^2 \end{pmatrix}.$$

By means of a suitable base change in $\mathfrak{c}^\perp$ we wish to have M as simple as possible. Now $\mathfrak{a}, \mathfrak{b} \in \mathfrak{c}^\perp$ define a matrix $A \in \mathrm{Mat}_3(\mathfrak{O})$ such that

$$\bar{A}^{\mathrm{T}} M A = \begin{pmatrix} 3 & 0 \\ 0 & 3 \end{pmatrix}, \qquad \det A \in (1 - \omega)\mathfrak{O}^*, \qquad \mathfrak{O}^* = \{\text{units of } \mathfrak{O}\},$$

hence

$$M = \bar{B}^{\mathrm{T}} B, \qquad \det B \in (1 - \omega)\mathfrak{O}^*, \qquad B \in \mathrm{Mat}_3(\mathfrak{O}).$$

By means of the Gauss algorithm for matrices and the euclidean algorithm in the ring $\mathfrak{O}$ one finds an element $X \in \mathrm{Gl}_2(\mathfrak{O})$ such that

$$BX \in \left\{ \begin{pmatrix} 1 & 0 \\ \eta & \omega - 1 \end{pmatrix}; \quad \eta \in \mathfrak{O}^* \cup \{0\} \right\}.$$

So we can assume that

$$M = \begin{pmatrix} 2 & 1 - \omega \\ 1 - \omega^2 & 3 \end{pmatrix} \quad \text{or} \quad \begin{pmatrix} 1 & 0 \\ 0 & 3 \end{pmatrix}.$$

It is easy to see that $\mathbb{U}(M, \mathfrak{O})$ consists of diagonal matrices. The fact $\gamma|_{\mathfrak{c}^\perp} \in \mathbb{U}(M, \mathfrak{O})$ implies that $\mathfrak{a}', \mathfrak{b}'$ are eigenvectors of $\gamma|_{\mathfrak{c}^\perp}$. Therefore $\mathfrak{a}' = \xi\mathfrak{a}$, $\mathfrak{b}' = \eta\mathfrak{b}$, $\xi, \eta \in \mathfrak{O}^*$. This leads to the contradiction $3 = \mathfrak{a}^2 = \mathfrak{a}'^2 = 2$ or 1 and proves 1.4.5.

Taking into account 1.4.3, 1.4.4, 1.4.5, $\partial_\Gamma \mathfrak{B} = \Gamma \cdot \varkappa_1$ (Feustel) and the faithful representation of $\Gamma/\Gamma' \cong S_4$ on $\{\varkappa_1, \varkappa_2, \varkappa_3, \varkappa_4\}$ (see 1.3) we obtain a Γ'-fixed point formation on $\mathfrak{B}^* = \mathfrak{B} \cup \partial_\Gamma \mathfrak{B}$ consisting of four cusps and three elliptic points (see Figure 1.4.A).

1.4.A.

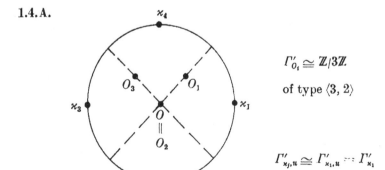

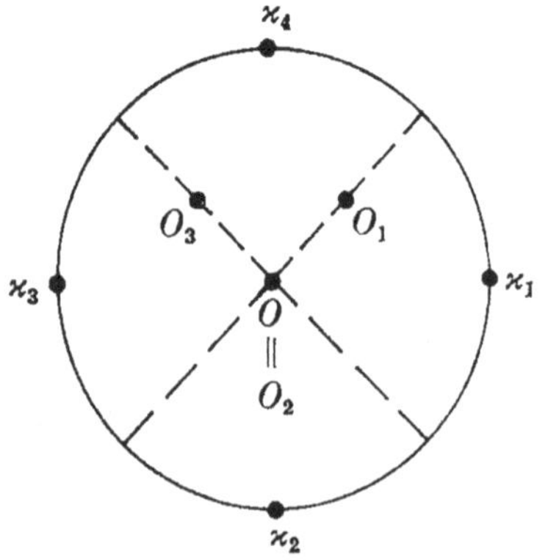

$\Gamma'_{O_1} \cong \mathbb{Z}/3\mathbb{Z}$

of type $\langle 3, 2 \rangle$

$\Gamma'_{\varkappa_j, u} \cong \Gamma'_{\varkappa_1, u} = \Gamma'_{\varkappa_1}$

We say that a cyclic stationary group acting (effectively) on a two-dimensional complex manifold is of *type* $\langle d, e\rangle$ in the corresponding fixed point if its representation in the tangent space at this point is generated by an element with eigenvalues $e^{2\pi i/d}$ and $e^{2\pi ie/d}$.

1.5. Selfintersection of D_{ij} on $\widetilde{\mathfrak{B}/\Gamma'}$

In [33] I proved a proportionality formula for the selfintersection numbers of arithmetic curves $\widetilde{\mathfrak{D}/\Omega}$ on the *filled resolution of singularities* $(\widetilde{\mathfrak{B}/\Omega})_{\mathfrak{D}}$ *of* $\widehat{\mathfrak{B}/\Omega}$ *along* $\mathfrak{D}$. Here $\mathfrak{D}$ is an Ω-rational disc. This means that $\mathfrak{D}$ is the intersection of $\mathfrak{B} \subset \mathbb{P}^2$ and a projective line of $\mathbb{P}^2$ such that $\Omega_{\mathfrak{D}} = N_{\Omega}(\mathfrak{D})/Z_{\Omega}(\mathfrak{D})$ is an arithmetic sublattice of $\mathrm{Aut}_{\mathrm{hol}}\,\mathfrak{D}$, where

$$N_{\Omega}(\mathfrak{D}) = \{\gamma \in \Omega; \gamma\mathfrak{D} = \mathfrak{D}\}, \qquad Z_{\Omega}(\mathfrak{D}) = \{\gamma \in \Omega; \gamma|_{\mathfrak{D}} = \mathrm{id}_{\mathfrak{D}}\}.$$

The filled resolution of singularities is a special smooth model of $\widehat{\mathfrak{B}/\Omega}$, which resolves simultaneously the singularities of $\widehat{\mathfrak{B}/\Omega}$ and of the closed curve $\widehat{\mathfrak{D}/\Omega}$ in $\widehat{\mathfrak{B}/\Omega}$. The conception of filled resolution of singularities is easily generalized to a finite set of Ω-rational discs. In our case I proved in [33] that the filled resolution of singularities of $\widehat{\mathfrak{B}/\Gamma'}$ along $\bigcup D_{ij}$ is nothing else than the minimal resolution of singularities $\widetilde{\mathfrak{B}/\Gamma'}$ of $\widehat{\mathfrak{B}/\Gamma'}$. The main part of the proportionality formula is the Euler-Poincaré volume $c_1(\mathfrak{D}/\Gamma'_{\mathfrak{D}})$ of a fundamental domain of $\Gamma'_{\mathfrak{D}}$. It is equal to $[\Gamma_{\mathfrak{D}} : \Gamma'_{\mathfrak{D}}]\, c_1(\mathfrak{D}/\Gamma_{\mathfrak{D}}) = 2c_1(\mathfrak{D}/\Gamma_{\mathfrak{D}})$. The group $\Gamma_{\mathfrak{D}}$ is isomorphic to $\mathbb{PU}((1, 1), \mathfrak{O}) = \mathbb{PSU}((1, 1), \mathfrak{O})$. Translating this group to the Siegel upper half plane $\mathfrak{H} \simeq \mathfrak{D}$ Feustel found in [17] by the geometric construction of a fundamental domain

$$c_1(\mathfrak{D}/\Gamma_{\mathfrak{D}}) = -\frac{1}{3}.$$

For the sake of completeness we give the idea for another proof using the isometry circle method of Ford described in [41]. The isometry circle of the element $\gamma = \begin{pmatrix} 2 & \omega^2 - 1 \\ \omega - 1 & 2 \end{pmatrix} \in \mathbb{U}((1, 1), \mathfrak{O})$ is

$$K(\gamma)\colon \left| z - \frac{2}{3}(1 - \omega^2) \right| = \frac{1}{\sqrt{3}}.$$

The circle $K(\gamma)$ cuts out a fundamental domain in the unit disc, which is drawn in Figure 1.5.A.

The classical arc formula yields

$$c_1(\mathfrak{D}/\Gamma_{\mathfrak{D}}) = -\frac{1}{2\pi}\left(\pi - \frac{\pi}{3} - 0 - 0\right) = -\frac{1}{3}.$$

From $c_1(\mathfrak{D}/\Gamma'_{\mathfrak{D}}) = -\frac{2}{3}$ and the knowledge of the correction part of the proportionality formula for the selfintersection number (see [33], Theorem 5.1) we deduced

$$(D^2_{ij,\widetilde{\mathfrak{B}/\Gamma'}}) = -1 \qquad \text{(see [33], IV, (2.13))}.$$

We also calculated the Euler number

$$e(D_{ij,\widetilde{\mathfrak{B}/\Gamma'}}) = 2 \qquad \text{(see [33], IV, (2.14))},$$

hence $D_{ij,\widetilde{\mathfrak{B}/\Gamma'}} \cong \mathbb{P}^1$. This is also clear from the fundamental domain in Figure 1.5.A.

1.5.A.

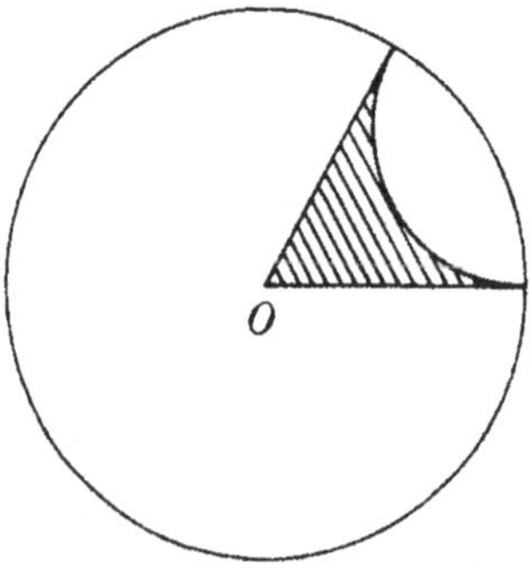

1.6. The Rough Classification of $\widetilde{\mathfrak{B}/\Gamma'}$

Let T_i, $i = 1, \ldots, 4$, be the elliptic curves resolving the cusp singularities $\mathfrak{k}_i$ of $\widehat{\mathfrak{B}/\Gamma'}$. We denote the images of the elliptic points O_i, $i = 1, 2, 3$, lying on $\mathfrak{D}_{i4}$ by O'_i (see Figure 1.4.A). These are the only singularities on $\mathfrak{B}/\Gamma'$. Their minimal resolutions of singularity are two normally crossing smooth rational curves with selfintersection number -2. Altogether we obtain the curve constellation on $\widetilde{\mathfrak{B}/\Gamma'}$ sketched in Figure 1.6.A. This curve will play an important role in our further considerations. We call it the *fundamental curve* on $\widetilde{\mathfrak{B}/\Gamma'}$ and denote it by $\mathfrak{F}$. The selfintersection number of T_i is the same as that of $T'_\varkappa$ in the cusp bundle $\mathbb{F}'_\varkappa$ (see 1.2.a). It was calculated in 1.2.2. A rough version of 1.6.A was found already in [33] (V, (2.15)). It was sufficient for the rough classification of $\widetilde{\mathfrak{B}/\Gamma'}$. The main steps were the following ones: In [32] I proved general proportionality formulas for the Chern invariants c_1^2, c_2 of ball quotient surfaces $\widetilde{\mathfrak{B}/\Omega}$. The main part is the volume $c_2(\mathfrak{B}/\Omega)$ of an Ω-fundamental domain with respect to the Bergmann metric on $\mathfrak{B}$. For $\Omega = \Gamma^{(d)}$ this volume was calculated by number theoretic methods. The correction parts come from Ω-fixed point formations and the corre-

1.6.A.

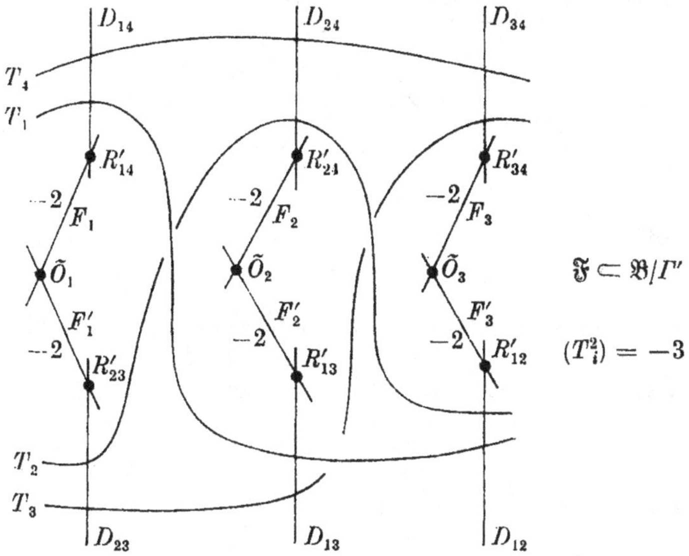

sponding isotropy groups. In our special situation I found

$$c_1^2(\widetilde{\mathfrak{B}/\Gamma'}) = -9, \qquad c_2(\widetilde{\mathfrak{B}/\Gamma'}) = 9 \qquad \text{(see [33], V, (2.16))}.$$

Blowing down the curve

1.6.1. $(D_{14} + F_1 + F_1') + (D_{24} + F_2 + F_2') + (D_{34} + F_3 + F_3')$

we get a smooth surface with vanishing Chern invariants supporting a smooth rational curve, e.g. D_{12}, with selfintersection number 0. By surface classification theory this is only possible for ruled surfaces over an elliptic curve with D_{12} as a fibre (see [33], IV, Prop. 4, V, Prop. 2). So we have

1.6.2. PROPOSITION. $\widetilde{\mathfrak{B}/\Gamma'}$ *is a ruled surface over an elliptic curve* (up to birational equivalence).

§ 2. FINE CLASSIFICATION OF THE SURFACES $\overline{\mathfrak{B}/\Gamma'\cdot H}$

2.1. Trivialization of $\widetilde{\mathfrak{B}/\Gamma'}$

The curve constellation 1.6.A is a trivializing one. That means that we obtain after blowing down some of these curves the trivial elliptic ruled surface $T' \times \mathbb{P}^1$. Indeed, if we blow down the curve (1.6.1) the remaining curve constellation is found in Figure 2.1.A where Q_1', Q_2', Q_3' are the image points of the curve (1.6.1). The elliptic curves T_i are sections of the fibre projection π' onto the elliptic base

2.1.A.

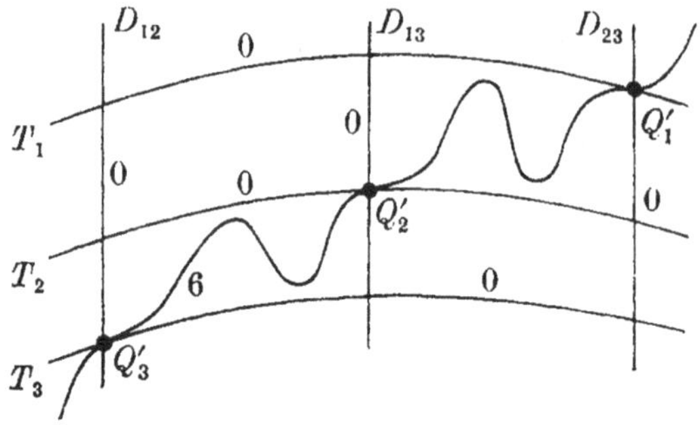

$(T_i \cdot T_4) = (T_i \cdot T_4)_{Q_i'} = 3$

curve T' because the intersection of T_i with a fibre, say D_{12}, is a simple point. So we have $T' \simeq T_i$, $i = 1, \ldots, 4$. The three sections T_1, T_2, T_3 are disjoint. Therefore our surface is nothing else than $T' \times \mathbb{P}^1$. Altogether we have a commutative diagram

2.1.a.

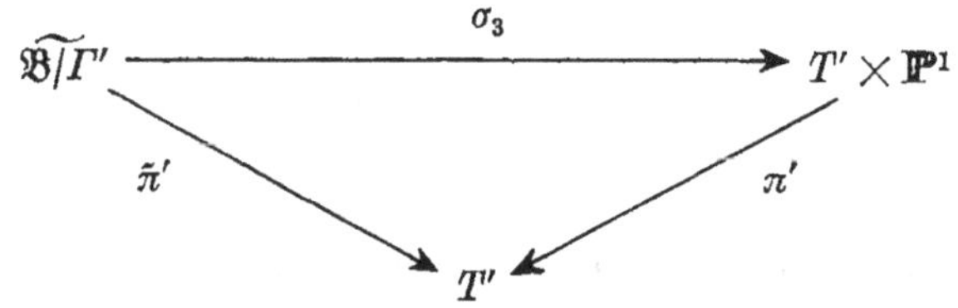

The three exceptional fibres of $\tilde{\pi}'$ are

$$\tilde{\pi}^{-1}\tilde{\pi}'\tilde{O}_i = D_{i4} + D_{jk} + F_i + F_i', \qquad i = 1, 2, 3; \qquad j, k \neq i, 4.$$

Later we will determine the position of Q_1', Q_2', Q_3' on $T' \times \mathbb{P}^1$. This will complete the fine classification of $\widetilde{\mathfrak{B}/\Gamma'}$.

2.2. S_4-Action on $\widetilde{\mathfrak{B}/\Gamma'}$

We give a three-dimensional picture of the dual graph of the fundamental curve (Figure 1.6.A) on $\widetilde{\mathfrak{B}/\Gamma'}$, which will be convenient for our further considerations. We let correspond the elliptic curves T_i to the vertices of a regular tetrahedron. The centers of the six edges of the tetrahedron are the vertices of a regular octahedron. We let the rational curves D_{ij} correspond to them. The points corresponding to F_i, F_i' we place on the three diagonals of the octahedron. We join two of the points corresponding to our curves with a segment if the curves intersect each other on $\widetilde{\mathfrak{B}/\Gamma'}$. In addition we join the points corresponding to D_{ij} and D_{jk}, $i < j < k$. So we obtain the beautiful Figure 2.2.A.

It is easy to deduce from Figures 1.3.A and 1.6.A that the action of $S_4 = \Gamma/\Gamma''$ (see the remark at the end of 1.3) on the set of irreducible components of the fundamental curve $\mathfrak{F}$ in 1.6.A is represented by the action of the group of motions of the tetrahedron in 2.2.A.

2.2.A.

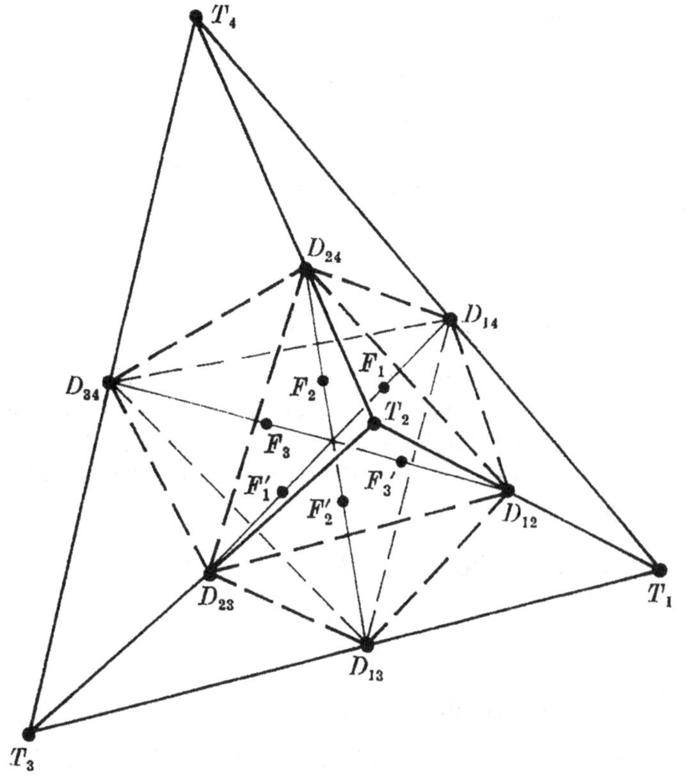

2.3. S_4-Fixed Point Formation on $\widetilde{\mathfrak{B}/\Gamma'}$

The action of S_4 on $\widetilde{\mathfrak{B}/\Gamma'}$ commutes with the projection $\tilde{\pi}' : \widetilde{\mathfrak{B}/\Gamma'} \to T'$ because the fibers and their S_4-images are rational curves. But a rational curve cannot cover the elliptic base curve T'. So we have for each $g \in S_4$ a commutative diagram

$$\begin{array}{ccc} \widetilde{\mathfrak{B}/\Gamma'} & \xrightarrow[g]{\sim} & \widetilde{\mathfrak{B}/\Gamma'} \\ {\scriptstyle \tilde{\pi}'} \downarrow & & \downarrow {\scriptstyle \tilde{\pi}'} \\ T' & \xrightarrow[\bar{g}]{\sim} & T' \end{array}$$

2.3.1. $\quad 1 \to K_4 \to S_4 \to \operatorname{Aut} T'$ *is exact,*

$$g \overset{\psi}{\longmapsto} \bar{g}.$$

Proof. Let $H \subset S_4$ be the kernel of ψ. The subgroups 1, K_4, A_4, S_4 of S_4 are the only normal ones. The kernel H is not equal to A_4 or S_4 because (123) acts effectively on the set of exceptional fibres of $\tilde{\pi}'$ and hence also on T'. On the other hand $H \cap K_4$ cannot be trivial because we have an exact commutative diagram for $i \in \{1, 2, 3\}$:

$$\begin{array}{ccccccc} 1 \to & H & \to & S_{4,\tilde{O}_i} & \to & \psi(S_4)_{\bar{\pi}'(\tilde{O}_i)} & \\ & \cup & & \cup & & \cup & \\ 1 \to & H \cap K_4 & \to & K_4 & \to & \psi(K_4) & \to 1 \end{array}$$

and $\psi(S_4)_{\bar{\pi}'(\tilde{O}_i)}$ is cyclic as an isotropy group of a curve. Therefore $H \cap K_4 \neq 1$ and $H \neq 1$.

2.3.2. LEMMA. *It holds that*

$$\mathrm{Fix}_{\widetilde{\mathfrak{B}/\Gamma'}}(12)\,(34) = E_{(12)(34)} \sqcup \{\tilde{O}_3, R'_{12}, R'_{34}\}.$$

Proof. Let $g = (12)\,(34)$ and $\mathrm{Fix}\, g = (C_1 \sqcup \cdots \sqcup C_r) \sqcup \{R_1, \ldots, R_s\}$ be the fixed point set of g on $\widetilde{\mathfrak{B}/\Gamma'}$, where the C_i's are irreducible (smooth) curves and the R_j's are isolated fixed points of g. The curve C_i cannot be a fibre component. Indeed, the action of g on the components of the exceptional fibres of $\bar{\pi}'$ is (look at the octahedron 2.2.A):

2.3.3. $$D_{14}, F_1, F'_1, D_{23} \mapsto D_{23}, F'_1, F_1, D_{14},$$
$$D_{24}, F_2, F'_2, D_{13} \mapsto D_{13}, F'_2, F_2, D_{24},$$
$$D_{34}, F_3, F'_3, D_{12} \mapsto D_{34}, F_3, F'_3, D_{12}.$$

The action of g on every other fibre $F \cong \mathbb{P}^1$ is effective because

$$g: T_1 \cap F, T_2 \cap F, T_3 \cap F, T_4 \cap F \mapsto T_2 \cap F, T_1 \cap F, T_4 \cap F, T_3 \cap F,$$

and $T_i \cap F$ is a point (see 2.1.A). If F_3 were a component of $\mathrm{Fix}\, g$, then for $h = (14)\,(23)$ we would find

2.3.4. $$F'_3 = hF_3 \subseteq h(\mathrm{Fix}\, g) = \mathrm{Fix}\, hgh^{-1} = \mathrm{Fix}\, g.$$

But F_3 and F'_3 cannot be simultaneous fixed curves of g because they intersect each other. The same conclusion holds with F'_3 instead of F_3. Further D_{34}, and analogeously D_{12}, is not a component of $\mathrm{Fix}\, g$ because the points $T_3 \cap D_{34}$ and $T_4 \cap D_{34}$ are interchanged by the action of g.

The group element g has only one fixed point in the exceptional fibres $\bar{\pi}'^{-1}\bar{\pi}'(\tilde{O}_1)$, $\bar{\pi}'^{-1}\bar{\pi}'(\tilde{O}_2)$, namely $\tilde{O}_1, \tilde{O}_2$. Therefore $r \leq 1$. $\mathrm{Fix}\, g$ is not a finite set of points because it acts effectively on the non-exceptional fibres $F \cong \mathbb{P}^1$, and an element of finite order acting on $\mathbb{P}^1$ has exactly two fixed points. Therefore

$$\mathrm{Fix}\, g = E_{(12)(34)} \sqcup \{R_1, R_2, \ldots, R_s\}.$$

Obviously, $E_{(12)(34)}$ is the projection of a reflection disc $\mathfrak{E}_{(12)(34)} = \mathrm{Fix}_{\mathfrak{B}}(12)\,(34)_{\mathfrak{B}}$. We know the intersections $\tilde{O}_1, \tilde{O}_2$ of $E_{(12)(34)}$ with two of the exceptional fibres. The intersection number is 2. Therefore

$$(E_{(12)(34)} \cdot F) = 2, \qquad F \text{ a non-exceptional fibre}.$$

Hence $\operatorname{Fix} g \cap F = E_{(12)(34)} \cap F$ and the points R_i must lie on the third exceptional fibre. On each of its four components we find two fixed points of g. By (2.3.1) three of the fixed points on the special fibre are $\tilde{O}_3$, R'_{34}, R'_{12}. The other two, sitting on D_{12} and D_{34}, respectively, are denoted by R_{12} and R_{34}. The point R'_{34} does not lie on $E_{(12)(34)}$. Indeed, assume that $R'_{34} \in E_{(12)(34)}$. The element h acts on $E_{(12)(34)}$ as in the right part of (2.3.3). Therefore we would have also $R'_{12} = h(R'_{34}) \in E_{(12)(34)}$. But then

$$(E_{(12)(34)} \cdot \bar{\pi}'^{-1}\bar{\pi}'\tilde{O}_3) = \big(E_{(12)(34)} \cdot (D_{34} + F_3)\big) + \big(E_{(12)(34)} \cdot (D_{12} + F'_3)\big) \geqq 4.$$

On the other hand we know that the intersection number of $E_{(12)(34)}$ and each fibre is 2. This is a contradiction. We can also exclude that $\tilde{O}_3 \in E_{(12)(34)} : \tilde{O}_3 \in E_{(13)(24)}, E_{(14)(23)}$. The group K_4 acts on the tangent space $T_{\tilde{O}_3}$ of $\widetilde{\mathfrak{B}/\Gamma'}$ in $\tilde{O}_3$. The two eigenlines of the non-trivial elements of the abelian group K_4 in $T_{\tilde{O}_3}$ are represented by the curves $E_{(13)(24)}$, $E_{(14)(23)}$ through $\tilde{O}_3$. There is no place for $E_{(12)(34)}$. Therefore the intersection points of $E_{(12)(34)}$ and the exceptional fibre are R_{12} and R_{34}. Lemma 2.3.2 is proved.

By S_4-action we get the curve constellation of Figure 2.3.A.

2.3.5. REMARK. $E_{(ij)(kl)}$ does not intersect the *compactification divisor* $\mathfrak{T} = T_1 + T_2 + T_3 + T_4$ because $(ij)\,(kl)$ acts by index permutation on the set $\{T_1, T_2, T_3, T_4\}$ without fixed element and T_1, T_2, T_3, T_4 are mutually disjoint curves.

2.3.A.

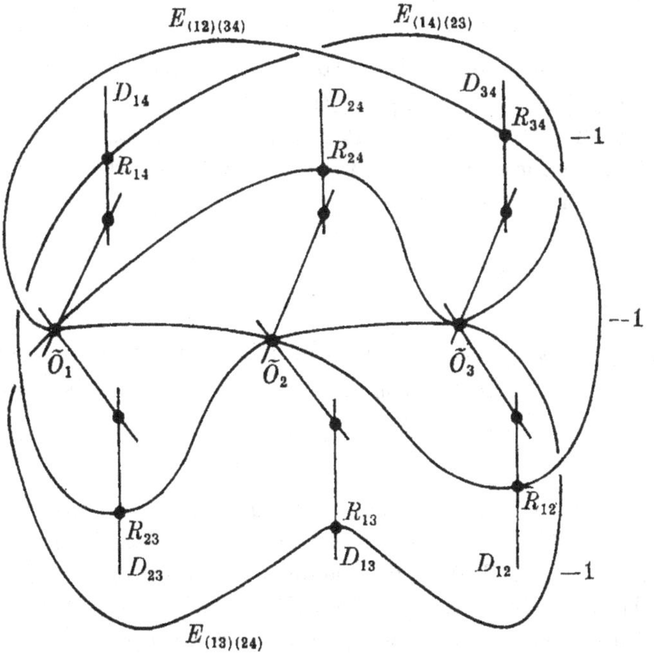

2.3.6. REMARK. K_4 acts on $E_{(ij)(kl)}$ with ineffective kernel $\langle (ij)\,(kl)\rangle \simeq \mathbb{Z}/2\mathbb{Z}$. This results from the right hand part of (2.3.4) and from the action on $\{R_{ij}, R_{kl}\}$.

2.3.7. LEMMA. *It holds that*

$$\text{(i)} \quad g(E_{(ij)(kl)}) = 2, \qquad \text{(ii)} \quad (E^2_{(ij)(kl)}) = -1.$$

Proof. (i) We use the *Hurwitz genus formula* (see [34]):

$$2.3.8. \qquad 2g(C) - 2 = [C : \bar{C}]\Big(2g(\bar{C}) - 2 + \sum_{\bar{P} \in \bar{C}} (e_{\bar{P}} - 1)/e_{\bar{P}}\Big)$$

for Galois or prime degree covering maps $p\colon C \to \bar{C}$ of smooth curves of degree $[C : \bar{C}]$ with ramification index $e_{\bar{P}}$ over $\bar{P}$. Apply (2.3.8) to $C = E = E_{(12)(34)}$ and $p = \bar{\pi}'|_E\colon E \to T'$ of degree 2 with the only ramification points $\tilde{O}_1, \tilde{O}_2$ (see Figure 2.3.A). Then one gets

$$2g(E) - 2 = 2\left(0 + \frac{1}{2} + \frac{1}{2}\right).$$

(ii) We apply the *adjunction formula*

$$2.3.9. \qquad 2g(C) - 2 = \big(C \cdot (K + C)\big)$$

for smooth curves C on a smooth surface with canonical divisor K. A canonical divisor on $\widetilde{\mathfrak{B}/\Gamma'}$ is

$$2.3.10. \qquad K = 3D_{34} + 2F_3 + F_3' - T_1 - T_2.$$

Indeed, $-T_1 - T_2$ is a canonical divisor on $T' \times \mathbb{P}^1$. We get from $T' \times \mathbb{P}^1$ to $\widetilde{\mathfrak{B}/\Gamma'}$ by nine blowing up procedures with centres over Q_1', Q_2', Q_3' (see 2.1.A). So we obtain K in (2.3.10) from $-T_1 - T_2$ by taking inverse images and by adding the exceptional curve for each blowing up. Figure 2.3.A shows that $(E \cdot K) = 3$. And (2.3.4) yields together with (i)

$$2 = (E^2) + 3.$$

2.3.11. LEMMA.

(i) $\{\mathfrak{D}_{14}, \mathfrak{E}_{(14)(23)}\}$ *is the one-dimensional part of a Γ-fixed point formation on* $\mathfrak{B}$.
(ii) $\{D_{14}, E_{(14)(23)}\}$ *is the one-dimensional part of an S_4-fixed point formation on*

$$\widetilde{\mathfrak{B}/\Gamma'} \setminus \left(\sum_{i=1}^{4} T_i + \sum_{j=1}^{3} (F_j + F_j')\right).$$

Proof. First we exclude the possibility that T_i is an S_4-reflection curve. By examination of the octahedron we see that no element of $S_4 \setminus \{1\}$ acts trivially on T_i. If $\gamma \in \Gamma \setminus \{1\}$ is a reflection, then, acting on $\mathbb{C}^3$, it has exactly two different eigenvalues. These eigenvalues are roots of unity of algebraic degree at most 3 over the field $K = \mathbb{Q}(\omega)$. Taking into account that $\operatorname{Tr}\gamma \in K$, $\det\gamma = 1$,

$\omega\gamma$, $\omega^2\gamma \in \Gamma$, we can assume that without loss of generality γ has eigenvalues 1, -1, -1. Hence γ is of order 2. Since

$$\mathfrak{B} \setminus \Gamma' \cdot \{O_1, O_2, O_3\} \to \mathfrak{B}/\Gamma' \setminus \{O_1', O_2', O_3'\}$$
$$= \widetilde{\mathfrak{B}/\Gamma'} \setminus \left(\sum T_i + \sum (F_j + F_j')\right)$$

is unramified (see 1.4) and $\Gamma \to S_4$ is surjective (see 1.3.1), it is easily seen that $g \in S_4$ has a reflection curve on $\widetilde{\mathfrak{B}/\Gamma'}$ iff one of its preimages $\gamma \in \Gamma$ is a reflection on $\mathfrak{B}$ (see [32], Lemma 2.5.6). Therefore, if g has a fixed curve, then g is of order 2. The elements of order 2 are just those of the conjugacy classes C_2, C_5 (see (1.3.2)). The one-dimensional components of Fix g, $g \in C_2$, are the curves $E_{(ij)(kl)}$ by Lemma 2.3.2. Now it suffices to prove that D_{14} is the only one-dimensional component of $\mathrm{Fix}_{\widetilde{\mathfrak{B}/\Gamma'}}(23)$. By 2.3.1 $\overline{(23)} = \psi(23)$ acts effectively on T'. Therefore each irreducible curve of Fix (24) is a $\tilde{\pi}'$-fibre component. If (23) acts on a non-exceptional fibre F, then it acts effectively on F because it interchanges the intersection points $F \cap T_3$, $F \cap T_2$. The fibre $D_{14} + F_1 + F_1' + D_{23}$ is the only exceptional fibre, where (23) fixes pointwise an irreducible component (look at the octahedron 2.2.A). The curve D_{14} is a component of Fix (23) because $\mathfrak{D}_{14}$ is the reflection disc of $(23)_{\mathfrak{B}}$ on $\mathfrak{B}$, but D_{23} is not a component because (23) interchanges $T_2 \cap D_{23}$ and $T_3 \cap D_{23}$. This proves (ii) and also (i).

2.3.12. Fix (234) $= \emptyset$.

Proof. $\overline{(234)} = \psi(234)$ acts effectively on T' by 2.3.1. It has no fixed point on T' because otherwise there would be a fixed point of (234) on T_1. Now consider $\mathbb{Z}/3\mathbb{Z} \cong \Gamma_{\varkappa,u}/\Gamma_\varkappa' \hookrightarrow S_3$, $\varkappa = \varkappa_1$, to see that (234) is represented by a unipotent element $(234)_{\mathfrak{B}} \in \Gamma_{\varkappa,u}$. The action of (234) on T_1 is the same as the action of $(234)_{\mathfrak{B}}$ on $T_\varkappa'$ in the cusp bundle $\mathbb{F}_\varkappa'$ because $\mathbb{F}_\varkappa'$ and $\widetilde{\mathfrak{B}/\Gamma'}$ are locally isomorphic near $T_\varkappa'$ and T_1, respectively (see 1.2.a, $G_\varkappa = 1$). The Diagram 1.2.b shows that $(234)_{\mathfrak{B}}$ has no fixed point on $T_\varkappa'$. The same is true for (234) on T_1. Hence $\overline{(234)}$ acts without fixed points on T'. Going back to $\widetilde{\mathfrak{B}/\Gamma'}$ by means of π' we get result 2.3.12.

In order to determine all isolated fixed points of elements of S_4 acting on $\widetilde{\mathfrak{B}/\Gamma'} \setminus \sum_{i=1}^{3} (F_i + F_i')$ we introduce special fibres of $\tilde{\pi}'$

$$F_{ji} = F_{ij} = F_{kl}, \qquad (ijkl) \in S_4,$$

in the following manner: (ij) acts on the exceptional fibre $D_{ij} + \cdots + D_{kl}$ and interchanges the two other exceptional fibres. The Hurwitz genus formula (2.3.8) applied to $T' \to T'/\langle\overline{(ij)}\rangle$ yields

$$0 = 2\left(2g(T'/\langle\overline{(ij)}\rangle) - 2 + \frac{1}{2} + \cdots + \frac{1}{2}\right).$$

So $\overline{(ij)}$ has exactly four fixed points on T': $\bar{\pi}'(D_{ij} + \cdots + D_{kl})$, P_{ij}, P'_{ij}, P''_{ij}.

We denote the corresponding (non-exceptional) fibres by F_{ij}, F'_{ij}, F''_{ij}. Since $\overline{(ij)} = \overline{(kl)}$ (see 2.3.1) we have $F_{ij} = F_{kl}$. So we have found nine special fibres

2.3.13.
$$\begin{matrix} F_{14}, F'_{14}, F''_{14}; \\ F_{24}, F'_{24}, F''_{24}; \\ F_{34}, F'_{34}, F''_{34} \end{matrix}$$

which are different from each other. For example we know that $\overline{(24)}$ and $\overline{(34)}$ have no common fixed point on T' because $\overline{(34)} \cdot \overline{(24)} = \overline{(234)}$ has no fixed point on T'. Observe that

2.3.14. (123) *acts transitively on each column of* (2.3.13).

2.3.15.
$$\mathrm{Fix}_{\widetilde{\mathfrak{B}/\Gamma'} \setminus \sum_{n=1}^{3}(F_n+F'_n)}(i, j) = \{D_{kl}\} \sqcup \{R_{ij}, F_{ij} \cap T_k, F_{ij} \cap T_l, F'_{ij} \cap T_k, F'_{ij} \cap T_l, F''_{ij} \cap T_k, F''_{ij} \cap T_l\}.$$

Proof. Take for example $(i, j) = (2, 4)$. On $D_{24} + D_{13}$ we find the isolated fixed points on D_{24}. The permutation (24) acts on $E_{(13)(24)}$. Therefore $R_{24} = E_{(13)(24)} \cap D_{24}$ is a fixed point. An element of order 2 acting effectively on $\mathbb{P}^1$ has exactly two fixed points. The point R'_{24} is the second one on D_{24}. In the same manner one finds the other isolated fixed points of (24) as intersection points of F_{24}, F'_{24}, F''_{24} and T_1, T_3.

2.3.16.
$$\mathrm{Fix}_{\widetilde{\mathfrak{B}/\Gamma'}}(ijk4) = \{E_{(ik)(j4)} \cap F_{ik}, E_{(ik)(j4)} \cap F'_{ik}, E_{(ik)(j4)} \cap F''_{ik}, \tilde{O}_j\}.$$

Proof. $(ijk4)$ acts on $E_{(ik)(j4)}$. On the other hand $\overline{(ijk4)} = \overline{(ik)}$ by 2.3.1. So $(ijk4)$ acts (effectively) on F_{ik}, F'_{ik}, F''_{ik} and there are no other non-exceptional fibres with $(ijk4)$-action. By 2.3.2 it holds that

$$\mathrm{Fix}\,(ijk4) \subset \mathrm{Fix}\,(ijk4)^2 = \mathrm{Fix}\,(ik)\,(j4) = E_{(ik)(j4)} \cup \{\tilde{O}_j, R'_{ik}, R'_{j4}\}.$$

Therefore the only possible fixed points on the exceptional fibres are $\tilde{O}_j$, R'_{ik}, R'_{j4}, R_{ik}, R_{j4}. But $(ijk4)$ properly moves all of them except $\tilde{O}_j$.

2.3.B.

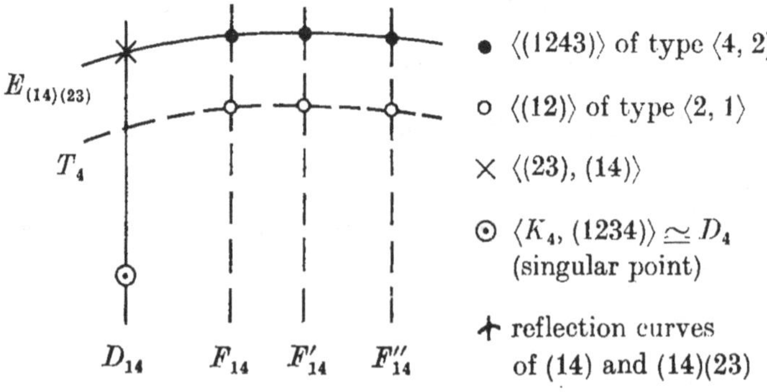

Altogether we found the S_4-fixed point formation on $\overline{\mathfrak{B}/\Gamma'}$ together with the corresponding isotropy groups described in Figure 2.3.B.

If we pull back the situation to $\mathfrak{B}$ taking into consideration that $\mathfrak{B} \to \mathfrak{B}/\Gamma'$ is unramified outside of O_1', O_2', O_3', then we get Feustel's *special fixed point formation* on $\mathfrak{B}$ (see [17]):

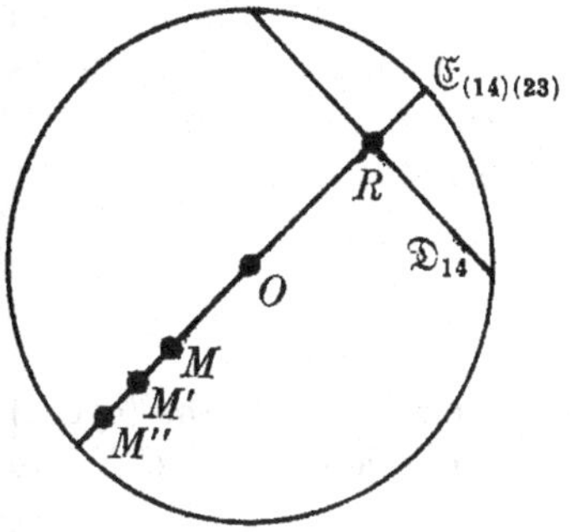

$$\Gamma_0 = \left\langle \begin{pmatrix} \omega^2 & 0 & 0 \\ 0 & \omega & 0 \\ 0 & 0 & 1 \end{pmatrix}, (1243)_{\mathfrak{B}}, (12)(34)_{\mathfrak{B}} \right\rangle$$

$\Gamma_R \cong K_4$

$\Gamma_M, \Gamma_{M'}, \Gamma_{M''}$ of type $\langle 4, 2\rangle$

$Z_\Gamma(\mathfrak{D}_{14}), Z_\Gamma(\mathfrak{C}_{(14)(23)})$ of type $\langle 2, 0\rangle$

2.4. *Classification of* $\overline{\mathfrak{B}/\Gamma'\cdot K_4}$

First we blow up the nine points R_{ij}', $\tilde{O}_l$ on $\widetilde{\mathfrak{B}/\Gamma'}$. We denote the surface thus obtained by $\widetilde{\widetilde{\mathfrak{B}/\Gamma'}}$ and the corresponding projection onto T' by $\tilde{\tilde{\pi}}'$. Then we have the configuration around the exceptional fibre $\tilde{\tilde{\pi}}'^{-1}\tilde{\pi}'\tilde{O}_1$ described in Figure 2.4.A.

2.4.A.

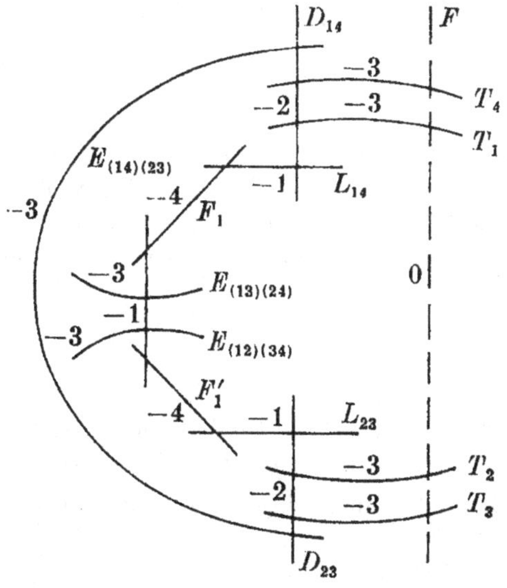

We denote the reduced curve underlying to the inverse image of $\mathfrak{F}$ by $\widetilde{\mathfrak{F}}$ and call it the *extended fundamental curve* (on $\widetilde{\mathfrak{B}/\Gamma'}$). The irreducible components of $\widetilde{\mathfrak{F}}$ are the S_4-images of the irreducible curves in 2.4.A (without the non-exceptional fibre F). We can take the quotient of the surface $\widetilde{\mathfrak{B}/\Gamma'}$ by K_4. The isotropy group $K_{4,P}$ of any point $P \in \widetilde{\mathfrak{B}/\Gamma'}$ is generated by reflections or trivial (see Lemma 2.3.2). Therefore the quotient surface $\widetilde{\mathfrak{B}/\Gamma'} \cdot K_4 = \widetilde{\mathfrak{B}/\Gamma'}/K_4$ is smooth. Furthermore, no of the irreducible curves C in 2.4.A has a K_4-contact point. This means that $gC \cap C = \emptyset$ or $gC = C$ for $g \in K_4$. So the *selfintersection formula* (see [33], II, Theorem 8)* for quotient curves on quotient surfaces is reduced in our case to

2.4.1. $(C/G)^2 = \dfrac{|Z|^2}{|N|}\,(C^2) - \sum\limits_{C \ni P \bmod N} e_P/d_P, \qquad G = K_4,$

where $Z = \{g \in G; \ g|_C = \mathrm{id}_C\}$, $N = \{g \in G; \ gC = C\}$, and $\langle d_P, e_P \rangle$ is the (cyclic) type of N_P/Z_P. Using the old notation C also for the quotient curve C/K_4 we obtain Figure 2.4.B for the image of $\widetilde{\mathfrak{F}}$ on $\widetilde{\mathfrak{B}/\Gamma'} \cdot K_4$. The curves $E_{(ij)(kl)}$ are sections of the projection $\widetilde{\mathfrak{B}/\Gamma'} \cdot K_4 \to T'$. Therefore they are isomorphic to the elliptic curve T'. So we have the same situation as in Figure 1.6.A. This yields the following commutative diagram of $\langle(123)\rangle$-equivariant maps:

2.4.a.

$$
\begin{array}{ccc}
 & \widetilde{\mathfrak{B}/\Gamma'} \cdot K_4 & \\
\sum\limits_{i=1}^{3}(L_i + F_i + L_{i4}) \swarrow & & \searrow \sum\limits_{i=1}^{3}(F_i + D_{i4} + L_{i4}) \\
\overline{\mathfrak{B}/\Gamma'} \cdot K_4 = \overline{\mathfrak{B}/\Gamma' \cdot K_4} & & T' \times \mathbb{P}^1 \\
 & \searrow \quad \swarrow & \\
 & T' &
\end{array}
$$

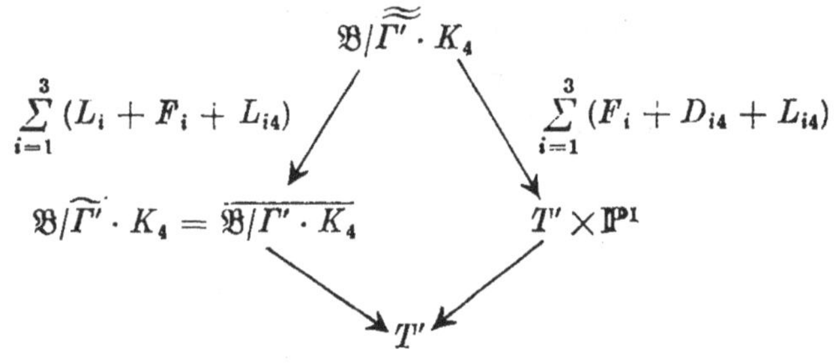

The divisors which mark the arrows consist of the exceptional curves of the corresponding birational morphisms. The notations $\overline{\mathfrak{B}/\Gamma'} \cdot K_4$, $\mathfrak{B}/\Gamma' \cdot K_4$ agree with those of 1.2.b because $L_i + F_i + L_{i4}$ is the result of the successive blowing up of the image point of $O_i \in \mathfrak{B}$ along $\mathfrak{B} \to \mathfrak{B}/\Gamma' \cdot K_4$, and we see that it is a regular point.

2.5. Classification of $\overline{B/\Gamma' \cdot A_4}$

From 2.3.12 we know that $\langle(123)\rangle$ acts freely on the base curve T' of the ruled surfaces in 2.4.a. Hence it acts freely on $\overline{\mathfrak{B}/\Gamma' \cdot K_4}$. The group $\langle(123)\rangle$ acts transitively on $\{E_{(14)(23)}, E_{(13)(24)}, E_{(12)(34)}\}$. Thus, using (2.4.1) again (for $G = \langle(123)\rangle$),

---

* See (3.6.2), p. 71.

2.4.B.

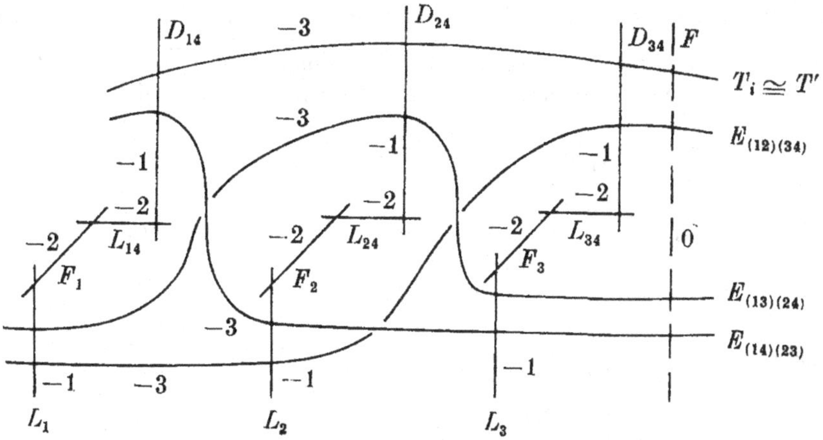

the image of our curve $\tilde{\tilde{\mathfrak{F}}}$ together with the special fibres (2.3.13) on the (smooth) quotient surface $\widetilde{\widetilde{\mathfrak{B}/\Gamma'}} \cdot A_4 = \widetilde{\widetilde{\mathfrak{B}/\Gamma'}} \cdot K_4/\langle(123)\rangle$ over $T = T'/\langle(123)\rangle$ is 2.4.C where we omitted some indices. Blowing down $D + L_{14} + F_1$ one gets $T' \times \mathbb{P}^1/\langle(123)\rangle = T \times \mathbb{P}^1$. Blowing down $L_1 + F_1 + L_{14}$ one gets $\overline{\mathfrak{B}/\Gamma' \cdot A_4}$. Indeed, we can take the quotient by $\langle(123)\rangle$ of the whole Diagram 2.4.a.

2.4.C.

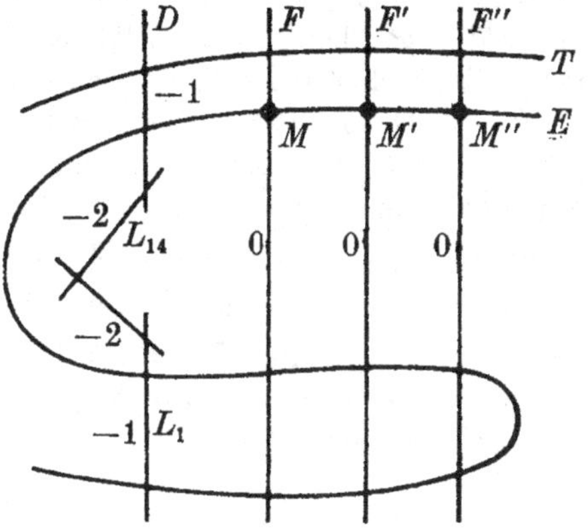

2.6. Classification of $\widehat{\mathfrak{B}/\Gamma'}$

We take the quotient of $\overline{\mathfrak{B}/\Gamma' \cdot A_4}$ by $\langle(12)\rangle$. The resulting surface is $\overline{\mathfrak{B}/\Gamma}$. From 2.3.15 it follows that

$$\mathrm{Fix}_{\overline{\mathfrak{B}/\Gamma \cdot A_4}}(12) = D \cup \{M, M', M'', F \cap T, F' \cap T, F'' \cap T\}.$$

Therefore $\overline{\mathfrak{B}/\Gamma}$ supports exactly six singularities of type $\langle 2, 1\rangle$. The minimal resolution of these singularities yields $\widetilde{\mathfrak{B}/\Gamma}$. Forgetting E we obtain with obvious notations the Figure 2.6.A. The surface $\widetilde{\mathfrak{B}/\Gamma}$ is fibred over $\mathbb{P}^1 \simeq \overline{T} = T/\langle(12)\rangle$ with exactly three exceptional fibres. Blowing down 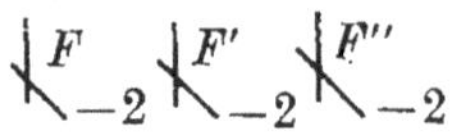

we obtain the trivial ruled surface $\mathbb{P}^1 \times \mathbb{P}^1$. Therefore

$$c_1^2(\widetilde{\mathfrak{B}/\Gamma}) = c_1^2(\mathbb{P}^1 \times \mathbb{P}^1) - 6 = 8 - 6 = 2.$$

Blowing down 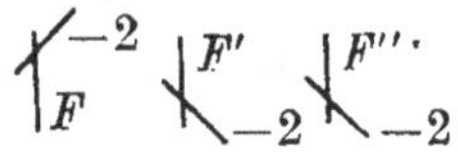

and $\overline{T}$ we get a smooth rational surface with $c_1^2 = 9$. This is only possible for $\mathbb{P}^2$.

2.6.A.

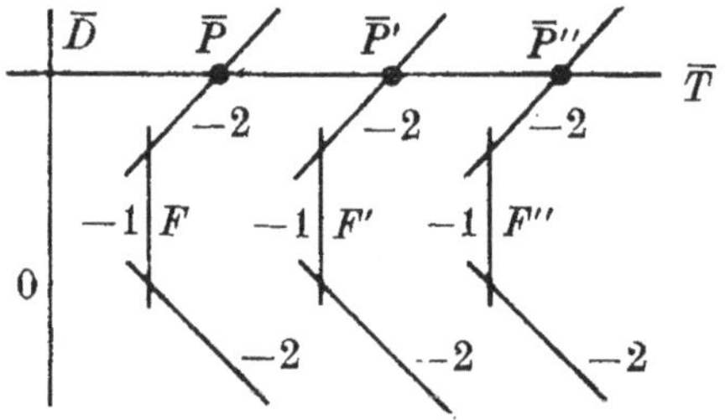

Inversely choose three points Q, P', P'' on $\mathbb{P} = \mathbb{P}^2$ in general position. Blow up one of these points, say Q, and denote the resulting exceptional curve by $\overline{T}$. The resulting surface is $\mathbb{F}_1$ fibred over $\overline{T} \simeq \mathbb{P}^1$. The points P', P'' lie on different fibres over $\overline{T}$. Choose a point $\overline{P}$ on $\overline{T} \subset \mathbb{F}_1$ such that $\overline{P}, P', P''$ project onto three different points on $\overline{T}$. Blow up $\overline{P}, P', P''$. Denote the proper transforms of the fibres through $\overline{P}, P', P''$ on $\mathbb{F}_1$ by Φ, Φ', Φ'' and the exceptional curves by F_1, F_2, F_3. Now blow up the intersection points $\Phi \cap F_1$, $\Phi' \cap F_2$, $\Phi'' \cap F_2$. We denote the resulting surface by $\tilde{\mathbb{P}}$. The construction of $\tilde{\mathbb{P}}$ is unique up to isomorphism. We have the curve constellation on $\tilde{\mathbb{P}}$ described in Figure 2.6.B. Comparing this with 2.6.A we see that $\widetilde{\mathfrak{B}/\Gamma} = \tilde{\mathbb{P}}$. Define $\hat{\mathbb{P}}$ and $\dddot{\mathbb{P}}$ by Diagram 2.6.a of birational morphisms, where, as usual in this book, the divisors marking the arrows indicate the curves being contracted:

2.6.a.

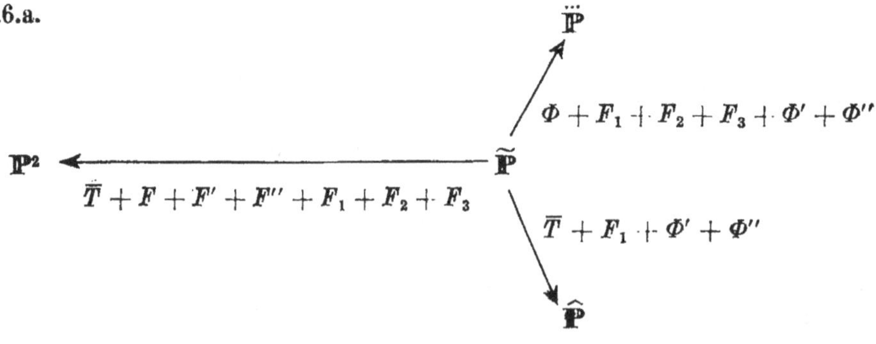

2.6.B.

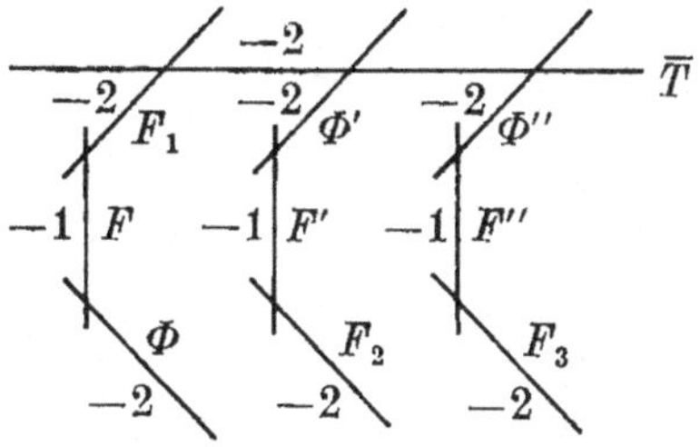

Altogether we have the following classification result:

2.6.1. PROPOSITION. *It holds that*

$$\widetilde{\mathfrak{B}/\Gamma} = \tilde{\mathbb{P}}, \qquad \overset{\cdots\cdots}{\mathfrak{B}/\Gamma} = \dddot{\mathbb{P}}, \qquad \widehat{\mathfrak{B}/\Gamma} = \hat{\mathbb{P}}.$$

2.6.2. COROLLARY ([29], I.9.6) *The cusp singularity of* $\widehat{\mathfrak{B}/\Gamma}$ *is the quaternion singularity.*

The quaternion singularity is the surface quotient singularity isomorphic to the singularity of $\mathbb{C}^2/Q$, where Q is the *quaternion group* $\left\langle \begin{pmatrix} i & 0 \\ 0 & -i \end{pmatrix}, \begin{pmatrix} 0 & 1 \\ -1 & 0 \end{pmatrix} \right\rangle$.
Its minimal resolution is isomorphic to the curve $\overline{T} + F_1 + \Phi' + \Phi''$ in 2.6.B.

2.6.3. REMARK. We can determine the exact position of $\overline{D}$ in 2.6.A. The curve $\overline{D}$ is related to $\overline{P}, \overline{P}', \overline{P}'' \in \overline{T} \simeq \mathbb{P}^1$ in the following manner: $\overline{D} \cap \overline{T}, \overline{P}, \overline{P}', \overline{P}''$ are the branch points of the covering map $T \to \overline{T}$ in the commutative diagram

$$\begin{array}{ccc}
\overline{\mathfrak{B}/\Gamma'' \cdot A_4} & \to & \overset{\cdots\cdots}{\mathfrak{B}/\Gamma} \\
\uparrow & & \uparrow \\
T = T_\varkappa & \to \overline{T} = & T_\varkappa/G_\varkappa \simeq \mathbb{P}^1 \qquad G_\varkappa \simeq \mathbb{Z}/2\mathbb{Z} \\
\downarrow & & \downarrow \\
\mathbf{F}_\varkappa & \to & \overset{\cdots\cdots}{\mathbf{F}_\varkappa/G_\varkappa}
\end{array}$$

Taking into consideration that $T = \mathbb{C}/\mathfrak{D}$ (see (1.2.1)) and that the ramification points are the 2-torsion points on T one can easily calculate the double ratio of the four branch points on $\overline{T} \simeq \mathbb{P}^1$. It is $-\omega$.

2.7. Summary

There is an interesting global connection between our surfaces $\overline{\mathfrak{B}/\Gamma' \cdot H}$ and the cusp bundles $\mathbf{F}_\varkappa, \mathbf{F}'_\varkappa$. If $\mathfrak{L}$ is a geometric line bundle over a curve C corresponding to the locally free sheaf $\mathscr{L}$ of rank 1 over C, then we denote the ruled surface $\mathbb{P}(\mathcal{O}_C \oplus \mathscr{L})$ by $\overline{\mathfrak{L}}$. On a ruled surface over C there is at most one section with negative selfintersection number (see e.g. [44], II, § 3). If it exists, then we

2.7.a.

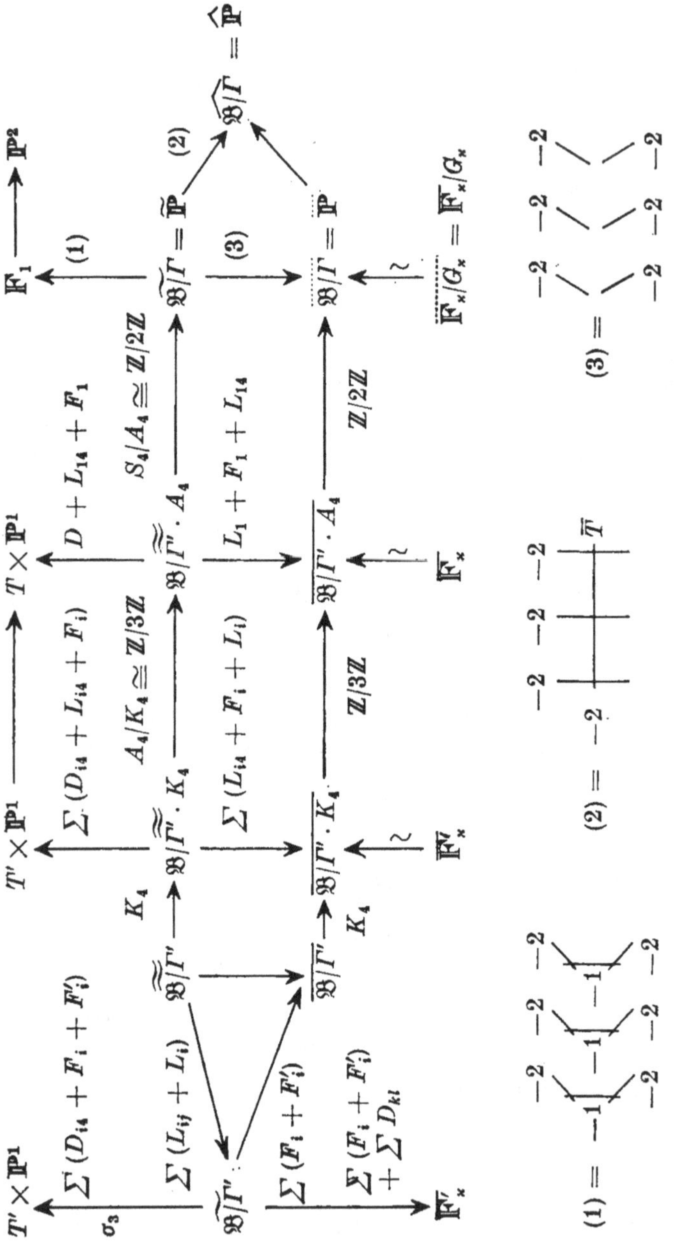

T' × P¹
σ₃
Σ (D_{i4} + F_i + F'_i)
Σ (L_{ij} + L_i)
Σ (F_i + F'_i)
Σ (F_i + F'_i) + Σ D_{kl}
F'_x
K₄
T' × P¹
Σ (D_{i4} + L_{i4} + F_i)
A₄/K₄ ≅ Z/3Z
Σ (L_{i4} + F_i + L_i)
Z/3Z
F'_x
T × P¹
D + L_{14} + F_1
S₄/A₄ ≅ Z/2Z
L_1 + F_1 + L_{14}
Z/2Z
F_x
F_1
T̄
P²
(1)
(2)
(3)
F_x/G_x = F_x/G_x
P̂
(1) =
−1 −1 −1
−2 −2 −2
−2 −2 −2
(2) =
−2 −2 −2 −2
T̄
(3) =
−2 −2 −2
−2 −2 −2

call it the 0-section and denote it also by C. The following lemma is an easy consequence of the classification results for elliptic and rational ruled surfaces (see e.g. [27]).

2.7.1. LEMMA. *If X is a ruled surface over an elliptic or rational curve with negative 0-section C, then it is isomorphic to $\overline{\mathbb{N}_{X/C}}$, where $\mathbb{N}_{X/C}$ is the (geometric) normal bundle of $C \subset X$.*

2.7.2. COROLLARY. *If X and Y are two ruled surfaces over an elliptic or rational curve with negative 0-sections C and if they are locally isomorphic near C, then they are globally isomorphic.*

For our proper normal subgroups H of S_4 we have local isomorphisms of $\widetilde{\mathfrak{B}/\Gamma'} \cdot H$ and $\overline{\mathbb{F}}_\varkappa$ or $\overline{\mathbb{F}}'_\varkappa$ near $T = T_\varkappa$ and $T' = T'_\varkappa$, respectively (see Diagram 1.2.b). Therefore we can find a global isomorphism from $\mathbb{F}_\varkappa$ and $\mathbb{F}'_\varkappa$, respectively, onto a suitable minimal (ruled) model of $\widetilde{\mathfrak{B}/\Gamma'} \cdot H$. A complete picture of the trivializations of our surfaces $\widetilde{\mathfrak{B}/\Gamma'} \cdot H$ and the connection with the cusp bundles is given in the summarizing Diagram 2.7.a (see p. 52).

2.8. *Fine Classification of $\widetilde{\mathfrak{B}/\Gamma'}$*

Let T' be the elliptic curve defined by the equation

$$WY^2 = X^3 - W^3$$

in the projective plane $\mathbb{P}^2$ with homogeneous coordinates $(w : x : y)$. Let $\varphi\colon T' \to \mathbb{P}^1$ be a 3-sheeted cyclic covering map. Such maps exist. For example one can take the restriction to T' of the rational projection $(w : x : y) \mapsto (w : y)$. By the Hurwitz genus formula (2.3.8) there are exactly three ramification points $t_1, t_2, t_3 \in T'$. Let $T_i = T' \times c_i$, $c_i \in \mathbb{P}^1$, be three different horizontal sections of the projection map $T' \times \mathbb{P}^1 \to T'$ onto the first factor. Now we define

$$D_{12} = t_3 \times \mathbb{P}^1, \qquad D_{13} = t_2 \times \mathbb{P}^1, \qquad D_{23} = t_1 \times \mathbb{P}^1, \qquad Q'_k = (t_k, c_k), \quad k = 1, 2, 3.$$

Then we have a situation as in 2.1.A. The construction of the divisor $T_1 + T_2 + T_3 + D_{12} + D_{13} + D_{23}$ together with Q'_1, Q'_2, Q'_3 on $T' \times \mathbb{P}^1$ is unique up to isomorphism. Indeed $\mathbb{P}\mathrm{Gl}_2(\mathbb{C})$ acts transitively on the set of triples of simple points on $\mathbb{P}^1$. Therefore our curve constellation does not depend on the choice of $c_1, c_2, c_3 \in \mathbb{P}^1$. On the other hand there exists up to isomorphism exactly one 3-sheeted cover over $\mathbb{P}^1$ with branch locus consisting of three different simple points. This is a one-dimensional special case of a general result (see 3.3.1 below). So T', φ and t_1, t_2, t_3 are uniquely determined up to isomorphism by the 3-sheeted cyclic cover condition over $\mathbb{P}^1$. The next step is the construction of T_4. Without loss of generality we can assume that $c_1 = \infty$. $c_2 = 0 \in \mathbb{A}^1 \subset \mathbb{P}^1$. By the Riemann-

Roch theorem we know that

$$H^0(T', 3t_1 - 3t_2) \subseteq H^0(T', 3t_1 - 2t_2) \simeq \mathbb{C}.$$

The first space is not 0 because there is a rational function f on $\mathbb{P}^1$ with divisor $(f) = \varphi(t_1) - \varphi(t_2)$; hence $\varphi^*(f) = \mathfrak{p}'$ lies in $H^0(T', 3t_1 - 3t_2) \simeq \mathbb{C}$. The function $\mathfrak{p}'$ defines a 3-sheeted covering map $T' \to \mathbb{P}^1$ $(t \mapsto \mathfrak{p}'(t))$; $\mathfrak{p}' \in H^0(T', 3t_1 - 3t_2)$ is uniquely determined by the condition $\mathfrak{p}'(t_3) = c_3$. By Hurwitz' genus formula (2.3.8) $\mathfrak{p}': T' \to \mathbb{P}^1$ has exactly three ramification points. Obviously, t_1 and t_2 are two of them. In order to see that t_3 is the third ramification point we return to the special situation at the beginning of this section. There we find $\mathfrak{p}' = \dfrac{Y}{W} - \mathrm{i}$, $t_1 = (0:0:1)$, $t_2 = (1:0:\mathrm{i})$, $t_3 = (1:0:-\mathrm{i})$, $c_3 = -2\mathrm{i}$ and the covering map defined by $\mathfrak{p}'$ is nothing else than φ. Going back isomorphically to the general situation we define T_4 as the graph of $\mathfrak{p}': T' \to \mathbb{P}^1$ embedded in $T' \times \mathbb{P}^1$. Then T_4 is a section of $\pi': T' \times \mathbb{P}^1 \to T'$ with $(T_k \cdot T_4) = (T_k \cdot T_4)_{Q'} = 3$ as sketched in Figure 2.1.A. Each section of π' defines a rational function on T'. The intersection condition is another formulation of the ramification condition above for the corresponding function. So T_4 is uniquely determined by the conditions in 2.1.A.

Now by nine blowing up procedures — three over each of the points Q'_k, $k = 1, 2, 3$ — we can resolve in a unique manner the singularities of the curve $T_1 + T_2 + T_3 + T_4$ on $T' \times \mathbb{P}^1$. We denote the resulting surface by $T' \widetilde{\times} \mathbb{P}^1$ and the morphism $T' \widetilde{\times} \mathbb{P}^1 \to T' \times \mathbb{P}^1$ by σ'_3.

2.8.1. PROPOSITION. *Up to isomorphism the surfaces $T' \widetilde{\times} \mathbb{P}^1$ and $\widetilde{\mathfrak{B}/\Gamma'}$ and the morphisms σ'_3 and σ_3 are the same. The reduced inverse image of $\sum_{i=1}^{4} T_i + \sum_{k<l} D_{kl} \in \mathrm{Div}\,(T' \times \mathbb{P}^1)$ by σ'_3 is the fundamental curve $\mathfrak{F}$ on $T' \widetilde{\times} \mathbb{P}^1 = \widetilde{\mathfrak{B}/\Gamma'}$.*

Proof. After the consideration above it suffices to show that the points $\pi' Q'_k = \tilde{\pi}' \tilde{O}_k$ (see Diagram 2.1.a) are the ramification points of a 3-sheeted cyclic covering map $T' \to \mathbb{P}^1$. Consider the element $\varrho = \begin{pmatrix} \omega & 0 & 0 \\ 0 & 1 & 0 \\ 0 & 0 & 1 \end{pmatrix}$. It defines morphisms

2.8.2. $\hat{\varrho}: \widehat{\mathfrak{B}/\Gamma'} \to \widehat{\mathfrak{B}/\Gamma'}$, $\quad \tilde{\varrho}: \widetilde{\mathfrak{B}/\Gamma'} \to \widetilde{\mathfrak{B}/\Gamma'}$.

By consideration of the rational fibres of $\tilde{\pi}'$ we see that $\tilde{\varrho}$ induces an isomorphism $\bar{\varrho} = \psi(\tilde{\varrho})$ on T' of order 3. In order to see that this action is not trivial we look at the cusp bundle $\mathbb{F}'_{\varkappa_2}$. The element $\varrho \in \tilde{\Gamma}'_{\varkappa_2}$ is not unipotent. Hence it acts effectively on the 0-section $T_2 = T'_{\varkappa_2}$ of $\mathbb{F}'_{\varkappa_2} = \mathbb{F}_{\varkappa_2}(\Gamma')$. By local isomorphy $\tilde{\varrho}$ acts effectively on $T_2 \in \mathrm{Div}\,(\widetilde{\mathfrak{B}/\Gamma'})$. So it cannot act trivially on the set of fibres of $\tilde{\pi}'$. The element ϱ acts trivially on the disc $\mathfrak{D}_{24}$, hence $\tilde{\varrho}$ acts trivially on D_{24}. Now D_{24} is a component of an exceptional fibre of $\tilde{\pi}'$. So $\tilde{\varrho}'(\tilde{O}_2)$ is a fixed point of $\bar{\varrho}$ and $\tilde{\varrho}$ acts on the set of three exceptional fibres. The $\bar{\varrho}$-image of $\pi'(\tilde{O}_1)$ is not equal to $\pi'(\tilde{O}_3)$ because $\bar{\varrho}$ has order 3. So the points $\pi'(\tilde{O}_k)$, $k = 1, 2, 3$, are the three ramification points of $T' \to T'/\langle\bar{\varrho}\rangle = \mathbb{P}^1$, q.e.d.

2.8.3. REMARK. The notation $\mathfrak{p}'$ for the function defining the section T_4 comes from the derivative of the classical *Weierstrass $\wp$-function*. Indeed, after a coordinate change T' becomes of equation type $Y^2 = 4X^3 - 4$. This equation is satisfied by $X = \wp$ and $Y = \wp'$. So we can regard T_4 in 2.1.A as the graph of the Weierstrass $\wp'$-function. It determines the whole curve constellation 2.1.A.

§ 3. THE LOGARITHMIC CANONICAL MAP Φ_1

3.1. The Image of the Logarithmic Canonical Map

3.1.1. DEFINITION. Let $\Omega = \Omega^2_{\widetilde{\mathfrak{B}/\Gamma'}}$, be the *canonical sheaf* on $\widetilde{\mathfrak{B}/\Gamma'}$.

$$\Omega(\mathfrak{T}) = \Omega \otimes \mathcal{O}(\mathfrak{T}), \qquad \mathcal{O} = \mathcal{O}_{\widetilde{\mathfrak{B}/\Gamma'}},$$

is called the *logarithmic canonical sheaf* of $\widetilde{\mathfrak{B}/\Gamma'}$ *along the compactification divisor* $\mathfrak{T} = T_1 + T_2 + T_3 + T_4$. The *m-th logarithmic pluricanonical sheaf* $\Omega^m(m\mathfrak{T})$ is the m-th tensor power of $\Omega(\mathfrak{T})$.

Each divisor $D > 0$ on a smooth surface X defines a rational map

$$\Phi_D\colon X \dashrightarrow \mathbb{P}^{\dim|D|},$$

where $|D| = \{E \in \operatorname{Div} X;\ E \equiv D,\ E \geqq 0\}$ is the complete linear system of the divisor class of D. The rational map defined by the class of $\Omega^m(m\mathfrak{T})$, $m > 0$, is denoted by Φ_m:

$$\Phi_m\colon \widetilde{\mathfrak{B}/\Gamma'} \dashrightarrow \mathbb{P}^{\dim H^0(\Omega^m(m\mathfrak{T}))-1}.$$

We call it the mth *logarithmic pluricanonical map of* $\widetilde{\mathfrak{B}/\Gamma'}$ (*with respect to* $\mathfrak{T}$). For $m = 1$ we call it the *logarithmic canonical map*.

3.1.2. LEMMA. *Let K be a canonical divisor of $\widetilde{\mathfrak{B}/\Gamma'}$. Then for $m \geqq 1$ the following condition are satisfied:*

(a) *$\Omega^m(m\mathfrak{T})$ is generated by its global sections.*

(b) *$|mK + m\mathfrak{T}|$ has no base points.*

(c) *Φ_m is a morphism.*

Proof. By (2.3.10) we know that

$$T_3 + T_4 + F_3' + 2F_3 + 3D_{34}$$

is a logarithmic canonical divisor. As

$$T_1 + T_2 + F_3 + 2F_3' + 3D_{12} \quad \text{and} \quad T_2 + T_4 + F_2' + 2F_2 + 3D_{24}$$

enjoy the same properties we see that they also are logarithmic canonical divisors (see 1.6.A). These three curves have no common point. This proves (b) for $m = 1$. Multiplying the three divisors by $m \geq 1$ we see that (b) holds in general. The equivalence of (a) and (b) is an easy and well-known consequence of the definitions. Property (c) is a consequence of (b).

3.1.3. PROPOSITION. *It holds that*

(a) $\dim H^0\big(\widetilde{\mathfrak{B}/\Gamma''}, \Omega(\mathfrak{T})\big) = 3$,

(b) $\Phi_1(\widetilde{\mathfrak{B}/\Gamma'}) = \mathbb{P}^2$,

(c) $\deg \Phi_1 = 3$.

The proof consists of a chain of six intermediate steps.

3.1.4. *Let C be an irreducible curve on $\widetilde{\mathfrak{B}/\Gamma''}$. Then $\Phi_1(C)$ is a point if and only if $\big((K + \mathfrak{T}) \cdot C\big) = 0$.*

This is a well-known general fact for morphisms $\Phi_D \colon X \to \mathbb{P}^N$ corresponding to a divisor D on a surface X. Indeed, if $\Phi_D(C)$ is a point, then we can choose a hyperplane H of $\mathbb{P}^N$ not containing $\Phi_D(C)$ ($N \geq 2$, see 3.1.9). The inverse image of H is an element of $|D|$. Now C and $\Phi_D^{-1}(H)$ are disjoint, hence $(D \cdot C) = 0$. If $\Phi_D(C)$ is a curve, then any hyperplane $H\big(\neq \Phi_D(C)\big)$ intersects $\Phi_D(C)$ properly in at least one point. Going back to X by means of Φ_D we find $(D \cdot C) > 0$.

3.1.5. *If $\big((K + \mathfrak{T}) \cdot C\big) = 0$, then C is a component of the fundamental curve $\mathfrak{F}$.*

Let C be an irreducible curve on $\widetilde{\mathfrak{B}/\Gamma'}$, which is not a component of $\mathfrak{F}$. Then $\sigma_3(C)$ is a curve on $T' \times \mathbb{P}^1$ (see 2.1). The curve $\sigma_3(C)$ intersects $T_3 + D_{12}$ properly. Going back to $\widetilde{\mathfrak{B}/\Gamma'}$ we see that C intersects $T_3 + D_{34} + D_{12} + F_3 + F_3'$ properly. But then it properly intersects one of the logarithmic canonical divisors

$$T_3 + T_4 + F_3' + 2F_3 + 3D_{34}, \qquad T_1 + T_2 + F_3 + 2F_3' + 3D_{12}.$$

Thus $\big((K + \mathfrak{T}) \cdot C\big) > 0$.

3.1.6. *$\Phi_1(\widetilde{\mathfrak{B}/\Gamma'})$ is a surface.*

This follows immediately from 3.1.4 and 3.1.5, because there are only a finite number of curves on $\widetilde{\mathfrak{B}/\Gamma'}$ which are contracted to a point by Φ_1.

3.1.7. $\big((K + \mathfrak{T}) \cdot D_{ij}\big) = 1, \qquad \big((K + \mathfrak{T}) \cdot T_l\big) = 0,$

$$\big((K + \mathfrak{T}) \cdot F_k\big) = \big((K + \mathfrak{T}) \cdot F_k'\big) = 0, \qquad (K + \mathfrak{T})^2 = 3.$$

This follows immediately from the adjunction formula (2.3.9), Figure 1.6.A and $K^2 = -9$ (see 1.6).

3.1.8. *$\Phi_1(C)$ is a point if and only if C is a component of* $\mathfrak{T} + \sum_{i=1}^{3} (F_i + F'_i)$.

This follows easily from 3.1.4, 3.1.5 and 3.1.7.

3.1.9. $3 \leqq \dim H^0\big(\widetilde{\mathfrak{B}/\Gamma'}, \mathcal{O}(K + \mathfrak{T})\big) \leqq 4.$

Consider the long exact sequence

$$0 \to H^0(K) \to H^0(K + \mathfrak{T}) \to H^0\big(\mathfrak{T}, (K + \mathfrak{T})|_{\mathfrak{T}}\big) \to H^1(K) \to \cdots$$

of the exact sequence of sheaves

$$0 \to \mathcal{O}(K) \to \mathcal{O}(K + \mathfrak{T}) \to \mathcal{O}_{\mathfrak{T}}\big(\mathcal{O}(K + \mathfrak{T})|_{\mathfrak{T}}\big) \to 0.$$

Now $\widetilde{\mathfrak{B}/\Gamma''}$ is birationally equivalent to a ruled surface over an elliptic curve. Therefore $H^0(K) = 0$, $H^1(K) \cong \mathbb{C}$. Furthermore

$$\mathcal{O}(K + \mathfrak{T})|_{\mathfrak{T}} = \bigoplus_{i=1}^{4} \mathcal{O}(K + \mathfrak{T})|_{T_i} = \bigoplus_{i=1}^{4} \mathcal{O}(K + T_i)|_{T_i} = \bigoplus_{i=1}^{4} \omega_{T_i} = \bigoplus_{i=1}^{4} \mathcal{O}_{T_i}.$$

Therefore $H^0\big(\mathfrak{T}, (K + \mathfrak{T})|_{\mathfrak{T}}\big) \cong \mathbb{C}^4$. The rest is clear.

Proof of 3.1.3. By 3.1.9 Φ_1 can be realized by four sections $f_0, f_1, f_2, f_3 \in H^0(K + \mathfrak{T})$. At least three of them are linearly independent. We can write

$$\Phi_1 : \widetilde{\mathfrak{B}/\Gamma'} \xrightarrow[(f_0:f_1:f_2:f_3)]{} \mathbb{P}^3.$$

Intersecting $\Phi_1(\widetilde{\mathfrak{B}/\Gamma''})$ with a hyperplane H of $\mathbb{P}^3$, going back to $\widetilde{\mathfrak{B}/\Gamma''}$ by taking inverse images and using the degree formula for intersection numbers of two curves we find

$$3 = (K + \mathfrak{T})^2 = \deg \Phi_1 \cdot \deg \Phi_1(\widetilde{\mathfrak{B}/\Gamma''}).$$

$\deg \Phi_1 = 1$ and $\deg \Phi_1(\widetilde{\mathfrak{B}/\Gamma'}) = 3$ is not possible. Indeed, in this case $\Phi_1(\widetilde{\mathfrak{B}/\Gamma''})$ would be a surface of degree 3 in $\mathbb{P}^3$, hence a rational surface (see e.g. [27], V, Remark 4.7.1). On the other hand Φ_1 would then be a birational morphism from the "elliptic ruled" surface $\widetilde{\mathfrak{B}/\Gamma'}$ onto the rational surface $\Phi_1(\widetilde{\mathfrak{B}/\Gamma'})$, which is impossible. Therefore $\deg \Phi_1 = 3$ and $\deg \Phi_1(\widetilde{\mathfrak{B}/\Gamma'}) = 1$. So f_0, f_1, f_2, f_3 are linearly dependent and $\Phi_1(\widetilde{\mathfrak{B}/\Gamma'}) \cong \mathbb{P}^2$.

3.2. The Branch Locus of Φ_1

By 3.1.8 we know the exceptional curve of the morphism Φ_1. It is the part $\mathfrak{T} + \sum_{i=1}^{3} (F_i + F'_i)$ of the fundamental curve $\mathfrak{F}$ coming from the resolution of singularities of $\widehat{\mathfrak{B}/\Gamma'}$. Therefore Φ_1 factorizes through $\widehat{\mathfrak{B}/\Gamma'}$:

3.2.a.

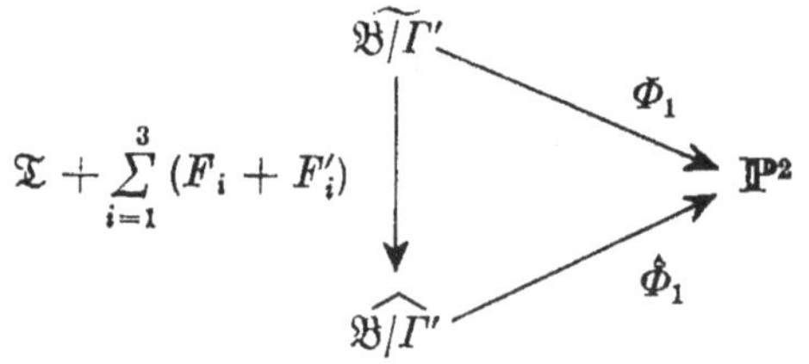

3.2.1. $\hat{\Phi}_1$ *is a finite morphism.*

Obviously $\hat{\Phi}_1$ is quasifinite. Then it has to be finite by the Zariski main theorem.

Now we consider the automorphisms $\hat{\varrho}$, $\tilde{\varrho}$ of order 3 on $\widetilde{\mathfrak{B}/\Gamma''}$ and $\widehat{\mathfrak{B}/\Gamma''}$, respectively, which were introduced in (2.8.2). It is easy to see that ϱ commutes with the elements of $\tilde{\Gamma}$ modulo Γ'. Therefore

3.2.2. $\tilde{\Gamma}/\Gamma' \cong \langle\tilde{\varrho}\rangle \times (\Gamma/\Gamma') \cong \langle\tilde{\varrho}\rangle \times S_4 = (\mathbb{Z}/3\mathbb{Z}) \times S_4.$

As we have seen in 2.8 $\tilde{\varrho}$ acts trivially on D_{24}. Applying S_4 to D_{24} we see that $\tilde{\varrho}$ acts trivially on each D_{ij}. Indeed, if $P \in D_{ij}$, $P = gQ$, $g \in S_4$, $Q \in D_{24}$, then

$$\tilde{\varrho}P = \tilde{\varrho}gQ = g\tilde{\varrho}Q = gQ = P.$$

The automorphism $\hat{\varrho}$ acts on the singular locus of $\widehat{\mathfrak{B}/\Gamma'}$. Because $\tilde{\varrho}$ acts trivially on D_{ij}, the automorphism $\hat{\varrho}$ does not move any singularity on $\widehat{\mathfrak{B}/\Gamma'}$ (see 1.6.A). It follows that $\tilde{\varrho}$ preserves and hence acts on each irreducible component of $\mathfrak{F}$. In particular $\tilde{\varrho}$ does not move the logarithmic canonical divisor (see (2.3.10))

$$B = T_3 + T_4 + F_3' + 2F_3 + 3D_{34}.$$

Consequently $\tilde{\varrho}$ is compatible with the logarithmic canonical map $\Phi_1 = \Phi_B$: $\widetilde{\mathfrak{B}/\Gamma'} \to \mathbb{P}^2$. Taking into account Diagram 3.2.a we can descend to $\widehat{\mathfrak{B}/\Gamma'}$; more precisely, we have the following commutative diagram:

3.2.b.

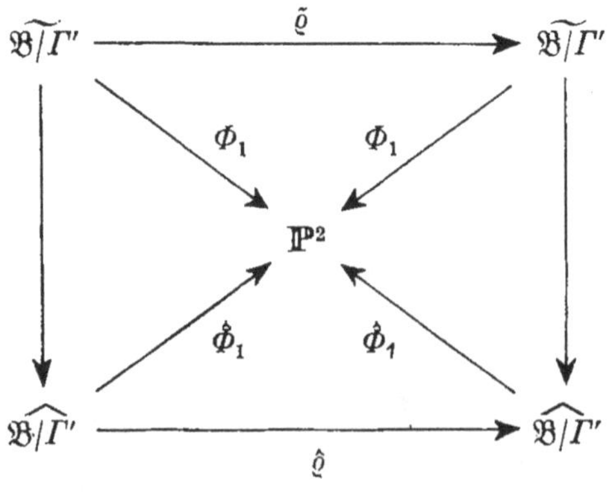

The universal property of quotient morphisms yields a commutative diagram

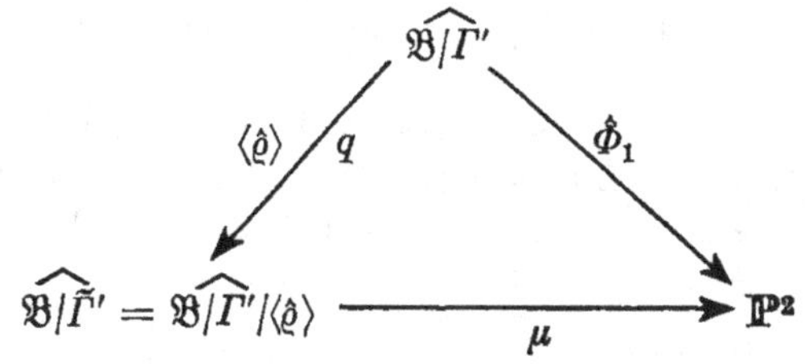

Observe that $\deg q = 3 = \deg \Phi_1 = \deg \hat{\Phi}_1$ by 3.1.3(c). Hence $\deg \mu = 1$. This means that μ is a birational morphism. Furthermore μ is quasifinite by 3.2.1. This is only possible if μ is an isomorphism. We have proved

3.2.3. PROPOSITION. *$\hat{\Phi}_1 : \widehat{\mathfrak{B}/\Gamma'} \to \mathbb{P}^2$ is the quotient morphism for the cyclic group $\langle\hat{\varrho}\rangle$ of order 3 acting on $\widehat{\mathfrak{B}/\Gamma'}$ and $\widehat{\mathfrak{B}/\tilde{\Gamma}'} = \mathbb{P}^2$.*

We are now going to determine the branch locus of $\hat{\Phi}_1$ or Φ_1.

3.2.4. *Φ_1 is ramified along $\sum_{i<j} D_{ij}$.*

This is an easy consequence of the fact that $\tilde{\varrho}$ acts trivially on D_{ij}, of (3.1.8), and of Proposition 3.2.3.

Now we will verify that Φ_1 is ramified nowhere else. More precisely it is true that

3.2.5. $\Phi_1 \mid \widetilde{\mathfrak{B}/\Gamma'} \setminus \mathfrak{F} = \hat{\Phi}_1 \mid \widehat{\mathfrak{B}/\Gamma'} \setminus \{\hat{\kappa}_1, \hat{\kappa}_2, \hat{\kappa}_3, \hat{\kappa}_4, O'_1, O'_2, O'_3\}$ *is unramified.*

Proof. Let P be a point of $\widetilde{\mathfrak{B}/\Gamma''} \setminus \mathfrak{F}$. We have to show that $\tilde{\varrho}(P) \neq P$. The automorphism $\bar{\varrho} : T' \xrightarrow{\sim} T'$ moves the point $\bar{\pi}'(P)$ properly because the only fixed points of $\bar{\varrho}$ are the projections of the three exceptional fibres of $\bar{\pi}'$. For this fact we refer the reader to the proof of 2.8.1 above. Now 3.2.4 follows immediately.

Summerizing what has been done so far, we have proved the main part of the following

3.2.6. PROPOSITION. *The fundamental curve $\mathfrak{F}$ in 1.6.A is the ramification locus of $\Phi_1 : \widetilde{\mathfrak{B}/\Gamma'} \to \mathbb{P}^2$. The branch locus $\Delta = \sum_{1 \le i < j \le 4} \Delta_{ij}$ is the image of $\sum_{i<j} D_{ij}$ and $\Delta_{ij} = \Phi_1(D_{ij})$ is the projective line on $\mathbb{P}^2$ going through the two points $P_i = \Phi_1(T_i)$, $P_j = \Phi_1(T_j)$. The points P_1, P_2, P_3, P_4 are in general position on $\mathbb{P}^2$. Let Q_i, $i = 1, 2, 3$, be the image point $\Phi_1(F_i + F'_i) = \hat{\Phi}_1(O'_i)$. The projective line going through Q_i, Q_j, $i \neq j$, is the image of $E_{(ij)(k4)}$.*

A picture of this situation is sketched in 3.4.A, p. 64.

Proof. It is easy to see that $\Phi_1(D_{ij})$ and $\Phi_1(E_{(ij)(kl)})$ have exactly one common point, namely $\Phi_1(R_{ij})$ (see 2.3.A). The image curves intersect transversally in this point. By Bezout's theorem Δ_{ij} and $\Phi_1(E_{(ij)(kl)})$ are curves of degree 1 on $\mathbb{P}^2$. These are the projective lines.

In order to see that P_1, P_2, P_3, P_4 are in general position, we must show, by definition, that no three of them lie on a projective line. Now assume that e.g. P_1, P_2, P_3 lie on a projective line L in $\mathbb{P}^2$. This line must then be equal to $\Delta_{12}, \Delta_{13}, \Delta_{23}$ at the same time. Then we have

$$\Phi_1(D_{12}) = \Phi_1(D_{13}) = \Phi_1(D_{23}) = L.$$

But Φ_1 is an isomorphism on $\sum_{i<j} D_{ij}$ by 3.2.3 and 3.2.4. Consequently $D_{12} = D_{13} = D_{23}$. This is a contradiction.

The last statement of the proposition comes from Figure 2.3.A.

3.3. Characterization of $\widehat{\mathfrak{B}/\Gamma'}$ as a Branched Cyclic Cover over $\mathbb{P}^2$

Next we state the well-known *cyclic cover theorem*. A proof can found in [42]. We will only prove part (a) by an explicit construction. This will be done later in 4.3 because there it will be needed directly. In the present section we just want to give an elegant formulation of the preceding results.

3.3.1. THEOREM. *Let V be a smooth algebraic variety, $d \geq 2$ a natural number, Δ a reduced effective divisor on V such that its class in* Pic V *is divisible by d. Then*

(a) *There exist d-sheeted cyclic covers $V(\delta) \to V$, totally branched over Δ and nowhere else.*

(b) *These covers $V(\delta)$ are in one-to-one correspondence with the dth roots of Δ in* Pic V, i.e., *divisor classes $\bar{\delta}$ satisfying $d \cdot \delta \equiv \Delta$ (linear equivalence), $\delta \in \bar{\delta}$.*

(c) *An automorphism $\varphi: V \to V$ lifts to $V(\delta)$ iff*

(i) $\varphi^*(\Delta) = \Delta$ *and* (ii) $\varphi^*(\bar{\delta}) = \bar{\delta}$.

If φ lifts, then it lifts in exactly d different ways.

We apply this theorem to our situation

$$V := \mathbb{P}^2, \qquad \Delta = \sum_{1 \leq i < j \leq 4} \Delta_{ij}, \qquad d = 3,$$

or more generally to the case where Δ is the reduced divisor consisting of the six projective lines obtained by taking all the lines going through exactly two of four given points P_1, P_2, P_3, P_4 in general position on $\mathbb{P}^2$. Since

$$\Delta \equiv 6H, \qquad H \text{ a projective line on } \mathbb{P}^2,$$

the assumptions of Theorem 3.3.1 are satisfied. The divisor class $\bar{\delta}$ in (b) is uniquely determined; namely $\delta \equiv 2H$, because Pic $\mathbb{P}^2 = \mathbb{Z} \cdot \bar{H}$. Therefore the cyclic cover $\mathbb{P}^2(\delta) \to \mathbb{P}^2$ is uniquely determined by Δ.

3.3.2. PROPOSITION. *The following covers are isomorphic:*

a) *The unique cyclic cover* $\mathbb{P}^2(2H) \to \mathbb{P}^2$ *branched along* Δ;

b) $\hat{\Phi}_1 : \widehat{\mathfrak{B}/\Gamma'} \to \mathbb{P}^2$ *in Diagram* 3.2.a;

c) *the quotient map* $\widehat{\mathfrak{B}/\Gamma'} \to \widehat{\mathfrak{B}/\tilde{\Gamma}'}$ *of* $\tilde{\Gamma}'/\Gamma' = \mathbb{Z}/3\mathbb{Z}$.

The equivalence of c) and b) is the statement of Proposition 3.2.3. The remaining part follows from Proposition 3.2.6 and the cyclic cover theorem 3.3.1. The independence from the choice of the four points $P_1, P_2, P_3, P_4 \in \mathbb{P}^2$ in general position follows from 3.3.3 below.

As a corollary we give an interpretation of $\Gamma/\Gamma' \subset \mathrm{Aut}\, \widehat{\mathfrak{B}/\Gamma'}$ in terms of the automorphism group of $\mathbb{P}^2$. We use the well-known

3.3.3. MAIN THEOREM of (elementary) projective geometry. *For two given sets of* $n + 2$ *points in general position on* $\mathbb{P}^n$

$$\mathsf{A} = \{A_1, A_2, \ldots, A_{n+2}\}, \qquad \mathsf{B} = \{B_1, B_2, \ldots, B_{n+2}\}$$

there is exactly one automorphism of $\mathbb{P}^n$ *extending the map*

$$A_1 \mapsto B_1,\ A_2 \mapsto B_2, \ldots, A_{n+2} \mapsto B_{n+2}.$$

Especially we have for $\mathsf{B} = \mathsf{A}$

3.3.4. $\mathrm{Aut}\,(\mathbb{P}^n; \{A_1, A_2, \ldots, A_{n+2}\}) \cong S_{n+2}$ (symmetric group).

The left-hand side consists of all (projective) automorphisms of $\mathbb{P}^n$, which permute the elements of the set $\{A_1, A_2, \ldots, A_{n+2}\}$.

3.3.5. COROLLARY. $\Gamma/\Gamma' \subset \mathrm{Aut}\, \widehat{\mathfrak{B}/\Gamma'}$ *is a lift of* $\mathrm{Aut}\,(\mathbb{P}^2; \{P_1, P_2, P_3, P_4\}) = S_4$ *along* $\hat{\Phi}_1 : \widehat{\mathfrak{B}/\Gamma'} \to \mathbb{P}^2$; *the full lift is* $\tilde{\Gamma}/\Gamma'$.

Proof. By interpretation c) in Proposition 3.3.2 of our 3-sheeted cover of $\mathbb{P}^2$ we have an action of $\Gamma/\Gamma' \cong \tilde{\Gamma}/\tilde{\Gamma}'$ on $\mathbb{P}^2$. Indeed, we have the commutative diagram of quotient morphisms

3.3.a.

$$\begin{array}{ccc} \widehat{\mathfrak{B}/\Gamma'} & \xrightarrow{\ \Gamma/\Gamma'\ } & \widehat{\mathfrak{B}/\Gamma} \\ {\scriptstyle\langle\hat{\varrho}\rangle}\big\downarrow{\scriptstyle\hat{\Phi}_1} & & \big\downarrow{\scriptstyle\langle\hat{\varrho}\rangle} \\ \mathbb{P}^2 = \widehat{\mathfrak{B}/\tilde{\Gamma}'} & \xrightarrow[\ \tilde{\Gamma}/\tilde{\Gamma}'\ ]{} & \widehat{\mathfrak{B}/\tilde{\Gamma}} \end{array}$$

The action of Γ/Γ' descends to $\mathbb{P}^2 = \widehat{\mathfrak{B}/\tilde{\Gamma}'}$ via the identification $\Gamma/\Gamma' = \tilde{\Gamma}/\tilde{\Gamma}'$. It acts effectively on $\mathbb{P}^2 = \widehat{\mathfrak{B}/\Gamma'}$. Each element of Γ/Γ' acts by permutations on

$$\{P_1, P_2, P_3, P_4\} = \{\hat{\Phi}_1(\hat{\mathfrak{K}}_1), \hat{\Phi}_1(\hat{\mathfrak{K}}_2), \hat{\Phi}_1(\hat{\mathfrak{K}}_3), \hat{\Phi}_1(\hat{\mathfrak{K}}_4)\} = \{\Phi_1(T_1)., \ldots, \Phi_1(T_4)\}$$

(see 2.2.A, 3.1.8). Therefore

$$\Gamma/\Gamma' = \tilde{\Gamma}/\tilde{\Gamma}' = \mathrm{Aut}\,(\mathbb{P}^2; \{P_1, P_2, P_3, P_4\}) = \mathrm{Aut}\,(\mathbb{P}^2; \Delta) = S_4.$$

Now each element of Aut $(\mathbb{P}^2; \Delta)$ lifts in exactly three different ways to $\widehat{\mathfrak{B}/\Gamma'}$ by Theorem 3.3.1(c). Using $\hat{\varrho}$ instead of $\tilde{\varrho}$ in (3.2.2) the last statement of Corollary 3.3.5 is also clear.

3.3.6. REMARK. It is important enough to point out specifically at this time that $\mathbb{P}^2 \setminus \{4 \text{ points } P_i$ in general position$\}$ *is covered by the ball* $\mathfrak{B}$. *The covering map is étale outside the curve* Δ. The situation is completely described by the commutative Diagram 3.3.b.

3.3.b.

3.3. b

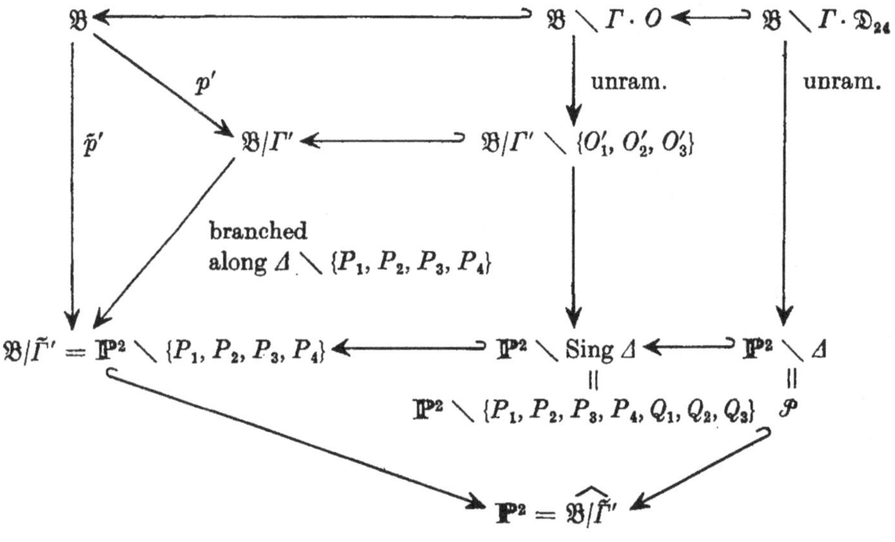

3.3.7. PROPOSITION. *It holds that*

$$\widehat{\mathfrak{B}/\tilde{\Gamma}} \cong \mathbb{P}^2/\mathrm{Aut}\,(\mathbb{P}^2; \{P_1, P_2, P_3, P_4\}) = \mathbb{P}^2/S_4.$$

Proof. This is immediately clear from 3.3.a and 3.3.5.

3.4. The Fixed Point Formation of the Full Eisenstein Lattice

3.4.1. DEFINITION. An S_{n+2}-*coordinate system on* $\mathbb{P}^n$ is an ordered $(n+2)$-tuple $(P_1, P_2, \ldots, P_{n+2})$ of $n+2$ points in *general position* on $\mathbb{P}^n$. This means that no $n+1$ points of this tuple lie in a hyperplane of $\mathbb{P}^n$.

The group S_{n+2} acts on $\mathbb{P}^{n+1}$ by permutation of the canonical homogeneous coordinates. It acts on $\mathbb{P}^n$ as $\operatorname{Aut}(\mathbb{P}^n; \{P_1, P_2, \ldots, P_{n+2}\})$ (see (3.3.4)).

$$P_1 \mapsto \big(-(n+1) : 1 : \ldots : 1\big), \quad P_2 \mapsto \big(1 : -(n+1) : 1 : \ldots : 1\big), \quad \ldots,$$
$$P_{n+2} \mapsto \big(1 : \ldots : 1 : -(n+1)\big)$$

extends uniquely to an S_{n+2}-equivariant embedding $\mathbb{P}^n \hookrightarrow \mathbb{P}^{n+1}$. The image is the hyperplane $\sum_{i=1}^{n+2} x_i = 0$. So we can let correspond to each point $P \in \mathbb{P}^n$ a coordinate $(n+2)$-tuple $\big(x_1(P) : \ldots : x_{n+2}(P)\big)$ with $\sum_{i=1}^{n+2} x_i(P) = 0$.

3.4.2. DEFINITION. $\big(x_1(P) : \ldots : x_{n+2}(P)\big)$ is called the *S_{n+2}-coordinate tuple* of P with respect to $(P_1, P_2, \ldots, P_{n+2})$.

3.4.3. LEMMA. *Let (P_1, P_2, P_3, P_4) be an (ordered) quadruple of points in general position on $\mathbb{P}^2$. Denote by $\overline{PQ}$ the projective line going through $P, Q \in \mathbb{P}^2$, $P \neq Q$, and by $\operatorname{Fix}^1 g$ the one-dimensional part of $\operatorname{Fix} g$, $g \in S_4 = \operatorname{Aut}(\mathbb{P}^2; \{P_1, P_2, P_3, P_4\})$. Using S_4-coordinates with respect to this quadruple we set:*

$$\Delta_{ij} = \overline{P_iP_j} = \operatorname{Fix}^1(kl) = \{(x_1 : x_2 : x_3 : x_4) \in \mathbb{P}^2;\ x_k = x_l\},\ i, j \neq k, l,$$
$$\Delta_{(ij)(kl)} = \operatorname{Fix}^1(ij)(kl)$$
$$= \{(x_1 : x_2 : x_3 : x_4) \in \mathbb{P}^2;\ x_i + x_j = x_k + x_l = 0\},$$
$$Q_1 = (1 : -1 : -1 : 1),\quad Q_2 = (1 : -1 : 1 : -1),\quad Q_3 = (1 : 1 : -1 : -1),$$
$$M = (1 : i : -i : 1),\quad S = (1 : \omega : \omega^2 : 0),\quad R = (1 : 0 : 0 : -1).$$

Then $\{\Delta_{14}, \Delta_{(14)(23)}\} \sqcup \{P_1, Q_2, R, M, S\}$ is an S_4-fixed point formation on $\mathbb{P}^2$. The isotropy groups are of the following types:

$$Z_{S_4}(\Delta_{ij}) = \langle (kl) \rangle \cong Z_{S_4}(\Delta_{(ij)(kl)}) = \langle (ij)(kl) \rangle \cong \mathbb{Z}/2\mathbb{Z},$$
$$S_{4,P_1} = \langle (34), (23), (24) \rangle \cong S_3, \quad \textit{generated by reflections},$$
$$S_{4,Q_2} = \langle (13), (24), (14)(23) \rangle \cong D_4, \quad \textit{generated by reflections},$$
$$S_{4,R} = \langle (13), (24) \rangle \cong K_4, \quad \textit{generated by reflections},$$
$$S_{4,M} = \langle (1243) \rangle \quad \textit{of type } \langle 4, 2 \rangle,$$
$$S_{4,S} = \langle (123) \rangle \quad \textit{of type } \langle 3, 2 \rangle.$$

Proof. We only have to check one representative of each of the four conjugacy classes C_2, C_3, C_4, C_5 (see (1.3.2)). Obviously

$$\operatorname{Fix}(23) = \Delta_{14} \sqcup \{(0 : 1 : -1 : 0)\}, \qquad \operatorname{Fix}(14)(23) = \Delta_{(14)(23)} \sqcup \{Q_1\},$$
$$\operatorname{Fix}(123) = \{P_4, S\}.$$
$$\operatorname{Fix}(1243) \subset \operatorname{Fix}(1243)^2 = \operatorname{Fix}(14)(23)$$

and

$$(1243): Q_1 \mapsto Q_1,\ Q_2 \mapsto Q_3,\ Q_3 \mapsto Q_2.$$

Therefore (1243) does not act trivially on $\Delta_{(14)(23)} = \overline{Q_2Q_3}$. Consequently (1243) has exactly three isolated fixed points, two of them lying on $\overline{Q_2Q_3}$, namely

$$M = (1:\mathrm{i}:-\mathrm{i}:-1) \quad \text{and} \quad (23)\,M = (1:-\mathrm{i}:\mathrm{i}:-1).$$

Therefore $\mathrm{Fix}(1243) = \{Q_1, M, (23)\,M\}$.

For the convenience of the reader and for later use we illustrate the situation by means of Figure 3.4.A.

3.4.A.

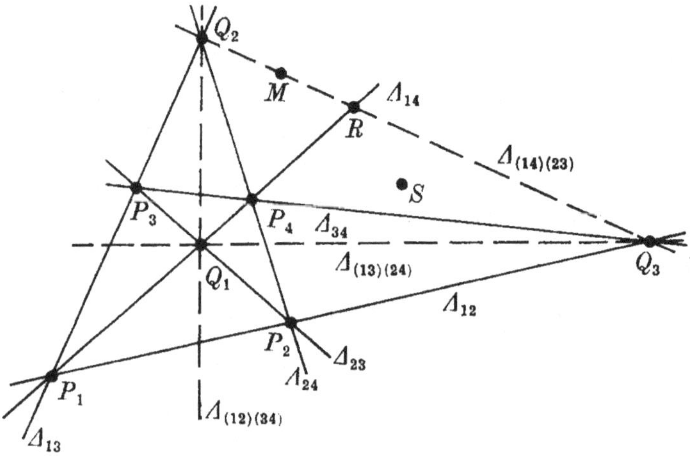

3.4.4. PROPOSITION (Feustel's fixed point formation, [17]). $\{\mathfrak{D}_{14}, \mathfrak{E}_{(14)(23)}\} \cup \{\varkappa_1, O, \tilde{R}, \tilde{M}, \tilde{S}\}$ *is a fixed point formation of the full Eisenstein lattice* $\tilde{\Gamma}$ *on* $\mathfrak{B}^* = \mathfrak{B} \cup \partial_\Gamma \mathfrak{B}$. (The relative position of the various points and discs is described by Figure 3.4.B. $\tilde{R}$, $\tilde{S}$, $\tilde{M}$ are preimages for $\tilde{p}'$ of R, S and M, respectively.) *The*

3.4.B.

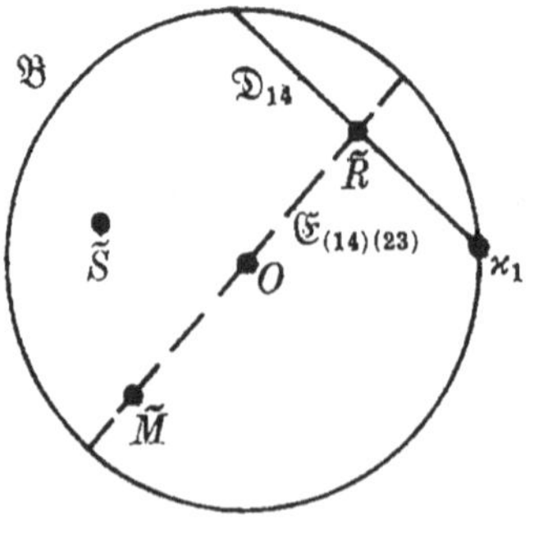

(effective) isotropy subgroups are of the following types:

$$Z_{\tilde{\Gamma}}(\mathfrak{D}_{14}) = \langle \varrho', (23)_{\mathfrak{B}} \rangle = \mathbb{Z}/6\mathbb{Z}, \qquad \varrho' = (12)_{\mathfrak{B}} \circ \varrho \circ (12)_{\mathfrak{B}}^{-1},$$

$$Z_{\tilde{\Gamma}}(\mathfrak{E}_{(14)(23)}) = \langle (14)\,(23)_{\mathfrak{B}} \rangle \simeq \mathbb{Z}/2\mathbb{Z},$$

$$\tilde{\Gamma}_O = \left\langle (14)\,(23)_{\mathfrak{B}}, (13)_{\mathfrak{B}}, \varrho, \begin{pmatrix} \omega & 0 & 0 \\ 0 & \omega^2 & 0 \\ 0 & 0 & 1 \end{pmatrix} \right\rangle,$$

generated by reflections, $\qquad |\tilde{\Gamma}_O| = 48,$

$\tilde{\Gamma}_{\tilde{R}} = \langle (23)_{\mathfrak{B}}, (14)\,(23)_{\mathfrak{B}}, \varrho \rangle \simeq K_4 \times (\mathbb{Z}/3\mathbb{Z})$, *generated by reflections,*

$\tilde{\Gamma}_{\tilde{M}} = \langle (1243)_{\mathfrak{B}} \rangle$ *of type* $\langle 4, 2 \rangle$, $\quad \tilde{\Gamma}_{\tilde{S}}$ *of type* $\langle 3, 2 \rangle$,

$\tilde{\Gamma}_{\varkappa_1}$ *generated by reflections.*

Proof. We lift the S_4-fixed point formation on $\mathbb{P}^2 \setminus \{P_1, P_2, P_3, P_4\}$ in Lemma 3.4.3 (omit P_1) in two steps along

$$\tilde{p}' : \mathfrak{B} \xrightarrow{p'} \mathfrak{B}/\Gamma' \xrightarrow{\langle \varrho \rangle} \mathfrak{B}/\tilde{\Gamma}' = \mathbb{P}^2 \setminus \{P_1, P_2, P_3, P_4\}.$$

The map p' is unramified outside of Δ (see Diagram 3.3.b). Therefore

$$\tilde{\Gamma}_P = (\tilde{\Gamma}/\tilde{\Gamma}')_{\tilde{p}'(P)} = S_{4,\tilde{p}'(P)} \qquad \text{for} \qquad P \in \mathfrak{B} \setminus \{\tilde{\Gamma} \cdot \mathfrak{D}_{14}\}.$$

So $\tilde{\Gamma}_{\tilde{M}}$ and $\tilde{\Gamma}_{\tilde{S}}$ are determined. The points R and Q_2 have uniquely determined preimages R', O_2' on $\mathfrak{B}/\Gamma'$ because they lie on the branch locus Δ. The isotropy groups of a preimage $\tilde{R} \in \mathfrak{B}$ of R and of R' coincide because p' is unramified around R' (see 3.3.b); that means $\tilde{\Gamma}_{\tilde{R}} = (\tilde{\Gamma}/\Gamma')_{R'}$. The latter group is obtained by a group extension of $(\tilde{\Gamma}/\tilde{\Gamma}')_{\tilde{R}}$ with $\langle \varrho \rangle$. Similarly we get $\tilde{\Gamma}_O$ by an additional extension with $\tilde{\Gamma}_O' = \langle \operatorname{diag}(\omega, \omega^2, 1) \rangle$. The reflection discs $\mathfrak{E}_{(ij)(kl)}$ or $\mathfrak{D}_{ij}$ on $\mathfrak{B}$ are found as S_4-reflection curves on $\mathbb{P}^2 \setminus \Delta$ and as $\langle \varrho \rangle$-reflection curves on $\widehat{\mathfrak{B}/\Gamma'}$, respectively. The points P_i correspond to the $\tilde{\Gamma}'$-cusps. Their common image point on $\mathbb{P}^2/S_4 = \widehat{\mathfrak{B}/\tilde{\Gamma}}$ is the only cusp "singularity" on $\widehat{\mathfrak{B}/\tilde{\Gamma}}$. It is a regular point because S_{4,P_i} is generated by reflections. Therefore $\tilde{\Gamma}_{\varkappa_1}$ is generated by reflections by a lemma of Švarčman ([65], see also [29], I).

3.5. Symmetric and Product Compactification of $\mathcal{P}$ and its 3-Sheeted Cyclic Covers

We call $\mathcal{P} = \mathbb{P}^2 \setminus \Delta \hookrightarrow \mathbb{P}^2$ the *symmetric compactification of* $\mathcal{P}$. We will see at once that $\mathcal{P}$ allows also the compactification $\mathbb{P}^1 \times \mathbb{P}^1$, which we call the *product compactification of* $\mathcal{P}$. First we define a surface X, which will interrelate $\mathbb{P}^2$ and $\mathbb{P}^1 \times \mathbb{P}^1$. The surface X is obtained by blowing up the four points $P_1, P_2, P_3, P_4 \in \mathbb{P}^2$. The resulting four exceptional curves are denoted by $\Theta_1, \Theta_2, \Theta_3, \Theta_4$. Then we blow down Λ_{14}, Δ_{24}, Δ_{34} on X. This is illustrated in Figure 3.5.A. The surface Y thus obtained on the right is $\mathbb{P}^1 \times \mathbb{P}^1$. Indeed, $c_1^2(\mathbb{P}^2) = 9$, $c_1^2(X) = c_1^2(\mathbb{P}^2) - 4 = 5$, $c_1^2(Y) = c_1^2(X) + 3 = 8$, and Y supports the constellation 

of two smooth rational curves. This is only possible for $Y = \mathbb{P}^1 \times \mathbb{P}^1$ (see e.g. [33], IV).

3.5.1. PROPOSITION. *The Diagram* 3.5.a *is commutative:*

3.5. a

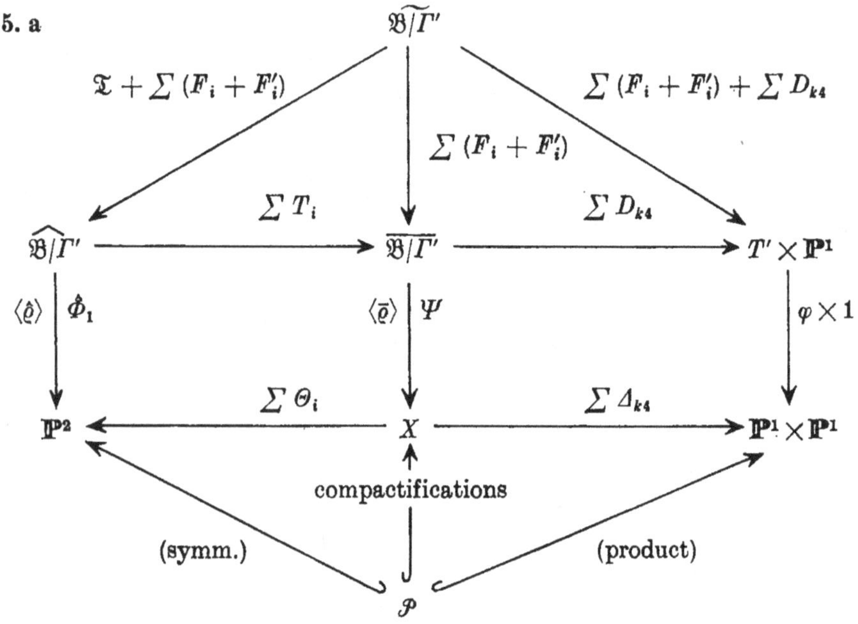

where $\varphi \times 1$ *is the 3-sheeted cyclic covering map branched along* $\Lambda_{12} + \Lambda_{13} + \Lambda_{23}$ *and* Ψ *is the quotient morphism* $\overline{\mathfrak{B}/\Gamma'} \to \overline{\mathfrak{B}/\Gamma''}/\langle \bar{\varrho} \rangle$.

Proof. The commutativity of the lower and upper triangles is clear from 3.5.A and from the curve constellation 1.6.A. The map φ is the 3-sheeted covering $T' \to \mathbb{P}^1$ defined in 2.8. Now $\overline{\mathfrak{B}/\Gamma'} \to \widehat{\mathfrak{B}/\Gamma''} \to \mathbb{P}^2$ contracts T_1, T_2, T_3, T_4 to P_1, P_2, P_3 and P_4, respectively and X is the result of blowing up of these four points. Therefore $\overline{\mathfrak{B}/\Gamma''} \to \mathbb{P}^2$ factorizes through X. For the same reason $\overline{\mathfrak{B}/\Gamma''}/\langle \bar{\varrho} \rangle \to \mathbb{P}^2$, defined by the universal property of quotients, factorizes through X. Obviously $\overline{\mathfrak{B}/\Gamma''}/\langle \bar{\varrho} \rangle \to X$ is quasifinite of degree 1, hence $X = \overline{\mathfrak{B}/\Gamma''}/\langle \bar{\varrho} \rangle$, and the left rectangle of 3.5.a is commutative. Also $\Psi\left(\sum_{k=1}^{3} D_{k4}\right) = \sum_{k=1}^{3} \Lambda_{k4}$. Therefore contractions and formations of surface quotients in the right rectangle commute with each other.

3.5.A.

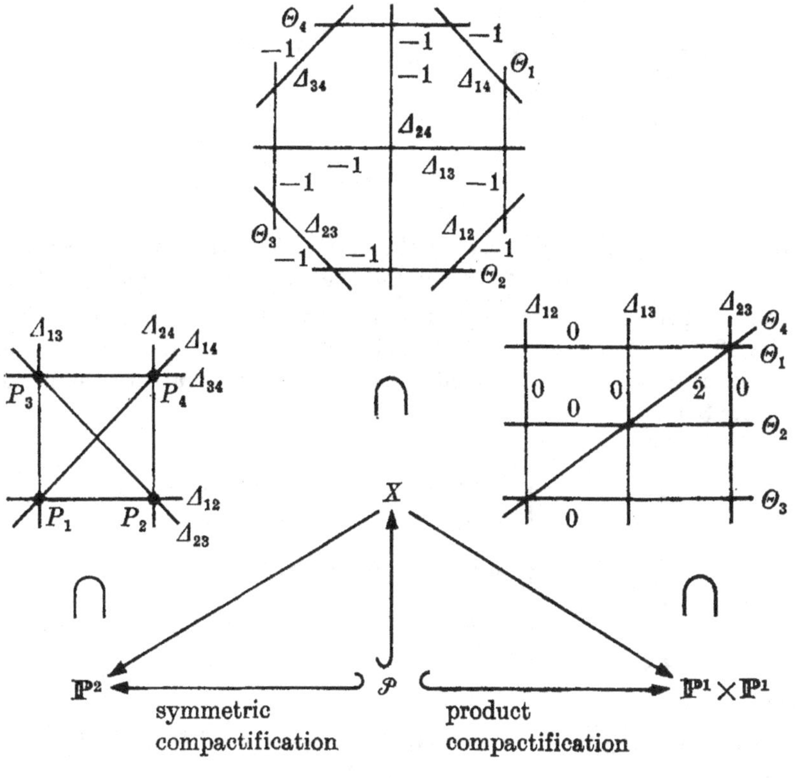

3.6. Fine Classification of the Surfaces $\widehat{\mathfrak{B}/\Gamma'} \cdot H$

We finish this section with the fine classification of $\widehat{\mathfrak{B}/\Gamma'} \cdot K_4$, $\widehat{\mathfrak{B}/\Gamma'} \cdot A_4$ and $\widehat{\mathfrak{B}/\Gamma}$ describing their "fibrations" over $\mathbb{P}^1$. As in the case of the special Eisenstein lattice we go down along the normal subgroup chain $1 \subset K_4 \subset A_4 \subset S_4$. We start with a simple construction of some rational surfaces with singularities.

Let $\hat{\Theta}$ be a point on $\mathbb{P}^2$. If we blow up $\hat{\Theta}$, we obtain the (non-minimal) Hirzebruch surface $\mathbb{F}_1$. It is characterized as the only rational ruled surface $\mathbb{F} \to \mathbb{P}^1$ with a section Θ with selfintersection number -1. Take two points S, M on $\mathbb{P}^2$ such that $\hat{\Theta}$, S, M do not lie on a projective line. Denote the fibres through S, M on $\mathbb{F}_1$ by $F(S)$ or $F(M)$, respectively. Now blow up twice the point M on $F(M)$ and three times the point S on $F(S)$. The resulting surface $\mathbb{F}_1(\widetilde{M}, S)$ projects onto $\mathbb{P}^1$ with two exceptional fibres. More precisely we have, including Θ, the following curve constellation with vertical fibres:

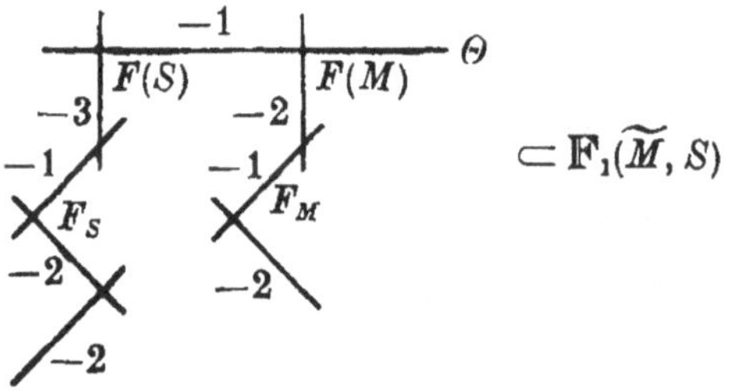

Here F_S, F_M denote the exceptional curves of first kind in the exceptional fibres. Now we contract the (−2)-curves and the (−3)-curve in the exceptional fibres. Then we get a surface $\mathbb{F}_1(\overline{M, S})$ with four singularities, two of type $\langle 2, 1\rangle$, one of type $\langle 3, 1\rangle$ and one of type $\langle 3, 2\rangle$. It is again "fibred" over $\mathbb{P}^1$ with the two exceptional fibres F_S and F_M containing the surface singularities as is sketched in Figure 3.6.A.

3.6.A.

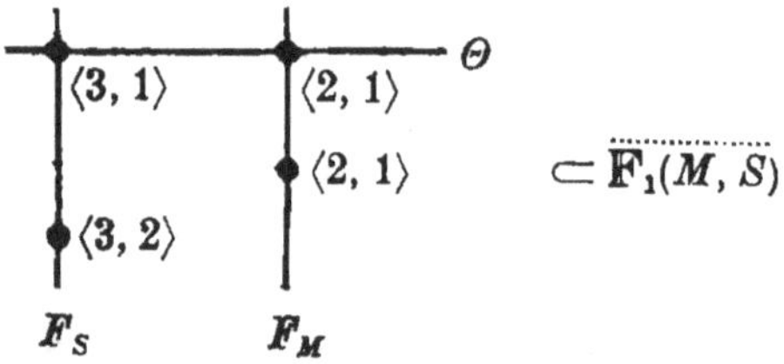

Returning to $\mathbb{F}_1(\widetilde{M}, S)$ we see that the contraction of Θ on $\mathbb{F}_1(\overline{M, S})$ is possible. We denote the resulting surface by $\mathbb{F}_1(\widehat{M}, S)$ and the contraction of Θ by $\Theta(\widehat{S, M})$. It is a regular point. So $\mathbb{F}_1(M, S)$ has exactly two singularities:

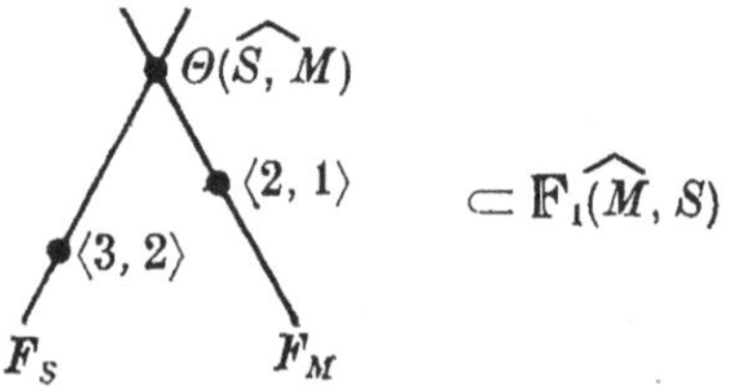

Next we start again with two points S_1, S_2 on $\mathbb{P}^2$ such that they are together with $\hat{\Theta} \in \mathbb{P}^2$ in general position on $\mathbb{P}^2$. Again we blow up $\hat{\Theta}$. But then we blow up three times the points S_1, S_2 on $\mathbb{F}(S_1)$ and $\mathbb{F}(S_2)$, respectively. We denote the resulting surface by $\mathbb{F}(\widetilde{S_1, S_2})$. We have a natural projection onto $\mathbb{P}^1$ with exactly two exceptional fibres:

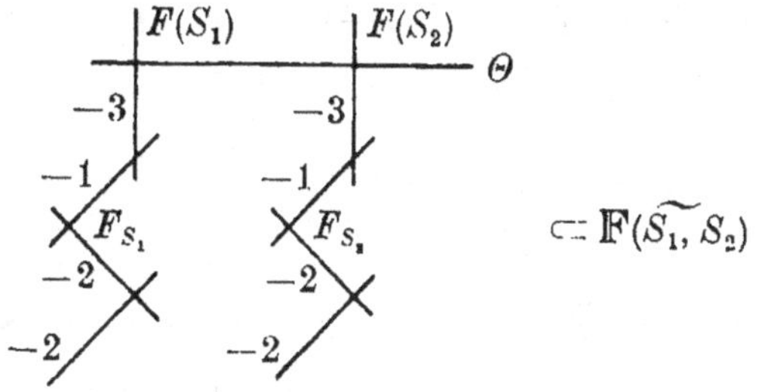

Now contract the (-3)- and (-2)-curves. We get a surface $\overline{\mathbb{F}(S_1, S_2)}$ "fibred" over $\mathbb{P}^1$ with two singularities of type $\langle 3, 1\rangle$ and two of type $\langle 3, 2\rangle$ sitting on the exceptional fibres F_{S_1} and F_{S_2}:

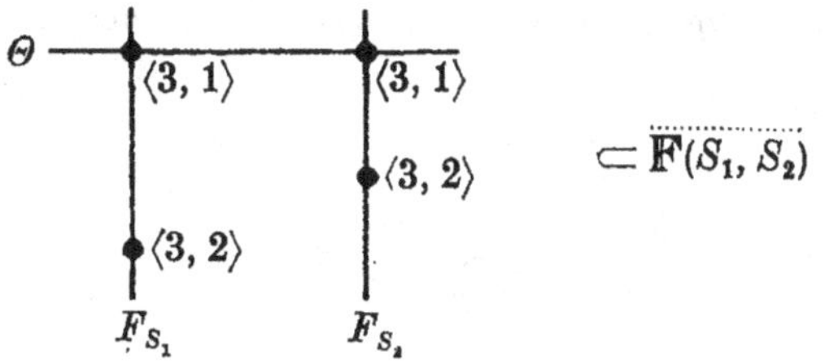

The contraction of Θ yields a surface $\mathbb{F}(\widehat{S_1, S_2})$ with three singularities $\Theta(\widehat{S_1, S_2})$, S_1', S_2' of type $\langle 3, 2\rangle$:

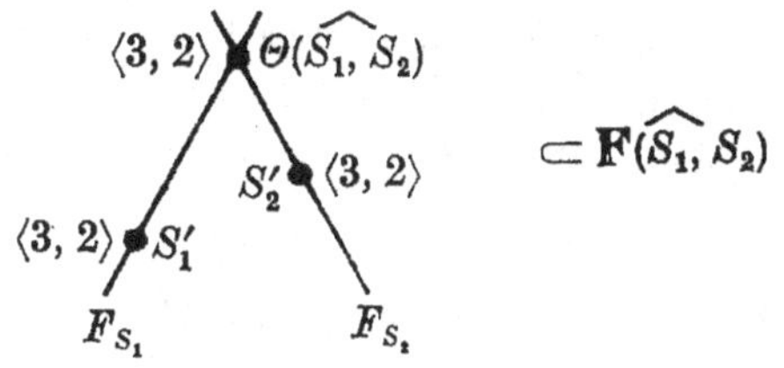

3.6.1. PROPOSITION. *It holds that*

$$\mathfrak{B}/\widehat{\tilde{\Gamma}' \cdot K_4} \cong \mathbb{P}^2, \qquad \mathfrak{B}/\widehat{\tilde{\Gamma}' \cdot A_4} \cong \mathbb{F}(\widehat{S_1, S_2}), \qquad \widehat{\mathfrak{B}/\tilde{\Gamma}} \cong \mathbb{F}_1(\widehat{M, S}).$$

Moreover we have the commutative Diagram 3.6.a, *where the vertical arrows represent the quotient morphisms with respect to the groups*

$$\tilde{\Gamma}'/\Gamma' \cong \tilde{\Gamma}' \cdot K_4/\Gamma' \cdot K_4 \cong \tilde{\Gamma}' \cdot A_4/\Gamma' \cdot A_4 \cong \tilde{\Gamma}/\Gamma \cong \mathbb{Z}/3\mathbb{Z},$$

and the horizontal arrows represent the quotient morphisms along the normal subgroup chain $1 \subset K_4 \subset A_4 \subset S_4$.

3.6.a.

$$
\begin{array}{ccccccc}
\overline{\mathfrak{B}/\Gamma'} & \longrightarrow & \overline{\mathfrak{B}/\Gamma'\cdot K_4} & \longrightarrow & \overline{\mathfrak{B}/\Gamma'\cdot A_4} & \longrightarrow & \overline{\mathfrak{B}/\Gamma} \\
\downarrow & & \downarrow & & \downarrow & & \downarrow \\
X = \overline{\mathfrak{B}/\tilde{\Gamma}'} & \longrightarrow & \overline{\mathfrak{B}/\tilde{\Gamma}'\cdot K_4} & \longrightarrow & \overline{\mathfrak{B}/\tilde{\Gamma}'\cdot A_4} & \longrightarrow & \overline{\mathfrak{B}/\tilde{\Gamma}} \\
\Sigma\Theta_i \swarrow \quad \searrow \Sigma\Delta_{i4} & & \| & & \| & & \| \\
 & & \tilde{\mathbb{F}}_1 & & \overline{\mathbb{F}(S_1, S_2)} & & \overline{\mathbb{F}_1(M, S)} \\
\mathbb{P}^2 = \widehat{\mathfrak{B}/\tilde{\Gamma}'} \qquad \mathbb{P}^1\times\mathbb{P}^1 & & & & & &
\end{array}
$$

After contraction of the compactification curves one obtains the commutative Diagram 3.6.b.

3.6.b.

$$
\begin{array}{ccccccc}
\widehat{\mathfrak{B}/\Gamma'} & \to & \widehat{\mathfrak{B}/\Gamma'\cdot K_4} & \to & \widehat{\mathfrak{B}/\Gamma'\cdot A_4} & \to & \widehat{\mathfrak{B}/\Gamma} \\
\downarrow & & \downarrow & & \downarrow & & \downarrow \\
\widehat{\mathfrak{B}/\tilde{\Gamma}'} & \to & \widehat{\mathfrak{B}/\tilde{\Gamma}'\cdot K_4} & \to & \widehat{\mathfrak{B}/\tilde{\Gamma}'\cdot A_4} & \to & \widehat{\mathfrak{B}/\tilde{\Gamma}} \\
\| & & \| & & \| & & \| \\
\mathbb{P}^2 & & \mathbb{P}^2 & & \mathbb{F}(\widehat{S_1, S_2}) & & \mathbb{F}_1(\widehat{M, S})
\end{array}
$$

Proof. S_4 acts on $\overline{\mathfrak{B}/\tilde{\Gamma}'} = X$. Projecting $\mathbb{P}^1\times\mathbb{P}^1$ onto the first factor we get a fibration $p\colon X \to \mathbb{P}^1\times\mathbb{P}^1 \xrightarrow[p_1]{} \mathbb{P}^1$ with exceptional fibres $\Delta_{12}+\Delta_{34}$, $\Delta_{13}+\Delta_{24}$, $\Delta_{23}+\Delta_{14}$. The action of S_4 on X is compatible with p. Indeed, the action of S_4 on $\overline{\mathfrak{B}/\Gamma''}$ is compatible with the projection $\pi\colon \overline{\mathfrak{B}/\Gamma''} \to T''\times\mathbb{P}^1 \to T'$ and $p\circ\Psi = \varphi\circ\pi$ by Proposition 3.5.1. Therefore, for $g\in S_4$ one finds

$$
\begin{aligned}
(p\circ g)\circ\Psi &= p\circ(g\circ\Psi) = p\circ(\Psi\circ g) = (p\circ\Psi)\circ g \\
&= (\varphi\circ\pi)\circ g = \varphi\circ(\pi\circ g) = \varphi\circ\pi = p\circ\Psi.
\end{aligned}
$$

From the universal property of the quotient morphism Ψ it follows that $p\circ g = p$. Consequently the quotient surfaces $\overline{\mathfrak{B}/\tilde{\Gamma}'\cdot K_4}$, $\overline{\mathfrak{B}/\tilde{\Gamma}'\cdot A_4}$, $\overline{\mathfrak{B}/\tilde{\Gamma}}$ are fibred over $\mathbb{P}^1$. So the classification of these surfaces is reduced to the following steps:

(i) find the singularities,

(ii) remove the singularities by minimal resolutions,

(iii) find a negative section over $\mathbb{P}^1$, such that it remains negatively after the next and last step:

(iv) Blow down the exceptional curves in the fibres.

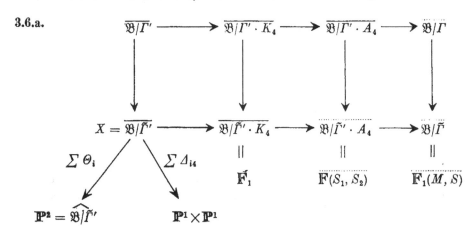

Then one obtains a Hirzebruch surface $\mathbb{F}_n$, $n \geqq 0$, where $-n$ is the selfintersection number of the section mentioned in (iii) after step (iv). A candidate for this section is the image curve Θ of the curves Θ_i because it has to be contractible. Indeed, the contraction is the cusp singularity. Hence it must be a negative section. In fact, we will see that after the steps (i), (ii), (iii), (iv) its selfintersection number is -1. Therefore we obtain in each case the Hirzebruch surface $\mathbb{F}_1$. Going backward through our four steps we obtain the classification of our surfaces. Tacitly we will use the *G-crosspoint formula for selfintersection numbers of quotient curves C/G on the filled resolution of singularities of a quotient surface Y/G* (see [33], II, Theorem 8):

3.6.2. $$(\widetilde{C/G})^2 = \frac{|Z_G(C)|^2}{|N_G(C)|}\big((C)^2 - \mathrm{cross}_G(C)\big) - \sum_{C \ni P \bmod N_G(C)} e_{C,P}/d_{C,P}.$$

This formula is also used for the determination of the singularity types $\langle d, e\rangle$ on Θ, using that the right-hand side has to be an integer. We will use the notations of the Figures 3.4.A, 3.5.A and of Diagram 3.5.a.

3.6.3. X/K_4:

The non-trivial isotropy groups of K_4 acting on X are generated by reflections. Therefore X/K_4 is smooth. The exceptional fibres $\Delta_{12} + \Delta_{34}$, $\Delta_{13} + \Delta_{24}$, $\Delta_{23} + \Delta_{14}$ of the fibration $p: X \dashrightarrow \mathbb{P}^1$ are mapped onto non-exceptional fibres Δ'_{12}, Δ'_{13}, Δ'_{23} on X/K_4 (over $\mathbb{P}^1$). The group K_4 acts transitively on the set of curves $\{\Theta_1, \Theta_2, \Theta_3, \Theta_4\}$. Therefore the common image Θ of the Θ_i's on X/K_4 is a section over $\mathbb{P}^1$ with selfintersection number -1. Thus $X/K_4 \cong \mathbb{F}_1$.

3.6.4. X/A_4:

We have to take the quotient of X/K_4 by the image of $\langle(123)\rangle$ in S_4/K_4. Denoting the two image points of the S_4-orbit of S by S'_1, S'_2 we find the following curve constellation on X/A_4:

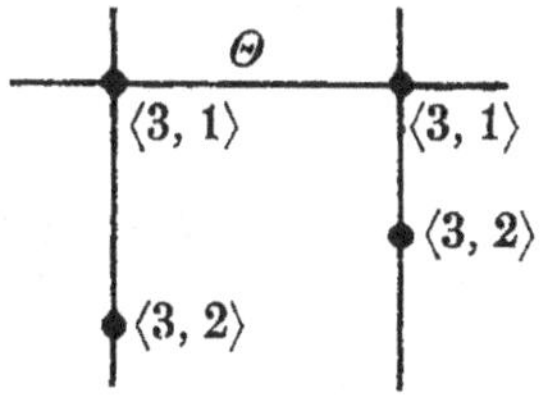

$\subset X/A_4$

The vertical lines correspond to fibres of X/A_4 over $\mathbb{P}^1$. Steps (ii), (iii), (iv) lead us to $\mathbb{F}_1$, thus

$$\overline{\mathfrak{B}/\tilde\Gamma'\cdot A_4} = X/A_4 = \overline{\mathbb{F}(S_1, S_2)}.$$

3.6.5. X/S_4:

We have to take the quotient of X/A_4 by $\langle(12)\cdot A_4\rangle = S_4/A_4 \cong \mathbb{Z}/2\mathbb{Z}$. The fixed points on Θ are the intersection points of Θ and the two fibres going through the image points of M and Q_i. The latter fibre is the image of Δ_{12}. It is a reflection curve. The element (12) acts effectively on the other fibre. It contains two fixed points of type $\langle 2, 1\rangle$. The other singularities of X/S_4 lie on the common image F_S of F_{S_1} and F_{S_2}. So we find after step (i) the following situation:

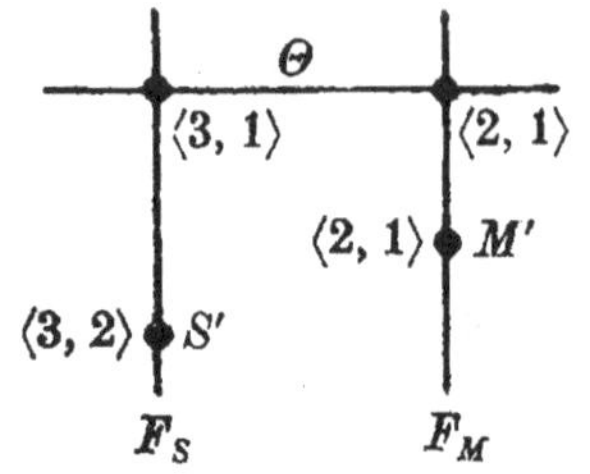

$\subset X/S_4 = \overline{\mathfrak{B}/\bar{\Gamma}}$

Steps (ii), (iii), (iv) lead us to $\mathbb{F}_1$. Therefore by the construction just above our proposition it follows that

$$\overline{\mathfrak{B}/\bar{\Gamma}'} = X/S_4 = \overline{\mathbb{F}(S, M)}.$$

§ 4. LOGARITHMIC PLURICANONICAL MAPS AND AUTOMORPHIC FORMS

4.1. Automorphic Forms and Logarithmic Pluricanonical Sheaves

For a divisor D on a smooth algebraic variety X and for $n \geqq m \geqq 1$ we have commutative diagrams of rational maps

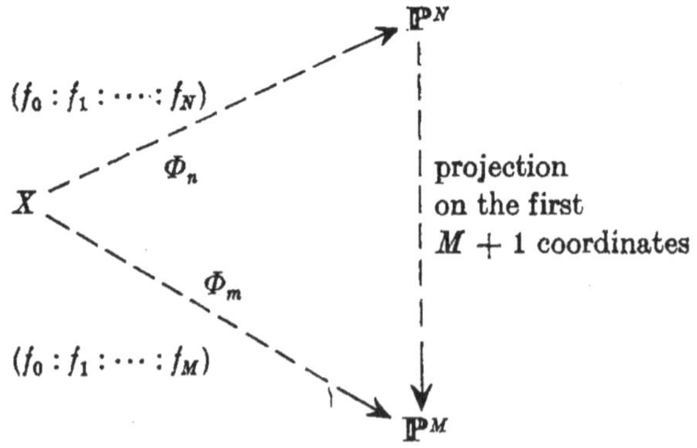

$N = h^0(nD) - 1$
$M = h^0(mD) - 1$

where $\Phi_k = \Phi_{kD}$ and $\{f_0, f_1, \ldots, f_N\}$ is a basis of $H^0(X, \mathcal{O}(D)^n)$ which extends the basis $\{f_0, f_1, \ldots, f_M\}$ of $H^0(X, \mathcal{O}(D)^m)$. If Φ_n and Φ_m are morphisms, then they define by restriction of the projection a morphism $\Phi_n(X) \to \Phi_m(X)$. In particular we have by Lemma 3.1.2 and Proposition 3.1.3 for the logarithmic canonical sheaf $\Omega(\mathfrak{T})$ on $X = \widetilde{\mathfrak{B}/\Gamma'}$ the following commutative diagrams of morphisms:

4.1.a.

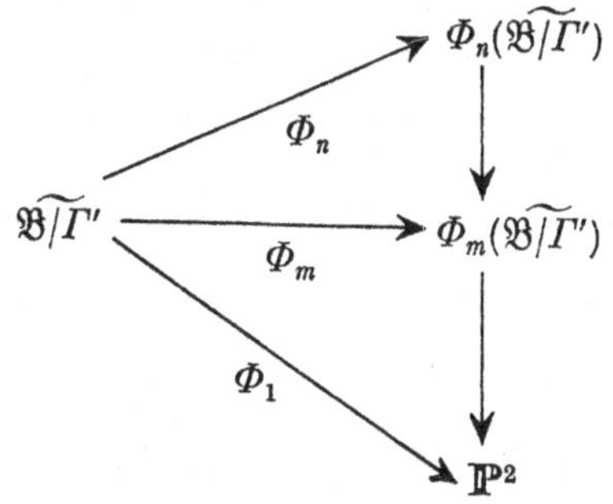

On the other hand consider the $\mathbb{C}$-vector spaces

$$[\Gamma', m] = H^0(\mathfrak{B}, \mathcal{O}_{\mathfrak{B}})^{\Gamma'}, \qquad m \geqq 0$$

of *Γ'-automorphic forms of weight m*, where the action of Γ' on the space $H^0(\mathfrak{B}, \mathcal{O}_{\mathfrak{B}})$ of holomorphic functions on $\mathfrak{B}$ is defined by

$$\Gamma' \ni \gamma : f \mapsto j_\gamma^m \cdot \gamma^*(f), \qquad j_\gamma \text{ the Jacobi determinant of } \gamma.$$

Therefore

$$[\Gamma', m] \cong \{f \in H^0(\mathfrak{B}, \mathcal{O}_{\mathfrak{B}});\ f(dz_1 \wedge dz_2)^m\ \Gamma'\text{-invariant}\} = H^0\big(\mathfrak{B}, (\Omega^2_{\mathfrak{B}})^m\big)^{\Gamma'}$$

with the action given by

$$\Gamma' \ni \gamma : \omega \mapsto \gamma^*(\omega), \qquad \omega \in H^0\big(\mathfrak{B}, (\Omega^2_{\mathfrak{B}})^m\big).$$

Factorizing $(\Omega^2_{\mathfrak{B}\setminus\Gamma\cdot O})^m$ by Γ' we obtain $(\Omega^2_{\mathfrak{B}/\Gamma'\setminus\{O_1',O_2',O_3'\}})^m$ because

$$\mathfrak{B} \setminus \Gamma \cdot O \to \mathfrak{B}/\Gamma' \setminus \{O_1', O_2', O_3'\}$$

is an unramified covering map (see Diagram 3.3.b). Thus each automorphic form $f \in [\Gamma', m]$ defines a section of $(\Omega^2_{\mathfrak{B}/\Gamma'\setminus\{O_1',O_2',O_3'\}})^m$ over $\mathfrak{B}/\Gamma' \setminus \{O_1', O_2', O_3'\}$.

4.1.1. PROPOSITION. *It holds that*

$$[\Gamma', m] \cong H^0\Big(\big(\Omega^2_{\widetilde{\mathfrak{B}/\Gamma'}} \otimes \mathcal{O}(\mathfrak{T})\big)^m\Big), \qquad m \geqq 0.$$

Proof. This result is well-known for neat arithmetic groups (see [50], Prop. 3.4). The additional thing to do is to show that the singular points O_1', O_2', O_3' on $\overline{\mathfrak{B}/\Gamma'}$ do not disturb things. Working outside $\{O_1', O_2', O_3'\}$ Mumford's proof yields

4.1.2. $$[\Gamma', m] \cong H^0\big(\mathfrak{B}, (\Omega^2_{\mathfrak{B}})^m\big)^{\Gamma'} = H^0\big(\mathfrak{B}, (\Omega^2_{\mathfrak{B}\setminus\Gamma\cdot O})^m\big)^{\Gamma'} = H^0(\overline{\mathfrak{B}/\Gamma'}, \mathcal{L}^m),$$

where

$$\mathcal{L} = i_*\big(\Omega^2_{\overline{\mathfrak{B}/\Gamma'}\setminus\{O_1',O_2',O_3'\}} \otimes \mathcal{O}_{\overline{\mathfrak{B}/\Gamma'}\setminus\{O_1',O_2',O_3'\}}(\mathfrak{T})\big)$$

with obvious notations and i denoting the open embedding

$$i\colon \overline{\mathfrak{B}/\Gamma'} \setminus \{O'_1, O'_2, O'_3\} \hookrightarrow \overline{\mathfrak{B}/\Gamma'}.$$

We prove that $i_*(\mathcal{L}^m)$ is an invertible $\mathcal{O}_{\overline{\mathfrak{B}/\Gamma'}}$-sheaf. The problem is local around $O' = O'_i$, $i \in \{1, 2, 3\}$. Take for example $i = 2$. Let U be a sufficiently small analytic Γ'_O-invariant open neighbourhood of $O \in \mathfrak{B}$. Then the projection $\mathfrak{B} \to \mathfrak{B}/\Gamma'$ induces an open embedding $j\colon (U/\Gamma'_O) \setminus \{O'\} \hookrightarrow \mathfrak{B}/\Gamma'$. Using local coordinates z_1, z_2 around $O \in U$ we find

$$\begin{aligned}\left(i_*(\mathcal{L}^m)\right)_{O'} &= j_*\left(\left(\mathcal{O}_{U\setminus\{O\}} \cdot (\mathrm{d}z_1 \wedge \mathrm{d}z_2)^m\right)^{\Gamma'_O}\right)_{O'} \\ &= j_*\left(\mathcal{O}_{U\setminus\{O\}}^{\Gamma'_O} \cdot (\mathrm{d}z_1 \wedge \mathrm{d}z_2)^m\right)_{O'} = \mathcal{O}_{\mathfrak{B}/\Gamma',O'} \cdot (\mathrm{d}z_1 \wedge \mathrm{d}z_2)^m\end{aligned}$$

because Γ'_O is generated by the element $(z_1 \mapsto \omega z_1, z_2 \mapsto \omega^2 z_2)$ with determinant 1. Let

$$\pi\colon \widetilde{\mathfrak{B}/\Gamma'} \to \overline{\mathfrak{B}/\Gamma'}$$

be the resolution of singularities. $\pi^*(\mathcal{L}^m)$ is an invertible sheaf, which coincides with $\left(\Omega(\mathfrak{T})\right)^m$ outside of $\sum_{i=1}^{3} (F_i + F'_i)$. In particular for $m = 1$ we can write

$$\Omega(\mathfrak{T}) = \pi^*(\mathcal{L})\left(\sum m_i F_i + \sum m'_i F'_i\right).$$

Let K be a canonical divisor and $\pi^*(\mathcal{L}) = \mathcal{O}(L)$ for a suitable divisor L on $\widetilde{\mathfrak{B}/\Gamma'}$. Then

4.1.3. $\quad K + \mathfrak{T} \equiv L + \sum m_i F_i + \sum m'_i F'_i.$

By the adjunction formula and 1.6.A one finds

$$(K \cdot F'_i) = (K \cdot F_i) = (K \cdot F_i) + (F_i^2) + 2 = -2 + 2 = 0, \quad i = 1, 2, 3.$$

Now intersect (4.1.3) with F_i, F'_i. Then one gets

$$\left.\begin{aligned} 0 &= (L \cdot F_i) - 2m_i + m'_i, \\ 0 &= (L \cdot F'_i) + m_i - 2m'_i, \end{aligned}\right\} \quad i = 1, 2, 3.$$

On the other hand

$$(L \cdot F'_i) = (L \cdot F_i) = \deg\left(\pi^*\mathcal{L}|_{F_i}\right) = \deg \mathcal{O}_{F_i} = 0$$

because $\mathcal{L}_{O'_i} \cong \mathcal{O}_{O'_i}$ and $\pi(F_i) = O'_i$. Therefore $m_i = m'_i = 0$, thus

$$\Omega^2_{\widetilde{\mathfrak{B}/\Gamma'}} \otimes \mathcal{O}(\mathfrak{T}) \cong \pi^*(\mathcal{L}) = \pi^*\left(i_*\left(\Omega^2_{\overline{\mathfrak{B}/\Gamma'}\setminus\{O'_1,O'_2,O'_3\}}(\mathfrak{T})\right)\right).$$

More generally we have

$$\big(\Omega(\mathfrak{X})\big)^m \cong \pi^*(\mathcal{L}^m).$$

It is well-known that

$$H^0\big(\overline{\mathfrak{B}/\Gamma'}, \pi_*\pi^*(\mathcal{L}^m)\big) = H^0\big(\widetilde{\mathfrak{B}/\Gamma'}, \pi^*(\mathcal{L}^m)\big)$$

(trivial lowest degree part of the π-spectral sequence). By the projection formula and the connectedness theorem it follows that

$$\pi_*\pi^*\mathcal{L}^m = \pi_*\mathcal{O}_{\widetilde{\mathfrak{B}/\Gamma'}} \otimes \mathcal{L}^m = \mathcal{O}_{\overline{\mathfrak{B}/\Gamma'}} \otimes \mathcal{L}^m = \mathcal{L}^m.$$

By substituting this in the formula above

$$H^0\big(\widetilde{\mathfrak{B}/\Gamma'}, \Omega(\mathfrak{X})^m\big) \cong H^0(\overline{\mathfrak{B}/\Gamma'}, \mathcal{L}^m)$$

and finally by (4.1.2)

$$[\Gamma', m] \cong H^0\big(\widetilde{\mathfrak{B}/\Gamma'}, \Omega(\mathfrak{X})^m\big).$$

REMARK. It was not necessary to make use of Mumford's general proof. Indeed, one can extend the considerations above also to the cusp singularities. This was done for Hilbert modular surfaces in [26].

The space $[\Gamma', m]$ defines a holomorphic map

$$\Phi_{[\Gamma',m]} : \mathfrak{B} \xrightarrow[f_0:f_1:\cdots:f_M]{} \mathbb{P}^{M-1}; \quad M = \dim\,[\Gamma', m],$$

where $f_0, f_1, \ldots, f_M$ is a basis of $[\Gamma', m]$. This map factorizes through $\mathfrak{B}/\Gamma'$. By extension we obtain a morphism

$$\phi_m : \widehat{\mathfrak{B}/\Gamma'} \to \phi_m(\widehat{\mathfrak{B}/\Gamma'}) \subset \mathbb{P}^{M-1}.$$

Obviously, we have commutative diagrams

4.1.b.

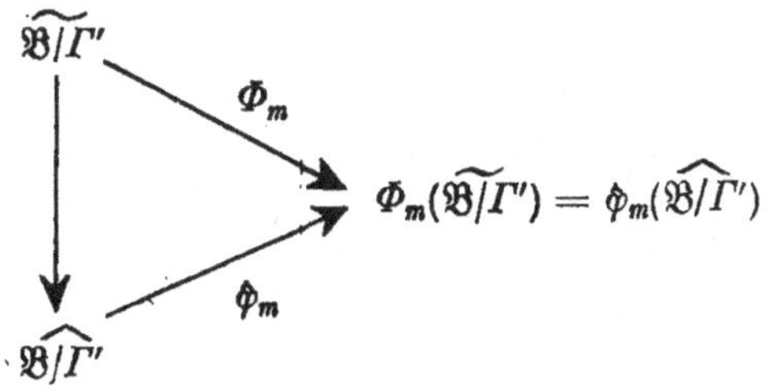

By a theorem of Baily-Borel in [4] ϕ_N is an isomorphism for $N \gg 0$. Taking into account 4.1.a we have for each pair n, m of natural numbers with $n \geqq m \geqq 2$ a commutative diagram

4.1.c.

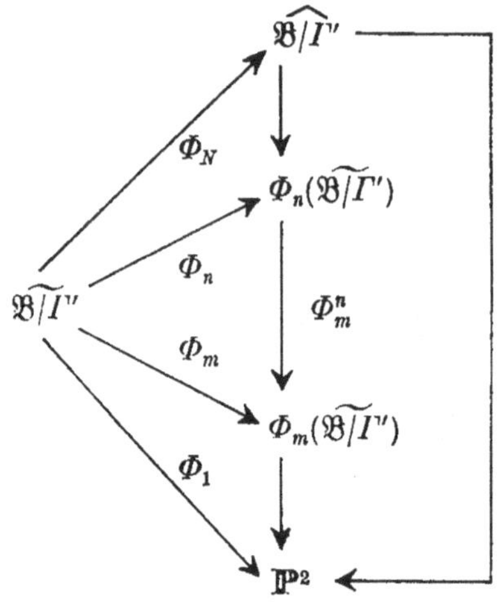

Later we will prove

4.1.4. PROPOSITION. *For* $m \geqq 2$ *it holds that* $\Phi_m(\widetilde{\mathfrak{B}/\Gamma'}) = \widehat{\mathfrak{B}/\Gamma'}$.

4.2. Dimension Formula for Spaces of Γ'-Automorphic Forms

In order to prove 4.1.4 we need

4.2.1. PROPOSITION. *Let K be a canonical divisor on $\widetilde{\mathfrak{B}/\Gamma'}$. Then*

$$\dim\,[\Gamma', m] = \dim H^0(\widetilde{\mathfrak{B}/\Gamma'}, mK + m\mathfrak{T}) = \begin{cases} 3, & m = 1 \\ \dfrac{3}{2}\, m(m-1) + 4, & m \geqq 2. \end{cases}$$

Proof. For $m = 1$ cf. 3.1.3 and 4.1.1. Assume that $m \geqq 2$. From Lemma 3.1.2 we know that $|mK + m\mathfrak{T}|$ has no base points. Furthermore $\dim \Phi_m(\widetilde{\mathfrak{B}/\Gamma'}) = 2$. In this situation we can apply a vanishing theorem of Mumford (see [49]):

$$H^1(-lK - l\mathfrak{T}) = 0 \qquad \text{for} \qquad l \geqq 1.$$

By Serre duality we find

$$H^1\big(mK + (m-1)\,\mathfrak{T}\big) = 0 \qquad \text{for} \qquad m \geqq 2.$$

$K + \mathfrak{T}$ is effective for a suitable K by (2.3.10). Thus

$$H^2\big(mK + (m-1)\,\mathfrak{T}\big) \cong H^0\big(-(m-1)\,(K + \mathfrak{T})\big) = 0.$$

If we denote the Euler-Poincaré characteristic by χ, then we can write

$$\begin{aligned} &\dim H^0\big(mK + (m-1)\,\mathfrak{T}\big) \\ &= \chi\big(\widetilde{\mathfrak{B}/\Gamma'}, mK + (m-1)\,\mathfrak{T}\big) \\ &= \frac{1}{2}\,\Big(\big(mK + (m-1)\,\mathfrak{T}\big) \cdot (m-1)\,(K + \mathfrak{T})\Big) + \chi(\mathcal{O}_{\widetilde{\mathfrak{B}/\Gamma'}}) \end{aligned}$$

by the Riemann-Roch-Hirzebruch theorem. We know the arithmetic genus

$$\chi(\mathcal{O}_{\widetilde{\mathfrak{B}/\Gamma'}}) = \chi(\mathcal{O}_{T'\times \mathbb{P}^1}) = 0.$$

Furthermore we know from 3.1.7 that

$$\big((K+\mathfrak{T})\cdot(K+\mathfrak{T})\big) = \big(K\cdot(K+\mathfrak{T})\big) = 3.$$

Therefore

4.2.2. $$\dim H^0\big(mK + (m-1)\,\mathfrak{T}\big) = \frac{3}{2}\, m(m-1) \quad \text{for} \quad m \geqq 2.$$

The exact sheaf sequence

$$0 \to \mathcal{O}\big(mK + (m-1)\,\mathfrak{T}\big) \to \mathcal{O}(mK + m\mathfrak{T}) \to \mathcal{O}(mK + m\mathfrak{T}|_{\mathfrak{T}}) \to 0$$

leads to the exact sequence of vector spaces

4.2.3. $$0 \to H^0\big(mK + (m-1)\,\mathfrak{T}\big) \to H^0(mK + m\mathfrak{T}) \to H^0\big(\mathfrak{T}, m(K + \mathfrak{T}|_{\mathfrak{T}})\big) \to 0.$$

Now $\mathcal{O}(K + T_i|_{T_i}) = \omega_{T_i}$ is isomorphic to the trivial sheaf on the elliptic curve T'_i. Therefore

$$\dim H^0\big(\mathfrak{T}, m(K + \mathfrak{T}|_{\mathfrak{T}})\big) = \dim \bigoplus_{i=1}^{4} H^0\big(T_i, m(K + T_i|_{T_i})\big) = 4.$$

The dimension formula in 4.2.1 now follows from (4.2.2) and (4.2.3).

4.2.4. REMARK. $H^0\big(mK + (m-1)\,\mathfrak{T}\big)$ is isomorphic to the space $[\Gamma', m]_{\text{cusp}}$ of Γ'-cusp forms of weight m. So we have proved

$$\dim\,[\Gamma', m]_{\text{cusp}} = \frac{3}{2}\, m(m-1) \quad \text{for} \quad m \geqq 2.$$

The next lemma prepares the proof of Proposition 4.1.4.

4.2.5. LEMMA. *$\Phi_1^2 \colon \Phi_2(\widetilde{\mathfrak{B}/\Gamma''}) \to \mathbb{P}^2$ is a finite morphism of degree* 3.

Proof. Diagram 4.1.c and the finiteness of $\hat{\Phi}_1$ (see 3.2.1) enforce the same property for Φ_1^2. Furthermore it is immediately clear that either Φ_1^2 is of degree 3 or it is a birational finite morphism, that means an isomorphism. Therefore it suffices to prove that $\Phi_2(\widetilde{\mathfrak{B}/\Gamma'})$ is not equal to $\mathbb{P}^2$. Asume that $\Phi_2(\widetilde{\mathfrak{B}/\Gamma''}) = \mathbb{P}^2$. We know that

$$|2K + 2\mathfrak{T}| = 6 \qquad \text{and} \qquad (2K + 2\mathfrak{T})^2 = 12$$

by Proposition 4.2.1 or by 3.1.7, respectively. Using a base $\{f_0, f_1, \ldots, f_6\}$ of $H^0(\widetilde{\mathfrak{B}/\Gamma''}, 2K + 2\mathfrak{T})$ we have the following situation:

$$\begin{array}{l} \Phi_2 \colon \widetilde{\mathfrak{B}/\Gamma'} \xrightarrow{(f_0:f_1:\cdots:f_6)} \Phi_2(\widetilde{\mathfrak{B}/\Gamma''}) \to \mathbb{P}^6 \\ \qquad\qquad\qquad\qquad\qquad \| \\ \qquad\qquad\qquad\qquad\quad \mathbb{P}^2 \end{array}$$

From the degree formula for intersection numbers of curves on surfaces we deduce

$$12 = \deg \Phi_2 \cdot \deg \Phi_2(\widetilde{\mathfrak{B}/\Gamma'}), \qquad \text{hence} \qquad \deg \Phi_2(\widetilde{\mathfrak{B}/\Gamma'}) = 4.$$

$\Phi_2(\widetilde{\mathfrak{B}/\Gamma'})$ is a non-degenerate surface in $\mathbb{P}^6$, that means there is no hyperplane of $\mathbb{P}^6$ containing $\Phi_2(\widetilde{\mathfrak{B}/\Gamma'})$. Otherwise $f_0, \ldots, f_6$ would not be linearly independent. It is well-known that the degree of a non-degenerate surface in $\mathbb{P}^n$ is not smaller than $n - 1$. This is a consequence of Clifford's theorem for projectively embedded curves. For the convenience of the reader we show how to reach this conclusion in our special case.

Let H be a hyperplane of $\mathbb{P}^6$ and $C = H \cap \Phi_2(\widetilde{\mathfrak{B}/\Gamma'})$ the corresponding hyperplane section. $(C^2)_{\mathbb{P}^2}$ is equal to the embedding degree 4. Therefore C is a curve of degree 2 on $\mathbb{P}^2 = \Phi_2(\widetilde{\mathfrak{B}/\Gamma'})$ by Bezout, hence $g = g(C) = 0$. On the other hand C is non-degenerate in $H \cong \mathbb{P}^5$ of degree 4. A corollary of Clifford's theorem (see e.g. [23], p. 251) says that $g \leqq d - n$ for a non-degenerated curve C in $\mathbb{P}^n$ of degree d. So we obtain in our situation the contradiction $0 \leqq 4 - 5$. Lemma 4.2.5 is proved.

4.3. *Structure of the Ring of Γ'-Automorphic Forms*

In order to determine the structure of $\bigoplus_{m=0}^{\infty} [\Gamma', m]$ we will first define a graded subring and subsequently compare the dimensions of the homogeneous parts of the same degree of both rings. It turns out that they are equal. Our result is

4.3.1. PROPOSITION. *It holds that*

$$\bigoplus_{m=0}^{\infty} [\Gamma', m] \cong \bigoplus_{m=0}^{\infty} H^0(\widetilde{\mathfrak{B}/\Gamma'}, mK + m\mathfrak{T}) = \mathbb{C}[Y_0, Y_1, Y_2, z],$$

where Y_0, Y_1, Y_2 are algebraically independent of weight 1, *and z is an element of weight* 2 *satisfying an irreducible equation*

4.3.2. $z^3 = az^2 + bz + c,$

with $a, b, c \in \mathbb{C}[Y_0, Y_1, Y_2]$ homogeneous of degree 2, 4 *and* 6, *respectively.*

Proof. The first isomorphism comes from Proposition 4.1.1. Indeed, the bi-unique correspondence between automorphic forms and global sections of logarithmic canonical sheaves is compatible with multiplication and degree. In order to prove the isomorphism on the right-hand side we set

$$V_0 = H^0(0 \cdot (K + \mathfrak{T})) = \mathbb{C} \cdot 1 = V^{(0)},$$

$$V_1 = H^0(1 \cdot (K + \mathfrak{T})) = \mathbb{C} \cdot 1 \oplus \mathbb{C} \cdot f_1 \oplus \mathbb{C} \cdot f_2 = V^{(1)},$$

$$V_2 = H^0(2 \cdot (K + \mathfrak{T})) = \mathbb{C} \cdot 1 \oplus \mathbb{C} \cdot f_1 \oplus \mathbb{C} \cdot f_2 \oplus \mathbb{C} \cdot f_1^2 \oplus \mathbb{C} \cdot f_1 f_2 \oplus \mathbb{C} \cdot f_2^2 \oplus \mathbb{C} \cdot g$$
$$= V^2 \oplus \mathbb{P} \cdot g,$$
$$V_3 = V^{(3)} + V^{(1)} \cdot g$$

and inductively

4.3.3. $$V_{m+1} = V^{(m+1)} + gV^{(m-1)} + g^2 V^{(m-3)}, \qquad m \geqq 3,$$

where

$$V^{(n)} = \{f \in \mathbb{C}[f_1, f_2];\ \deg f \leqq n \text{ with respect to } f_1, f_2\}.$$

We used the dimension formula (4.2.1) for the definition of f_1, f_2, g:

$$\dim H^0(K + \mathfrak{T}) = 3, \qquad \dim H^0(2(K + \mathfrak{T})) = 7.$$

Furthermore, f_1, f_2 are algebraically independent by 3.1.3(b). The degree of $\Phi_2(\widetilde{\mathfrak{B}/\Gamma''})$ over $\Phi_1(\widetilde{\mathfrak{B}/\Gamma''}) = \mathbb{P}^2$ is 3 by Lemma 4.2.5. The function fields are $\mathbb{C}(f_1, f_2, g)$ and $\mathbb{C}(f_1, f_2)$, respectively. Therefore g is algebraically dependent of order 3 over $\mathbb{C}(f_1, f_2)$. Consequently the sum in (4.3.3) is direct. Hence

$$\dim V_{m+1} = \dim V^{(m+1)} + \dim V^{(m-1)} + \dim V^{(m-2)}, \qquad m \geqq 3.$$

The dimension of $V^{(n)}$ is $\dfrac{(n+1)(n+2)}{2}$. By induction one finds

$$\dim V_m = \begin{cases} 1, & m = 0, \\ 3, & m = 1, \\ \dfrac{3}{2}\, m(m-1) + 4, & m \geqq 2. \end{cases}$$

Comparing this with the result of Proposition 4.2.1 we see that we have an isomorphism of graded $\mathbb{C}$-vector spaces

4.3.4. $$\bigoplus_{m=0}^{\infty} V_m \cong \bigoplus_{m=0}^{\infty} H^0(m(K + \mathfrak{T})).$$

If g did not satisfy an equation of type

$$g^3 = a'g^2 + b'g + c',$$

$a', b', c' \in \mathbb{C}[f_1, f_2]$ of degree 2, 4 and 6, respectively, then $\mathbb{C}g^3 + V_6$ would be a direct sum. Now $\mathbb{C}g^3 + V_6$ is a subspace of $H^0(6(K + \mathfrak{T}))$. Comparing dimensions we get the contradiction

$$50 = \dim(\mathbb{C}g^3 + V_6) \leqq \dim H^0(6(K + \mathfrak{T})) = 49.$$

It follows immediately that (4.3.4) is an isomorphism of graded rings. Setting $f_1 = Y_1/Y_0, f_2 = Y_2/Y_0, z = g/Y_0^2$ we see that $\bigoplus_{m=0}^{\infty} V_m$ is isomorphic to $\mathbb{C}[Y_0, Y_1, Y_2, z]$ and our proposition is proved.

Now we are ready for the

Proof of Proposition 4.1.4. It suffices to prove that $\Phi_2(\widetilde{\mathfrak{B}/\Gamma''}) = \widehat{\mathfrak{B}/\Gamma'}$. Indeed, Φ_2^m is finite of degree 1 by 4.1.c, 4.2.5 and 3.2.1. If we know that $\Phi_2(\widetilde{\mathfrak{B}/\Gamma'}) = \widehat{\mathfrak{B}/\Gamma''}$, then it follows that Φ_2^m is a finite birational morphism onto a normal surface. This is only possible if Φ_2^m is an isomorphism.

We know from the Baily-Borel *embedding theorem* that

4.3.5. $$\widehat{\mathfrak{B}/\Gamma''} = \operatorname{Proj} \bigoplus_{m \geq 0} [\Gamma', m] = \operatorname{Proj} \mathbb{C}[Y_0, Y_1, Y_2, z]$$
$$= \operatorname{Proj} \mathbb{C}[Y_0^2, Y_1^2, Y_2^2, Y_0Y_1, Y_0Y_2, Y_1Y_2, z] = V \subset \mathbb{P}^6.$$

Here we used the well-known isomorphism

$$\operatorname{Proj} S \simeq \operatorname{Proj} \bigoplus_{d=0}^{\infty} S_{2d}$$

for graded rings $S = \bigoplus_{d=0}^{\infty} S_d$, and in order to get the last inclusion we changed degrees by dividing them by 2:

$$S'_d = S_{2d}, \qquad \operatorname{Proj} \bigoplus_{d=1}^{\infty} S'_d \subset \mathbb{P}^6.$$

On the other hand, Φ_2 is the following morphism:

$$\Phi_2 \colon P \mapsto \big(1 : f_1^2(P) : f_2^2(P) : f_1(P) : f_2(P) : f_1f_2(P) : g(P)\big).$$

It is immediately clear that the image of Φ_2 lies on V, hence it is V. Therefore $\Phi_2(\widetilde{\mathfrak{B}/\Gamma'}) = \widehat{\mathfrak{B}/\Gamma'}$.

Next we will determine an exact relation (4.3.2) for a suitable g; this means that we will determine precisely the homogeneous polynomials a, b, c. The considerations above suggest us to use the *Veronese embedding*

$$v \colon \mathbb{P}^2 \hookrightarrow \mathbb{P}^5$$
$$(y_0 : y_1 : y_2) \mapsto (y_0^2 : y_1^2 : y_2^2 : y_0y_1 : y_0y_2 : y_1y_2)$$
$$= (z_0 : z_1 : z_2 : z_3 : z_4 : z_5).$$

An system of equations for $v(\mathbb{P}^2)$ on $\mathbb{P}^5 = \operatorname{Proj} \mathbb{C}[Z_0, Z_1, Z_2, Z_3, Z_4, Z_5]$ is

4.3.6. $$v(\mathbb{P}^2) \colon Z_3Z_4 = Z_0Z_5, \; Z_3Z_5 = Z_1Z_4, \; Z_4Z_5 = Z_2Z_3.$$

We will see that v lifts along branched 3-sheeted covering maps to the Baily-Borel embedding. The corresponding cover over $\mathbb{P}^5$ will be simple enough to allow an explicit description of the image of $\widetilde{\mathfrak{B}/\Gamma'}$ in it in terms of projective coordinates. Recall the situation on $\mathbb{P}^2$. The surface $\widehat{\mathfrak{B}/\Gamma'}$ is the cyclic 3-sheeted cover $\mathbb{P}^2(\delta)$ over $\mathbb{P}^2$ branched along $\Delta = \sum \Delta_{ij}$ (see Figure 3.4.A, Theorem

3.3.1, Proposition 3.3.2), where

$$\mathcal{O}_{\mathbb{P}^2}(\Delta) \simeq \mathcal{O}_{\mathbb{P}^2}(6), \qquad \mathcal{O}_{\mathbb{P}^2}(\delta) \simeq \mathcal{O}_{\mathbb{P}^2}(2).$$

The essential point is that $\mathcal{O}_{\mathbb{P}^2}(\delta)$ is the inverse image by v of the generating class $\mathcal{O}_{\mathbb{P}^5}(1)$ of Pic $\mathbb{P}^5 \simeq \mathbb{Z}$ and Δ is the inverse image of a divisor Δ' on $\mathbb{P}^5$, $\mathcal{O}_{\mathbb{P}^5}(\Delta') \simeq \mathcal{O}_{\mathbb{P}^5}(3)$. The space $\mathbb{P}^5(\delta')$ will allow a convenient explicit description.

We begin with an explicit equation for Δ on $\mathbb{P}^2$:

$$\Delta\colon (Y_0 + Y_1)(Y_0 + Y_2)(Y_1 + Y_2)(Y_2 - Y_1)(Y_2 - Y_0)(Y_1 - Y_0) = 0$$

$$\text{or} \qquad (Y_2^2 - Y_1^2)(Y_2^2 - Y_0^2)(Y_1^2 - Y_0^2) = 0.$$

We define on $\mathbb{P}^5$ the divisor

$$\Delta'\colon (Z_2 - Z_1)(Z_2 - Z_0)(Z_1 - Z_0) = 0.$$

Obviously, $v^*(\Delta') = \Delta$. We will use the following functorial property for cyclic covers:

4.3.7. PROPOSITION. *Let $\iota\colon X \hookrightarrow V$ be an embedding of smooth algebraic varieties, Δ' an effective divisor on V, $\delta' \in \operatorname{Pic} V$ a d-th root of the class of $\mathcal{O}(\Delta)$, $\Delta = \iota^*(\Delta')$ a reduced divisor on X and $\delta = \iota^*(\delta')$. Then ι lifts to the d-sheeted cyclic covers $X(\delta)$, $V(\delta')$ and gives an embedding $\tau\colon X(\delta) \hookrightarrow V(\delta')$. More precisely we have a cartesian diagram*

$$\begin{array}{ccc} X(\delta) & \xrightarrow{\ \tau\ } & V(\delta') \\ \Big\downarrow{\scriptstyle \pi} & & \Big\downarrow{\scriptstyle \pi'} \\ X & \xrightarrow{\ \iota\ } & V \end{array} \qquad X(\delta) = X \times_V V(\delta').$$

For the proof we need the

Construction of cyclic d-sheeted covers $V(\delta) \to V$ branched along Δ

Let $\{V_i\}_{i \in I}$ be a sufficiently small affine open covering of V such that $\{\varphi_{ij}\}_{(i,j)\in I \times I}$ represents $\delta \in \operatorname{Pic} V = H^1(V, \mathcal{O}_V^*)$, $\{(V_i, f_i)\}_{i\in I}$ represents Δ and

$$f_i = \varphi_{ij}^d \cdot f_j \quad \text{on} \quad V_i \cap V_j \qquad (\Delta \cap V_i\colon f_i = 0).$$

This choice is possible because $\{f_i/f_j\}_{i,j}/\{\varphi_{ij}^d\}$ is a coboundary $\{\varepsilon_i/\varepsilon_j\}$, $\varepsilon_i \in H^0(V_i, \mathcal{O}_{V_i}^*)$, and we can replace f_i by f_i/ε_i. The φ_{ij} are the *transition functions* of a line bundle $\mathbb{L}(\delta)$. Together with the projections onto the base we have the following commutative diagrams:

$$\begin{array}{ccc} \mathbb{L}(\delta) & \hookleftarrow & \mathbb{L}(\delta)_i = \mathbb{A}^1 \times V_i \\ \Big\downarrow & & \Big\downarrow \\ V & \hookleftarrow & V_i \end{array} \qquad \mathbb{A}^1 = \operatorname{Spec} \mathbb{C}[U_i].$$

The gluing isomorphisms are

$$\mathbb{L}(\delta)_j \mid V_i \cap V_j \xrightarrow{\sim} \mathbb{L}(\delta)_i \mid V_i \cap V_j, \qquad u_i = \varphi_{ij} \cdot u_j .$$

Locally $V(\delta)_i$ is defined by

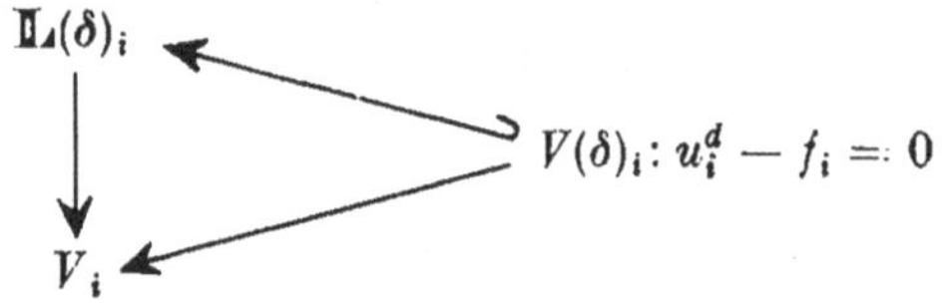

We illustrate the situation in the following figure using local coordinates t_1, t_2 on V_i:

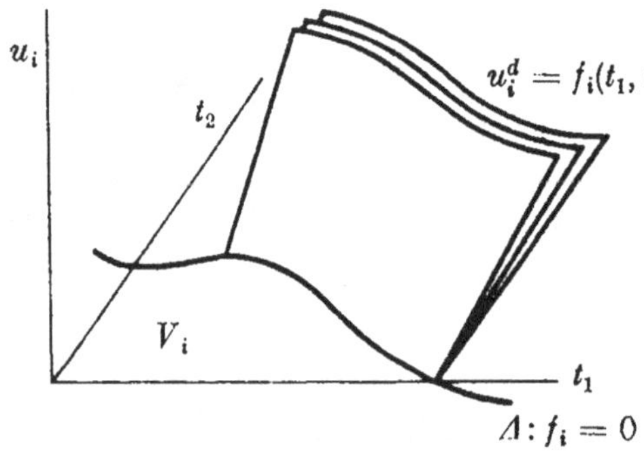

Clearly the $V(\delta)_i$'s can be glued together to a cyclic d-sheeted cover $V(\delta) \to V$ branched along Δ.

Changing notations we obtain the proof of 4.3.7 by restriction of the construction of $V(\delta')$ to $X \subset V$. Indeed, we obtain the following commutative diagram with cartesian quadrangles:

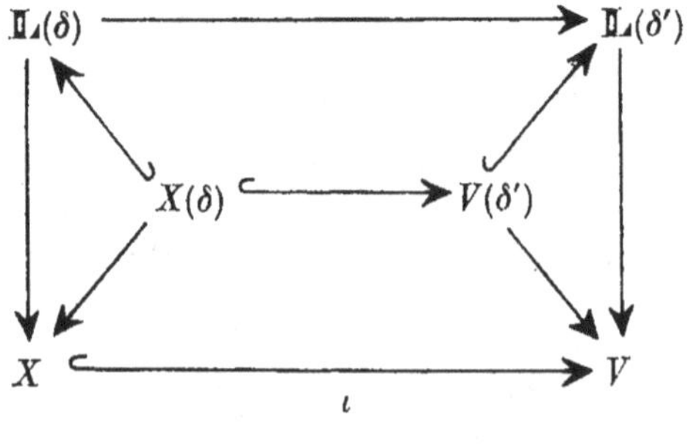

Now we apply the construction to our special situation

$$\iota = v\colon X = \mathbb{P}^2 \hookrightarrow V = \mathbb{P}^5, \qquad \Delta\colon \prod_{0 \le i < j \le 2} (Y_j^2 - Y_i^2) = 0, \qquad \Delta'\colon \prod_{0 \le i < j \le 2} (Z_j - Z_i) = 0 .$$

4.3.8. COROLLARY. *There is a cartesian diagram*

$$\begin{array}{ccc} \widehat{\mathfrak{B}/\Gamma'} = \mathbb{P}^2(\delta) & \overset{\tilde{v}}{\hookrightarrow} & \mathbb{P}^5(\delta') \\ {\scriptstyle \hat{\Phi}_1}\downarrow & & \downarrow{\scriptstyle \pi'} \\ \mathbb{P}^2 & \underset{v}{\hookrightarrow} & \mathbb{P}^5 \end{array}$$

Working on $V = \mathbb{P}^5$, $\mathbb{P}^5 \supset V_i \colon Z_i \neq 0$ with $\bar{\delta}' = \mathcal{O}_{\mathbb{P}^5}(1)$ we set

$$u_i = \frac{z}{z_i}, \qquad \varphi_{ij} = \frac{z_j}{z_i}, \qquad f_i = \frac{(z_2 - z_1)(z_2 - z_0)(z_1 - z_0)}{z_i^3},$$

$$\left(\frac{z}{z_i}\right)^3 = \frac{(z_2 - z_1)(z_2 - z_0)(z_1 - z_0)}{z_i^3} \qquad \text{on} \qquad V_i.$$

So $\mathbb{P}^5(\delta')$ is explicitly determined. Globally it is a subvariety of

$$\mathbb{P}^6 = \operatorname{Proj} \mathbb{C}[Z_0, \ldots, Z_5, Z],$$

defined by

$$\mathbb{P}^5(\delta') \colon Z^3 = (Z_2 - Z_1)(Z_2 - Z_0)(Z_1 - Z_0).$$

By Corollary 4.3.8 and (4.3.6) we see that

$$\begin{aligned} \mathbb{P}^6 \supset \widehat{\mathfrak{B}/\Gamma'} &= \operatorname{Proj} \mathbb{C}[Z_0, \ldots, Z_5, Z] \Big/ \Big(Z^3 - \prod_{0 \le i < j \le 2} (Z_j - Z_i),\ Z_3 Z_4 - Z_0 Z_5, \\ &\qquad\qquad Z_3 Z_5 - Z_1 Z_4,\ Z_4 Z_5 - Z_2 Z_3\Big) \\ &= \operatorname{Proj} \mathbb{C}[Y_0^2, Y_1^2, Y_2^2, Y_0 Y_1, Y_0 Y_2, Y_1 Y_2, Z] \Big/ \Big(Z^3 - \prod_{0 \le i < j \le 2} (Y_j^2 - Y_i^2)\Big) \\ &= \operatorname{Proj} \mathbb{C}[Y_0, Y_1, Y_2, Z] \Big/ \Big(Z^3 - \prod_{0 \le i < j \le 2} (Y_j^2 - Y_i^2)\Big). \end{aligned}$$

The last step comes from the identification $\operatorname{Proj} \bigoplus_{d=0}^{\infty} S_{2d} = \operatorname{Proj} \bigoplus_{d=0}^{\infty} S_d$ for graded rings $S = \bigoplus_{d=0}^{\infty} S_d$. We restrict $\widehat{\mathfrak{B}/\Gamma'} \to \mathbb{P}^2$ to the affine part $Y_0 \neq 0$. The surface $\widehat{\mathfrak{B}/\Gamma'}$ is the normalization of $\mathbb{P}^2$ in the function field of $\widehat{\mathfrak{B}/\Gamma'}$. Setting $f_1 = Y_1/Y_0$, $f_2 = Y_2/Y_0$, $z = Z/Y_0^2$ we see that

$$\mathbb{C}[f_1, f_2, z], \qquad z^3 = (f_2^2 - f_1^2)(f_2^2 - 1)(f_1^2 - 1),$$

is the affine coordinate ring of $\widehat{\mathfrak{B}/\Gamma'}$ over $Y_0 \neq 0$. On the other hand this ring was

$$\mathbb{C}[f_1, f_2, g], \qquad g^3 = a' g^2 + b' g + c'.$$

Now look again at the proof of 4.3.1 to see that one can choose $g = z$ there. Proposition 4.3.1 now yields the following central result:

4.3.9 THEOREM. *There is an isomorphism of graded rings*

$$\bigoplus_{m=0}^{\infty} [\Gamma', m] \cong \mathbb{C}[Y_0, Y_1, Y_2, z]/(Z^3 - (Y_2^2 - Y_1^2)(Y_2^2 - Y_0^2)(Y_1^2 - Y_0^2)),$$

where Y_0, Y_1, Y_2, Z are algebraically independent; Y_0, Y_1, Y_2 are of weight 1, *Z is of weight* 2.

4.4. *Structure of the Rings* $\bigoplus_{m=0}^{\infty} [\Gamma' \cdot H, m]_\chi$

The group $\tilde{\Gamma}/\Gamma'$ acts as group of automorphisms of the graded ring $\bigoplus_{m=0}^{\infty} [\Gamma', m]$. This action transfers to an action on the graded ring

$$R' = \mathbb{C}[Y_0, Y_1, Y_2, z], \qquad z^3 = \prod_{0 \leq i < j \leq 2} (Y_j + Y_i)(Y_j - Y_i).$$

By considering the generators Y_0, Y_1, Y_2, z we note that the action is uniquely determined by the action on the first two homogeneous parts

$$R_1' = \mathbb{C}Y_0 + \mathbb{C}Y_1 + \mathbb{C}Y_2, \qquad R_2' = \bigoplus_{0 \leq i < j \leq 2} \mathbb{C}Y_iY_j \oplus \mathbb{C}z.$$

We know the projective action of $\tilde{\Gamma}/\Gamma' \cong (\mathbb{Z}/3\mathbb{Z}) \times S_4$ (see (3.2.2)) on these parts by 3.3.5. Also $\mathbb{P}\bar{\varrho}$, $\langle\bar{\varrho}\rangle \cong (\mathbb{Z}/3\mathbb{Z})$, acts trivially on $\mathbb{P}R_1'$. In order to determine the action of $\mathbb{P}S_4$ on $\mathbb{P}R_1'$ we set

$$x_1 = \frac{1}{2}(Y_0 - Y_1 - Y_2), \qquad x_2 = \frac{1}{2}(-Y_0 + Y_1 - Y_2),$$

$$x_3 = \frac{1}{2}(-Y_0 - Y_1 + Y_2), \quad x_4 = \frac{1}{2}(Y_0 + Y_1 + Y_2).$$

Then

$$R_1' = \mathbb{C}x_1 + \mathbb{C}x_2 + \mathbb{C}x_3 + \mathbb{C}x_4, \qquad x_1 + x_2 + x_3 + x_4 = 0.$$

The inverse substitution is

$$Y_0 = x_1 + x_4, \qquad Y_1 = x_2 + x_4, \qquad Y_2 = x_3 + x_4.$$

The triple points of

$$\Delta: d = \prod_{0 \leq i < j \leq 2} (Y_j + Y_i)(Y_j - Y_i) = \prod_{1 \leq i < j \leq 4} (x_j - x_i) = 0$$

are in x-coordinates on $\mathbb{P}^2 = \mathbb{P}R_1'$ equal to

$$P_1 = (-3 : 1 : 1 : 1), \qquad P_2 = (1 : -3 : 1 : 1),$$
$$P_3 = (1 : 1 : -3 : 1), \qquad P_4 = (1 : 1 : 1 : -3).$$

So $(x_1 : x_2 : x_3 : x_4)$ are S_4-coordinates on $\mathbb{P}R_1' = \mathbb{P}^2$. Hence the action of S_4 on R_1' induces the symmetric action on $\mathbb{P}^2$ with respect to the x-coordinates. Two representations of S_4 on R_1' with the same projective action on $\mathbb{P}R_1'$ differ from each other by a character $\sigma\colon S_4 \to \mathbb{C}^*$. The only non-trivial character of S_4 is the signature sgn. So we have two possibilities, S_4^+ and S_4^-, of representations of S_4 on R_1' together with their natural extensions to R':

S_4^+: S_4 acts *symmetrically* on R_1' with respect to x_1, x_2, x_3, x_4;

S_4^-: S_4 acts *asymmetrically* on R_1' with respect to x_1, x_2, x_3, x_4:

$$S_4 \ni g\colon \mathbb{C}[x_1, x_2, x_3, x_4] \to \mathbb{C}[x_1, x_2, x_3, x_4],$$
$$x_1^i x_2^j x_3^k x_4^l \mapsto \big(\operatorname{sgn}(g)\big)^m \cdot x_{g(1)}^i x_{g(2)}^j x_{g(3)}^k x_{g(4)}^l,$$
$$m = i + j + k + l.$$

Extending the action to R' we find in both cases for $g \in S_4$

$$\big(g(z)\big)^3 = g(z^3) = g(d) = \operatorname{sgn}(g) \cdot d, \qquad g(z) = \operatorname{sgn}(g) \cdot z$$

because $d = z^3$, hence $g(z) = \tau(g)\, z$, τ a character of S_4, and

$$\operatorname{sgn}(g) \cdot d = g(z^3) = \big(g(z)\big)^3 = \big(\tau(g)\, z\big)^3 = \tau(g)\, z^3 = \tau(g) \cdot d.$$

Now $\bar{\varrho}$ acts trivially on $\mathbb{P}R_1'$ by Proposition 3.2.3. Therefore $\bar{\varrho} = \lambda \cdot \mathrm{id}_{R_1'}$, $\lambda^3 = 1$. On the other hand

$$\big(\bar{\varrho}(z)\big)^3 = \bar{\varrho}(z^3) = \bar{\varrho}(d) = \lambda^6 \cdot d = d, \qquad \bar{\varrho}(z) = \mu z, \qquad \mu^3 = 1.$$

Since $\bar{\varrho} \in \mathrm{Gal}(\widehat{\mathfrak{B}/\Gamma'}/\mathbb{P}^2)$ does not act trivially on $z/\xi_1\xi_2 \in \mathbb{C}(\widehat{\mathfrak{B}/\Gamma'})$ we have $\mu \cdot \lambda \neq 1$.

At the end of this section we will prove the following

4.4.1. LEMMA. *The action of $\tilde{\Gamma}/\Gamma' = (\mathbb{Z}/3\mathbb{Z}) \times S_4$ on $\bigoplus_{m=0}^{\infty} [\Gamma', m] \cong R'$ is the extension of the asymmetric action S_4^- of S_4 on R_1':*

$$(\mathbb{Z}/3\mathbb{Z}) \ni \bar{\varrho}\colon x_1, x_2, x_3, x_4, z \mapsto \omega x_1, \omega x_2, \omega x_3, \omega x_4, \mu z, \qquad (\mu\omega \neq 1)$$
$$S_4 \ni g\colon x_i, z \mapsto \operatorname{sgn}(g) \cdot x_{g(x_i)}, \operatorname{sgn}(g) \cdot z, \qquad i = 1, \ldots, 4.$$

Therefore there is a character $\chi\colon (\mathbb{Z}/3\mathbb{Z}) \times S_4 \to \mathbb{C}^*$ such that the twist of the original action of $\tilde{\Gamma}/\Gamma'$ with χ acts symmetrically on $\mathbb{C}[x_1, \ldots, x_4]$, $\bar{\varrho}_\chi(x_i) = x_i$, $\bar{\varrho}_\chi(z) = \omega z$, $g_\chi(z) = \operatorname{sgn}(g) \cdot z$. We define the $(\tilde{\Gamma}/\Gamma')_\chi$-action on $\mathbb{C}[x_1, \ldots, x_4, z]$ by

$$\gamma_\chi(f) = \chi(\gamma)^{\mathrm{weight}(f)} \cdot \gamma(f), \qquad \gamma \in (\mathbb{Z}/3\mathbb{Z}) \times S_4, \qquad f \textit{ homogeneous.}$$

If H is a normal subgroup of $\tilde{\Gamma}/\Gamma'$, then, by the definition of the group action on automorphic forms (see 4.1), we have

4.4.2. $$\bigoplus_m [\Gamma' \cdot H, m]_\chi = \Big(\bigoplus_m [\Gamma', m]\Big)^{H_\chi} \qquad (\chi \text{ indicates the twisted action}).$$

From 4.4.1 and (4.4.2) we deduce immediately

4.4.3. PROPOSITION. *It holds that*

$$\bigoplus_{m=0}^{\infty}[\tilde{\Gamma}', m]_\chi \cong \mathbb{C}[Y_0, Y_1, Y_2] \cong \mathbb{C}[X_1, X_2, X_3, X_4]/(X_1 + X_2 + X_3 + X_4),$$

with Y_0, Y_1, Y_2 *and* X_1, X_2, X_3, X_4 *algebraically independent of weight* 1. *In other words: The ring of* $\tilde{\Gamma}'_\chi$*-automorphic forms is generated by three algebraically independent forms of weight* 1.

Instead of 4.3.9 we write

4.4.4. $$\bigoplus_{m=0}^{\infty}[\Gamma', m] = \mathbb{C}[X_1, X_2, X_3, X_4, Z]/(X_1 + X_2 + X_3 + X_4, Z^3 - D),$$

$$D = \prod_{1 \le i < j \le 4} (X_j - X_i).$$

Substituting X_i and Z in 4.4.1 for x_i and z, respectively, we lift the action of $\tilde{\Gamma}/\Gamma'$ to the polynomial ring $\mathbb{C}[X_1, X_2, X_3, X_4, Z]$. Then the group action commutes with the factorization in (4.4.4). The same is true for the formation of H-invariant subrings. We denote the elementary symmetric functions of X_1, X_2, X_3, X_4 by

$$\Sigma_1 = -(X_1 + X_2 + X_3 + X_4),$$

$$\Sigma_2 = X_1X_2 + X_1X_3 + X_1X_4 + X_2X_3 + X_2X_4 + X_3X_4,$$

$$\Sigma_3 = -(X_1X_2X_3 + X_1X_2X_4 + X_1X_3X_4 + X_2X_3X_4)$$

$$\Sigma_4 = X_1X_2X_3X_4.$$

Their images in $\mathbb{C}[x_1, x_2, x_3, x_4] = \mathbb{C}[X_1, X_2, X_3, X_4]/(X_1 + X_2 + X_3 + X_4)$ are denoted by $G_1 = 0$, G_2, G_3 and G_4, respectively.

4.4.5. PROPOSITION. *There is an isomorphism of graded rings*

$$\bigoplus_{m=0}^{\infty}[\tilde{\Gamma}, m]_\chi \cong \mathbb{C}[G_2, G_3, G_4],$$

Proof. Use (4.4.2) with $H = \tilde{\Gamma}/\Gamma'$, (4.4.4) and 4.4.1. The element z is killed by the action of $\bar{\varrho}$. Therefore our ring of invariants is contained in $\mathbb{C}[x_1, x_2, x_3, x_4]$. Therefore 4.4.5 follows from classical elementary invariant theory for symmetric groups acting on polynomial rings.

4.4.6. PROPOSITION. *There is an isomorphism of graded rings*

$$\bigoplus_{m=0}^{\infty}[\Gamma, m]_\chi \cong \mathbb{C}[G_2, G_3, G_4, z^2].$$

The fundamental relation between the generators is

$$(z^2)^3 = д.$$

where

$$д = 16G_2^4G_4 - 128G_2^2G_4^2 - 4G_2^3G_3^2 + 144G_2G_4G_3^2 - 27(G_3^2)^2 + 256G_4^3.$$

Proof. The elements G_2, G_3, G_4 in the set of generators are found in the same manner as in the proof of 4.4.5. The action of S_4 kills z (see 4.4.1) but it does not kill z^2. With the notations of 4.4.4 we see that $Д = D^2$ is S_4-invariant. By invariant theory we know that

$$Д = D^2 = \prod_{i \neq j} (X_j - X_i)$$

can be expressed as a polynomial in $\Sigma_1, \Sigma_2, \Sigma_3, \Sigma_4$. This polynomial is explicitly known (see e.g. [59]):

4.4.7. $$27Д = 4(\Sigma_2^2 - 3\Sigma_1\Sigma_3 + 12\Sigma_4)^3 - (27\Sigma_1^2\Sigma_4 + 27\Sigma_3^2 + 2\Sigma_2^3 - 72\Sigma_2\Sigma_4 + 9\Sigma_1\Sigma_2\Sigma_3)^2,$$

hence

$$д = d^2 = 16G_2^4 - 128G_2^2G_4^2 - 4G_2^3G_3^2 + 144G_2G_3^2G_4 - 27G_3^4 + 256G_4^3.$$

The rest is clear.

There are no difficulties to obtain also the structure of the other rings $\bigoplus_{m=0}^{\infty} [\Gamma' \cdot H, m]_\chi$, $\bigoplus_{m=0}^{\infty} [\tilde{\Gamma}' \cdot H, m]_\chi$, $H = K_4$ or A_4. By invariant theory it follows that

$$\mathbb{C}[x_1, x_2, x_3, x_4]^{K_4} = \mathbb{C}[U, V, W, R]/(R^2 - UVW),$$

where

$$U = (x_1 + x_2)^2, \qquad V = (x_1 + x_3)^2, \qquad W = (x_1 + x_4)^2,$$
$$R = (x_1 + x_2)(x_1 + x_3)(x_1 + x_4).$$

On the other hand,

$$\mathbb{C}[X_1, X_2, X_3, X_4]^{A_4} = \mathbb{C}[\Sigma_1, \Sigma_2, \Sigma_3, \Sigma_4, Д]$$

with (4.4.7) as the fundamental relation.

The Table 4.4.α summarizes all our rings of automorphic forms. Since $\widehat{\mathfrak{B}/\Gamma' \cdot H} = \mathrm{Proj}\left(\bigoplus_m [\Gamma' \cdot H, m]\right)$ by the Baily-Borel theorem, the table is a finer version of Diagram 3.6.b in terms of automorphic forms, because we can change over to subrings of forms of weight 12 m, where the χ-twist is not perceptible.

Now we are ready to do the announced

4.4. α.

H	1	K_4	A_4	S_4
Proj R_{H_χ}	$\widehat{\mathfrak{B}/\Gamma'}$	$\mathfrak{B}/\widehat{\Gamma' \cdot K_4}$	$\mathfrak{B}/\widehat{\Gamma' \cdot A_4}$	$\widehat{\mathfrak{B}/\Gamma}$
R_{H_χ}	$\dfrac{\mathbb{C}[Y_0, Y_1, Y_2, Z]}{(Z^3 - d)}$	$\dfrac{\mathbb{C}[U, V, W, Z, R]}{(Z^3 - d, R^2 - UVW)}$	$\dfrac{\mathbb{C}[Z, G_2, G_3, G_4]}{(Z^6 - \text{д})}$	$\dfrac{\mathbb{C}[G_2, G_3, G_4]}{(S^3 - \text{д})}$
$\tilde{R}_{H_\chi}$	$\mathbb{C}[Y_0, Y_1, Y_2]$	$\dfrac{\mathbb{C}[U, V, W, R]}{(R^2 - UVW)}$	$\dfrac{\mathbb{C}[G_2, G_3, G_4, D]}{(D^2 - \text{д})}$	$\mathbb{C}[G_2, G_3, G_4]$
Proj $\tilde{R}_{H_\chi}$	$\widehat{\mathfrak{B}/\tilde{\Gamma}'}$	$\mathfrak{B}/\widehat{\tilde{\Gamma}' \cdot K_4}$	$\mathfrak{B}/\widehat{\tilde{\Gamma}' \cdot A_4}$	$\widehat{\mathfrak{B}/\tilde{\Gamma}}$

$$R_{H_\chi} = \bigoplus_{m=0}^{\infty} [\Gamma' \cdot H_\chi, m], \qquad \tilde{R}_{H_\chi} = \bigoplus_{m=0}^{\infty} [\tilde{\Gamma}' \cdot H_\chi, m]$$

$$d(Y_0, Y_1, Y_2) = (Y_2^2 - Y_1^2)(Y_2^2 - Y_0^2)(Y_1^2 - Y_0^2), \quad d(U, V, W) = (U - V)(U - W)(V - W)$$

$$\text{д}(G_2, G_3, G_4) = 16G_2^4G_4 - 128G_2^2G_4^2 - 4G_2^3G_3^2 + 144G_2G_3^2G_4 - 27G_3^4 + 256G_4^3$$

$\mathbb{C}[\ldots]$: graded ring generated by the algebraically independent elements listet between the brackets of the following weights:

element:	Y_0	Y_1	Y_2	Z	U	V	W	R	G_2	G_3	G_4	S
weight:	1	1	1	2	2	2	2	3	2	3	4	4

Proof of Lemma 4.4.1. Assume that the action of $S_4 = \tilde{\Gamma}/\tilde{\Gamma}'$ on $[\tilde{\Gamma}', 1]$ is symmetric. That means that there are four $\tilde{\Gamma}'$-automorphic forms of weight 1 $\xi_1, \xi_2, \xi_3, \xi_4$ such that $\gamma(\xi_i) = \xi_{g(i)}$, where $g = \gamma \bmod \tilde{\Gamma}' \in S_4$. In particular we find for $i = 1, 3$, $g = (2, 4)$, $\gamma = (2, 4)_{\mathfrak{B}} = \operatorname{diag}(-1, 1, -1)$, $\mathfrak{D}_{13} = \operatorname{Fix}_{\mathfrak{B}}\gamma$:

$$\xi_i = \gamma(\xi_i) = j_\gamma \cdot \gamma^*(\xi_i) = -\gamma^*\xi_i, \qquad \xi_i | \mathfrak{D}_{13} = 0.$$

Going down to $\widehat{\mathfrak{B}/\tilde{\Gamma}'} = \mathbb{P}^2$ we find for the corresponding S_4-coordinates x_1, x_3 that the points on Δ_{13} have coordinates $(0 : x_2 : 0 : +x_2)$. But this is only one point on $\mathbb{P}^2$.

This is a contradiction.

The same conclusion works for ξ_i, $\varrho = \operatorname{diag}(\omega, 1, 1)$, $\mathfrak{D}_{24} = \operatorname{Fig}_{\mathfrak{B}}\, \varrho$. It follows that $\varrho(\xi_i) = \omega\xi_i$. Lemma 4.4.1 is proved.

Let $\tilde{\Gamma} \to \mathbb{C}^*$ be the non-trivial character $\tilde{\Gamma} \to \tilde{\Gamma}/\Gamma'' = (\mathbb{Z}/3\mathbb{Z}) \times S_4 \xrightarrow[\chi]{} \mathbb{C}^*$. We denote it also by χ.

4.4.8. DEFINITION. A holomorphic function f on $\mathfrak{B}$ is called a $\tilde{\Gamma}_\chi$-*automorphic form* (*of Nebentypus*) of weight r, if f is $\tilde{\Gamma}$-invariant under the action

$$f \mapsto \chi(\gamma)^r \cdot j_\gamma^r \cdot \gamma^*(f), \qquad \gamma \in \tilde{\Gamma}.$$

Let $[\tilde{\Gamma}, m]_\chi$ be the vector space of $\tilde{\Gamma}_\chi$-automorphic forms of weight m. The ring $\bigoplus_{m=0}^{\infty} [\tilde{\Gamma}, m]_\chi$ is called the *ring of* $\tilde{\Gamma}_\chi$-*automorphic forms*.

From the definitions and the invariant theory of symmetric groups acting on polynomial rings we obtain the following result:

4.4.9. PROPOSITION. *The ring of* $\tilde{\Gamma}_\chi$-*automorphic forms is the weighted polynomial ring* $\mathbb{C}[G_2, G_3, G_4]$.

§ 5. ALGEBRAIC CLASSIFICATION OF PICARD CURVES

5.1. *Projective Equivalence*

For the sake of short and precise notations and conclusions we use categorial language. We start with the following definitions and notations:

Div$^+\mathbb{P}^2$ denotes the category of projective plane curves. The objects are the (effective, embedded) plane algebraic curves $C \hookrightarrow \mathbb{P}^2$. We have a biunique correspondence

$$\mathbf{Div}^+\, \mathbb{P}^2 = \operatorname{Ob} \mathbf{Div}^+\, \mathbb{P}^2 \Leftrightarrow \{\text{homogeneous polynomials in } W, X, Y\}/\mathbb{C}^*.$$

$$C(f) : f = 0 \leftarrow\!\!\shortmid f = f(W, X, Y)$$

A morphism $C \to C'$ is a projective transformation $\alpha \in \mathbb{P}\mathrm{Gl}_3(\mathbb{C}) = \mathrm{Aut}\,\mathbb{P}^2$ sending C to C'; or more precisely $\alpha^*(C') = C$ in $\mathrm{Div}\,\mathbb{P}^2$. For reduced curves C, C' a morphism is a commutative diagram

$$\begin{array}{ccc} C & \hookrightarrow & \mathbb{P}^2 \\ \downarrow \wr & & \downarrow \wr\,\alpha \\ C' & \hookrightarrow & \mathbb{P}^2 \end{array}$$

PiC denotes the complete subcategory of $\mathbf{Div}^+\,\mathbb{P}^2$ of all Picard curves

$$C\colon\ Y^3 - p_4(X) = 0, \qquad p_4(X) \text{ a polynomial of degree } 4,$$

more precisely C is the projective closure of the zero set defined above:

$$C\colon\ WY^3 - W^4 p_4\left(\frac{X}{W}\right) = 0.$$

Now we let correspond to each *Picard curve*

$$C\colon f = WY^3 - \lambda \prod_{i=1}^{4} (X - e_i W) = 0$$

a plane curve $\mathfrak{R}(C)$ consisting of a number of projective lines:

$$\mathfrak{R}(C) = L_Y + L_W + N_1 + N_2 + N_3 + N_4,$$

$$L_Y\colon Y = 0, \qquad L_W\colon W = 0, \qquad N_i\colon e_i W - X = 0, \qquad i = 1, \ldots, 4.$$

$\mathfrak{R}(C)$ is called the *rack* of C. The intersection points of C and $\mathfrak{R}(C)$ are denoted by B_i, more precisely

$$B_0 = (0:0:1), \qquad B_i = (1:e_i:0), \qquad i = 1, \ldots, 4.$$

Geometrically we can describe the curve constellation $C + \mathfrak{R}(C)$ in Figure 5.1.A.
Notice the following properties:

5.1.1.

(a) $C \cap N_i = \{B_0, B_i\}, \qquad i = 1, \ldots, 4$

(b) L_W is the (inflection) tangent of C through B_0.

(c) N_i is the (inflection) tangent line of C through the inflection point B_i, if B_i is a regular point of C. The line N_i is the only projective line through B_i with the property $(N_i \cdot C)_{B_i} = 3$ if B_i coincides with at most one B_j, $j \neq i$.

Proof. (a) is trivial. B_0 is in any case a regular point of C. This is easily checked by the well-known fact that the singular locus $\mathrm{Sg}\,(C)$ of C is determined by the following equations:

5.1.A.

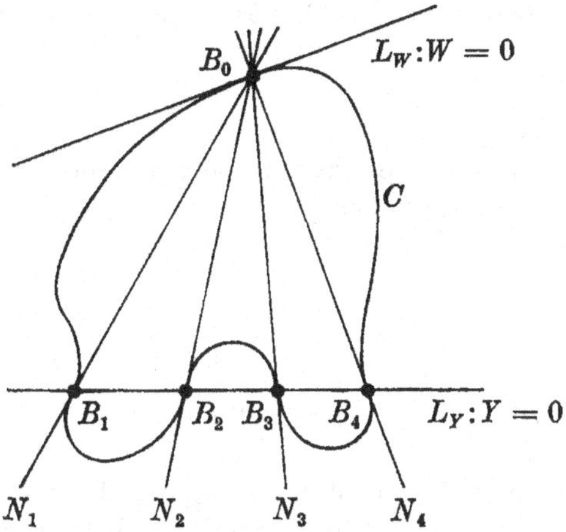

5.1.2. $\operatorname{Sg}(C) : f = \dfrac{\partial f}{\partial W} = \dfrac{\partial f}{\partial X} = \dfrac{\partial f}{\partial Y} = 0,$

$$\operatorname{Sg}(C) = \{B_i ; B_i = B_j \text{ for a } j \neq i\} \subseteqq \{B_1, B_2, B_3, B_4\}.$$

B_0 is the only intersection point of C and L_W. Therefore

$$(C \cdot L_W)_{B_0} = (C \cdot L_W) = 4$$

by Bezout's theorem. This implies (b). Since $N_i \neq L_W$, N_i cannot be the tangent line of C through B_0, hence $(N_i \cdot C)_{B_0} = 1$. By (a) and Bezout's theorem it follows that

$$4 = (N_i \cdot C) = (N_i \cdot C)_{B_0} + (N_i \cdot C)_{B_i} = 1 + (N_i \cdot C)_{B_i}.$$

Therefore $(N_i \cdot C)_{B_i} = 3$ and the first part of (c) is proved. The second part is an easy local coordinate calculation.

The essential reason for the use of categorial language is the presence of the *Hesse functor*

$$\mathfrak{h} : \mathbf{Div}^+ \mathbb{P}^2 \to \mathbf{Div}^+ \mathbb{P}^2 .$$

It is defined in the following manner: Let $f = f(W, X, Y)$ be a homogeneous polynomial. The *Hesse polynomial* of f is defined by

$$\mathfrak{h}(f) = \det \begin{pmatrix} \dfrac{\partial^2 f}{\partial W^2} & \dfrac{\partial^2 f}{\partial W\, \partial X} & \dfrac{\partial^2 f}{\partial W\, \partial Y} \\[2ex] \dfrac{\partial^2 f}{\partial X\, \partial W} & \dfrac{\partial^2 f}{\partial X^2} & \dfrac{\partial^2 f}{\partial X\, \partial Y} \\[2ex] \dfrac{\partial^2 f}{\partial Y\, \partial W} & \dfrac{\partial^2 f}{\partial Y\, \partial X} & \dfrac{\partial^2 f}{\partial W^2} \end{pmatrix}.$$

If $f = f_C$ is a polynomial corresponding to $C \in \mathrm{Div}^+\mathbb{P}^2$, then we define the *Hesse curve* of C as

$$\mathfrak{h}(C): \mathfrak{h}(f_C) = 0.$$

It follows easily from the chain rule for derivatives that $\mathfrak{h}$ is a functor.

Now let $\mathbf{PiC}_0$ denote the complete subcategory

$$\mathrm{i}_0: \mathbf{PiC}_0 \hookrightarrow \mathbf{PiC}$$

of curves of equation type

5.1.3. $$C: WY^3 - \prod_{i=1}^{4} (X - e_i W) = 0, \qquad \sum_{i=1}^{4} e_i = 0.$$

The embedding i_0 defines an equivalence of the categories $\mathbf{PiC}_0$ and $\mathbf{PiC}$. A corresponding inverse functor is realized by (*generalized*) *Tschirnhaus transformations*

5.1.4. $$\tau = \begin{pmatrix} * & 0 & 0 \\ * & * & 0 \\ 0 & 0 & * \end{pmatrix}: \mathbb{P}^2 \xrightarrow{\sim} \mathbb{P}^2$$

of Picard curves C. This sends C to a Picard curve $C_0 \in \mathbf{PiC}_0$.

The central point of this section is the following

5.1.5. LEMMA. $\mathfrak{R}: \mathbf{PiC} \to \mathbf{Div}^+ \mathbb{P}^2$ *is a functor.*

First we reduce the proof to the objects of $\mathbf{PiC}_0$. A Tschirnhaus transformation (5.1.4) $\tau: C \to C_0$ of a Picard curve C does not move B_0, L_W and L_Y. It sends $\{B_0, B_1, B_2, B_3\} = C \cap L_Y$ to $C_0 \cap L_Y = \{B_1^0, B_2^0, B_3^0, B_4^0\}$. Therefore we have with obvious notations

$$\tau(\mathfrak{R}(C)) = \tau\left(L_Y + L_W + \sum_{i=1}^{4} N_i\right) = L_Y + L_W + \sum_{i=1}^{4} N_i^0 = \mathfrak{R}(C_0).$$

The commutative Diagram 5.1.a with the two Tschirnhaus transformations τ, τ' of the Picard curves C and C', respectively, shows that it suffices to prove the Lemma for the restriction of the functor $\mathfrak{R}$ to $\mathbf{PiC}_0$.

So assume that $C \in \mathbf{PiC}_0$ is given by equation (5.1.3) written out as

5.1.6. $$f = WY^3 - (X^4 + aX^2W^2 + bXW^3 + cW^4) = 0.$$

5.1.7. L_Y is a linear component of $\mathfrak{h}(C)$. If $a \neq 0$ or $b \neq 0$, then L_Y is the only linear component of $\mathfrak{h}(C)$. If $a = b = 0$, $c \neq 0$, then the linear components of $\mathfrak{h}(C)$ are:

$$L_X: X = 0, L_Y, L_k: Y + 2 \cdot \omega^k \cdot \sqrt[3]{c}\, W = 0, \qquad k = 0, 1, 2,$$

with multiplicities 2 for L_X and 1 for L_Y and L_k.

Proof. We calculate the Hesse polynomial:

$$\mathfrak{h}(f) = \det\begin{pmatrix} 2aX^2 + 6bXW + 12cW^2 & 4aXW + 3bW^2 & 3Y^2 \\ 4aXW + 3bW^2 & 12X^2 + 2aW^2 & 0 \\ -3Y^2 & 0 & 6WY \end{pmatrix}$$

5.1.a.

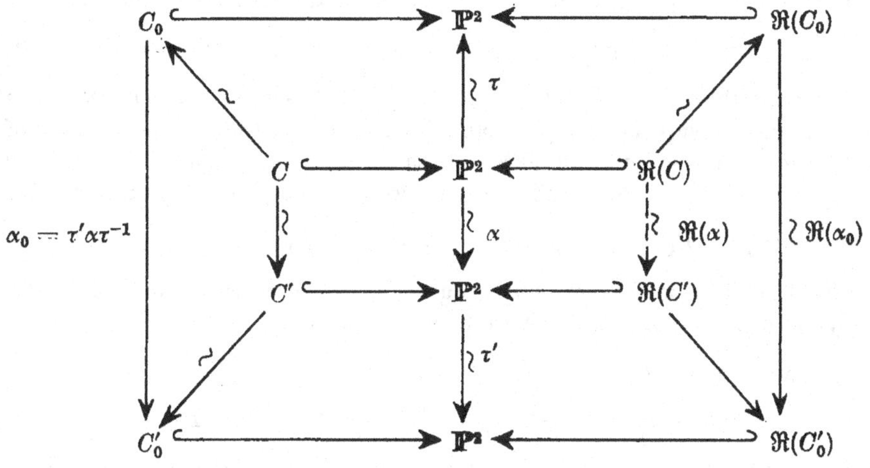

5.1.8. $\mathfrak{h}(f) = 6YG(W, X, Y)$, G homogeneous of degree 5,

$$G(1, X, Y) = g(X, Y)$$
$$= (3Y^3 + 4aX^2 + 12bX + 24c)(6X^2 + a) - (4aX + 3b)^2.$$

The first part of 5.1.7 is immediately clear. From (5.1.8) it follows that L_W is not a component of $\mathfrak{h}(C)$. So we have to find the linear factors of the inhomogeneous polynomial $g(X, Y)$. Up to a scalar factor they have the form $Y - (rX + s)$ or $X - t$. But the polynomials $g(X, rX + s)$, $g(t, Y)$ vanish identically for suitable fixed values $r, s, t \in \mathbb{C}$ if and only if $a = b = 0$. This proves the second part of 5.1.7. In order to prove the third part of 5.1.7 it suffices to check that

5.1.9. $\mathfrak{h}(f) = 108X^2Y(Y^3 + 8cW^3)$ if $a = b = 0$.

5.1.10. $\mathfrak{l}_Y$: $\mathbf{PiC}_0 \to \mathbf{Div}^+ \mathbb{P}^2$ is a functor.

$$C \mapsto L_Y\colon Y = 0.$$

Proof. Let $\alpha: C \to C'$ be a morphism in $\mathbf{PiC}_0$. Since $\mathfrak{h}$ is a functor we have a commutative diagram

$$\begin{array}{ccccc} C & \hookrightarrow & \mathbb{P}^2 & \hookleftarrow & \mathfrak{h}(C):\ \mathfrak{h}(f) = 0 \\ \downarrow \wr & & \alpha \downarrow \wr & & \downarrow \wr \\ C' & \hookrightarrow & \mathbb{P}^2 & \hookleftarrow & \mathfrak{h}(C'):\ \mathfrak{h}(f') = 0 \end{array}$$

We divide the curves of $\mathbf{PiC}_0$ given by equations (5.1.6) into three classes:

$$\text{(i) } a \neq 0 \text{ or } b \neq 0, \qquad \text{(ii) } a = b = 0,\ c \neq 0, \qquad \text{(iii) } a = b = c = 0.$$

The diagram above, 5.1.7 and (5.1.9) show that α preserves these classes. L_Y is the only linear component of $\mathfrak{h}(C)$, $\mathfrak{h}(C')$ in case (i) and the only component of multiplicity 4 in case (iii). Therefore in both of these cases $\alpha(L_Y) = L_Y$. Assume that $a = b = 0$, $c \neq 0$. For similar reasons we find with obvious notations that

$$\alpha(L_X) = L_X, \qquad \alpha(L_Y + L_0 + L_1 + L_2) = L_Y + L_0' + L_1' + L_2'.$$

$\mathbf{P} = (0:1:0)$ is a fixed point of α because it is the intersection point of L_Y, L_0, L_1, L_2 and of L_Y, L_0', L_1', L_2'. Taking into consideration

$$\alpha(C \cap L_X) = C' \cap L_X,$$

$$\alpha\big(C \cdot (L_Y + L_0 + L_1 + L_2)\big) = C' \cdot (L_X + L_0' + L_1' + L_2')$$

an easy calculation shows that $\alpha|_{L_X}$ is the identity. In particular $\alpha(Q) = Q$ for $Q = (1:0:0)$, hence

$$\alpha(L_Y) = \alpha(\overline{PQ}) = \overline{\alpha(P)\,\alpha(Q)} = \overline{PQ} = L_Y.$$

Proof of Lemma 5.1.5. From 5.1.10 it follows that

$$\alpha(\{B_1, B_2, B_3, B_4\}) = \{B_1', B_2', B_3', B_4'\}.$$

If B_i is a regular point of C, then B_0 is characterized by 5.1.1 as the intersection point of C and the tangent line of C through B_i. Now α preserves this property, hence we find successively

$$\alpha(B_0) = B_0,$$

$$\alpha(N_1 + N_2 + N_3 + N_4) = \alpha(\textstyle\sum \overline{B_0 B_i}) = \sum \overline{B_0 B_i'}$$
$$= N_1' + N_2' + N_3' + N_4'.$$

If C has exactly two singularities, say $B_1 = B_2 \neq B_3 = B_4$ (see (5.1.2)), then by 5.1.1(c) N_i is the only projective line through B_i such that $(G \cdot C)_{B_i} = 3$. Therefore

$$\alpha(\{N_1, N_3\}) = \{N_1', N_3'\}, \qquad \alpha(B_0) = \alpha(N_1 \cap N_3) = N_1' \cap N_3' = B_0.$$

In the remaining case $Y^3 = X^4$ we have $\alpha(N_i) = N_i$ because $L_X = N_i$, $i = 1, \ldots, 4$, and $\alpha(L_X) = L_X$ by (5.1.9). Intersecting L_X with C it follows that $\alpha(B_0) = B_0$. So we proved the last identity in all cases. L_W is the tangent line of C through B_0 and α preserves this property. Hence $\alpha(L_W) = L_W$. This finishes the proof of Lemma 5.1.5.

5.1.11. COROLLARY. *Two Picard curves in* $\mathbf{PiC}_0$

$$C: WY^3 - \prod_{i=1}^{4}(X - e_iW) = 0, \qquad C': WY^3 - \prod_{i=1}^{4}(X - e'_iW) = 0,$$

$$\sum_{i=1}^{4} e_i = \sum_{i=1}^{4} e'_i = 0$$

are projectively equivalent if and only if there exists a $c \in \mathbb{C}^*$ *such that* $e_i = c \cdot e'_i$ *for* $i = 1, \ldots, 4$ *up to a permutation of the* $e_1, \ldots, e_4$.

Proof. Let $C' \xrightarrow{\alpha} C$ be a morphism in $\mathbf{PiC}_0$. Lemma 5.1.5 implies that $\alpha: \mathbb{P}^2 \xrightarrow{\sim} \mathbb{P}^2$ is a Tschirnhaus transformation (5.1.4). We denote the polynomials in 5.1.11 by f and f'. From $\alpha^*(f) \in \mathbb{C}^* \cdot f'$ and the vanishing conditions in 5.1.11 one deduces easily that α is represented by a diagonal matrix

$$A = \begin{pmatrix} \nu^3/\mu^4 & 0 & 0 \\ 0 & \mu & 0 \\ 0 & 0 & \nu \end{pmatrix}, \qquad \mu, \nu \in \mathbb{C}^*,$$

and that $e'_i = \dfrac{1}{\mu} e_i$ for $i = 1, \ldots, 4$. Conversely the diagonal matrix above defines a morphism $C' \xrightarrow{\alpha} C$ in $\mathbf{PiC}_0$ with the proportionality property for the e_i's and e'_i's in 5.1.11.

We denote by $\mathbf{PiC}^*$ the complete subcategory of $\mathbf{PiC}$ consisting of all Picard curves which are not projectively equivalent to the unstable Picard curve $Y^3 = X^4$. Its class of objects is denoted by PiC^*. Furthermore we set

$$\mathbb{C}_0^{4*} = \left\{(e_1, e_2, e_3, e_4) \in \mathbb{C}^4;\ \sum_{i=1}^{4} e_i = 0\right\} \setminus \{(0, 0, 0, 0)\}.$$

$\mathbb{C}^* \times S_4$ acts on $\mathbb{C}_0^{4*}$ in the following way:

$$\mathbb{C}^* \times S_4 \ni (\lambda, g): (e_1, e_2, e_3, e_4) \mapsto \lambda \cdot (e_{g(1)}, e_{g(2)}, e_{g(3)}, e_{g(4)}).$$

The quotient space $\mathbb{C}_4^0/\mathbb{C}^*$ is identified with $\mathbb{P}^2$. With these notations we have the following biunique correspondence:

5.1.12. PROPOSITION. *It holds that*

$$\mathbb{P}^2/S_4 = \mathbb{C}_0^{4*}/\mathbb{C}^* \times S_4 \leftrightarrow \mathrm{PiC}^*/\textit{projective equivalence},$$

$$(e_1 : e_2 : e_3 : e_4) \mapsto C: Y^3 - (X - e_1)(X - e_2)(X - e_3)(X - e_4) = 0.$$

Proof. C on the right-hand side of the correspondence belongs to PiC_0^*. Taking into account the Tschirnhaus transformations (5.1.4) we see that PiC^* can be replaced by PiC_0^* in 5.1.12. Now the proposition is an immediate consequence of Corollary 5.1.11.

5.2. *Classification up to Isomorphism*

We start with the following well-known

5.2.1. LEMMA. *Two smooth projective plane curves C, C' of degree 4 are projectively equivalent if and only if they are isomorphic in the category of schemes over $\mathbb{C}$.*

We have to check the if-part. A projective line $H \subset \mathbb{P}^2$ restricts to a canonical divisor $h = \iota^*(H)$, $\iota\colon C \hookrightarrow \mathbb{P}^2$, on C because

$$\mathcal{O}(H) = \mathcal{O}(1) = \mathcal{O}(-3) \otimes \mathcal{O}(4) \cong \Omega^2_{\mathbb{P}^2} \otimes \mathcal{O}(C), \qquad \mathcal{O} = \mathcal{O}_{\mathbb{P}^2}.$$

Furthermore

5.2.2. $H^0(\mathbb{P}^2, \mathcal{O}(H)) \xrightarrow[\text{restr.}]{} H^0(C, \mathcal{O}_C(h))$ is an isomorphism.

Indeed, (5.2.2) is a part of the exact cohomology sequence of the exact sheaf sequence

$$0 \to \Omega^2_{\mathbb{P}^2} \to \Omega^2_{\mathbb{P}^2} \otimes \mathcal{O}(C) \to \mathcal{O}_C(h) \to 0$$

and $H^0(\mathbb{P}^2, \Omega^2_{\mathbb{P}^2}) = H^1(\mathbb{P}^2, \Omega^2_{\mathbb{P}^2}) = 0$. The morphism $\Phi_H\colon \mathbb{P}^2 \to \mathbb{P}^2$ is an isomorphism and $\Phi_h = \Phi_H \circ \iota$. Therefore ι is a canonical embedding. The same is true for $\iota'\colon C' \hookrightarrow \mathbb{P}^2$. An isomorphism $\alpha\colon C \xrightarrow{\sim} C'$ makes correspond canonical divisors to canonical divisors. On the other hand two embeddings of linearly equivalent divisors are projectively equivalent. The lemma is proved.

We denote by $\mathrm{PiC}^{\mathrm{sm}}$ the set of all smooth Picard curves. From (5.1.2) we know that $\mathrm{PiC}_0^{\mathrm{sm}}$ consists of all curves

$$C\colon Y^3 - \prod_{i=1}^{4} (X - e_i) = 0, \qquad \sum_{i=1}^{4} e_i = 0, \qquad \prod_{1 \le i < j \le 4} (e_j - e_i) \neq 0.$$

We see that $\mathcal{P} = \mathbb{P}^2 \setminus \Delta$ is the parameter space of "*labelled*" smooth Picard curves. The space of smooth Picard curves is $\mathcal{P}/S_4 = (\mathbb{P}^2 \setminus \Delta)/S_4$. Indeed, by 5.1.12 and 5.2.1 we have the following biunique correspondence:

5.2.3. PROPOSITION. *It holds that*

$$\mathcal{P}/S_4 = (\mathbb{P}^2 \setminus \Delta)/S_4 \leftrightarrow \mathrm{PiC}^{\mathrm{sm}}/\textit{isomorphy}.$$

Now we will investigate in more detail the singular Picard curves, that means the Picard curves with singularities. The set of these curves is denoted by $\mathrm{PiC}^{\mathrm{sg}}$. These are the Picard curves with a non-reduced rack. According to the type of the rack we divide PiC^{sg} into four classes:

$$\mathrm{PiC}^{sg} = \mathrm{PiC}^{h} \sqcup \mathrm{PiC}^{c} \sqcup \mathrm{PiC}^{r} \sqcup \mathrm{PiC}^{u}.$$

The notations become clear in Table 5.2.α. It is easy to see that the table is complete, that means that each embedded Picard curve with a singularity is projectively equivalent to one of the table. The geometric genus $g(C)$ is the genus of the normalization $\tilde{C}$ of C. For the calculation of the genus we use the 3-sheeted Galois covering map $C \to \mathbb{P}^1$ coming from the coordinate projection $\mathbb{P}^2 \dashrightarrow \mathbb{P}^1$, $(w : x : y) \mapsto (w : x)$. Geometrically it is the central projection of $C \subset \mathbb{P}^2$ onto L_Y with center B_0. The ramification locus is

$$\mathrm{Ram}\,(C) = \{B_0 = (0:0:1), B_1, B_2, B_3, B_4\}.$$

Now apply the Hurwitz genus formula (2.3.8) to $\tilde{C} \to C \to \mathbb{P}^1$ using the fact that the resolutions of singularities of type $Y^3 = X^2$, $Y^3 = X^3$ or $Y^3 = X^4$ consist of 1, 3 and 1 point(s), respectively.

From the singularity constellation in 5.2.α it is clear that two singular Picard curves of different classes cannot be isomorphic. The row listing the genus of the various curves shows that even the smooth models of two singular non-unstable Picard curves of different classes are not isomorphic. Two non-hyperelliptic singular Picard curves of the same class are projectively equivalent by 5.1.12, hence they are isomorphic.

In the remaining part of this section we prove that all hyperelliptic Picard curves are isomorphic to each other. The proof rests upon Igusa's classification of special curves of genus 2.

5.2.4. PROPOSITION. *Let C, C' be two hyperelliptic Picard curves and $\tilde{C}$, $\tilde{C}'$ their normalizations. Then $\tilde{C}$ and $\tilde{C}'$ are isomorphic.*

Proof. Aut $\tilde{C}$ contains an element of order 3 coming from the covering map $\tilde{C} \to \mathbb{P}^1$ described above. Therefore $\tilde{C}$ is a very special curve of genus 2. One can find a classification of all curves of genus two with higher automorphism groups, that means with automorphism group greater than $\mathbb{Z}/2\mathbb{Z}$, in Igusa's paper [35]. There are exactly three types with automorphism groups containing $\mathbb{Z}/3\mathbb{Z}$ as a subgroup. Their Rosenhain normal forms are:

$$\text{(I)} \qquad V^2 = U(U-1)(U-\lambda)\left(U - \frac{\lambda-1}{\lambda}\right)\left(U - \frac{1}{1-\lambda}\right),$$

$$\lambda \neq 0, 1, 2, \mathrm{i} = \sqrt{-1},$$

$$\text{(II)} \qquad V^2 = U(U-1)(U-2)\left(U-\frac{1}{2}\right)(U+1) \qquad (\lambda = 2),$$

$$\text{(III)} \qquad V^2 = U(U-1)(U-\mathrm{i})(U+\mathrm{i})(U+1) \qquad (\lambda = \mathrm{i}).$$

5.2. α. Table of singular Picard curves C

class of C	PiC^h	PiC^e	PiC^r	PiC^u
$C + \mathfrak{R}(C)$	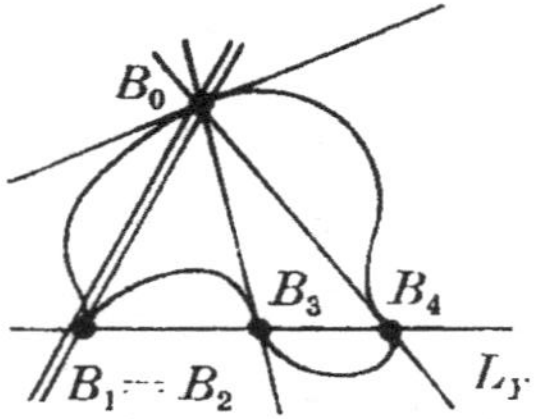	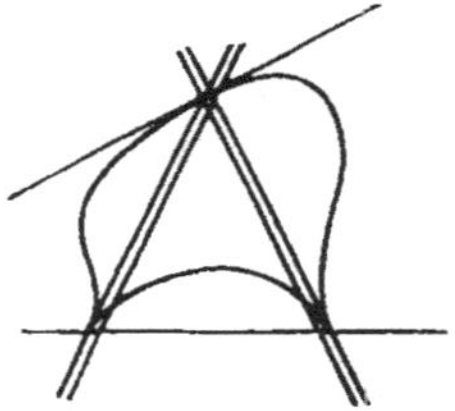	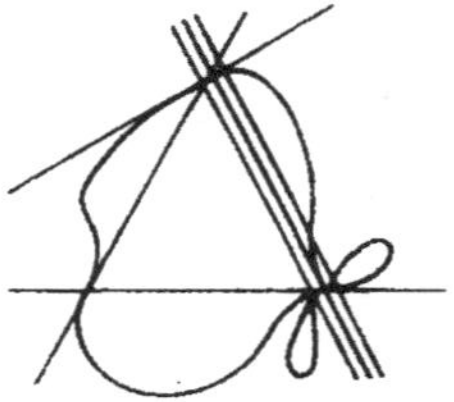	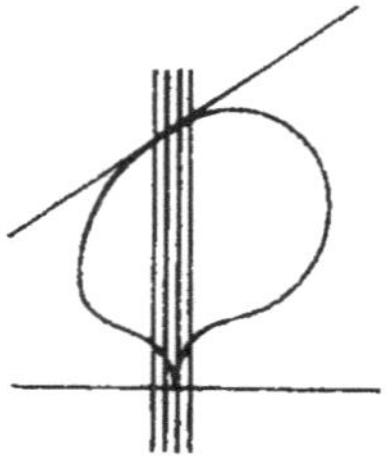
representative equation	$Y^3 - (X+1)^2\big(X-(1+c)\big)$ $\times \big(X-(1-c)\big) = 0$ $c \neq 0$	$Y^3 - (X-1)^2$ $\times (X+1)^2 = 0$	$Y^3 - (X+3)$ $\times (X-1)^3 = 0$	$Y^3 - X^4 = 0$
representing points of Λ	$\Delta \setminus \mathrm{Sg}\,\Delta$	$\{Q_1, Q_2, Q_3\}$	$\{P_1, P_2, P_3, P_4\}$	
singularities of type $Y^3 = X^2$	$B_1 = B_2$ (w.l.o.g.)	$B_1 = B_2$, $B_3 = B_4$		
singularities of type $Y^3 = X^3$			$B_2 = B_3 = B_4$	
singularities of type $Y^3 = X^4$				$B_1 = B_2 = B_3 = B_4$
geometric genus $g(C) = g(\tilde{C})$	2	1	0	0
name of the Picard curve	hyperelliptic	elliptic	rational	unstable

The automorphism groups are $2D_3$, $2D_6$ or $2S_4$, respectively, where D_n denotes the dihedral group of the regular n-gone and $2D_3$, $2D_6$, $2D_4$ are the unique full lifts of the groups $D_3, D_6, S_4 \subset \operatorname{Aut} \mathbb{P}^1$ acting on $\left\{\infty, 0, 1, \lambda, \frac{\lambda-1}{\lambda}, \frac{1}{1-\lambda}\right\}$ to the double covers (I), (II) and (III), respectively, branched on this 6-tuple (see 3.3.1). So we have the exact sequences

$$1 \to \mathbb{Z}/2\mathbb{Z} \to 2D_3 \to D_3 \to 1,$$

$$1 \to \mathbb{Z}/2\mathbb{Z} \to 2D_6 \to D_6 \to 1,$$

$$1 \to \mathbb{Z}/2\mathbb{Z} \to 2S_4 \to S_4 \to 1.$$

We will show that $\tilde{C}$ cannot be of type (I) or (III). Then it has to be isomorphic to the curve of type (II) and the proposition is proved.

5.2.5. *Each element $\gamma \in \operatorname{Aut} \tilde{C}$ has a fixed point on $\tilde{C}$.*

Proof. The Hurwitz genus formula applied to $\tilde{C} \to \tilde{C}/\langle\gamma\rangle$ yields

$$2 = |\langle\gamma\rangle| \left(2\bar{g} - 2 + \sum_i (v_i - 1)/v_i\right), \qquad \bar{g} = g(\tilde{C}/|\langle\gamma\rangle|) = 0, 1, 2.$$

Therefore the sum in the bracket cannot be equal to 0 for $\gamma \neq 1$. Hence γ does not act freely on $\tilde{C}$.

Now we assume that $\tilde{C}$ is of type (III). We take the quotient of $\tilde{C}$ by $2S_4$. The Hurwitz genus formula yields

$$2 = 48\left(-2 + a\frac{1}{2} + b\frac{2}{3} + c\frac{3}{4} + d\frac{5}{6} + e\frac{7}{8}\right),$$

$$49 = 12a + 16b + 18c + 20d + 21e, \qquad e = 1,$$

$$14 = 6a + 8b + 9c + 10d, \qquad c = d = 0, \qquad a = b = 1.$$

It follows that any $(\operatorname{Aut} \tilde{C})$-fixed point formation consists of three points P_2, P_3, P_8 with isotropy groups of order 2, 3 and 8, respectively, and that their orbits consist of 24, 16 and 6 points, respectively. Now add the quotient map coming from our element $g \in \operatorname{Aut} \tilde{C}$ of order 3 mentioned at the beginning of the proof of 5.2.4:

5.2.6. $\tilde{C} \xrightarrow{(\mathbb{Z}/3\mathbb{Z})} \tilde{C}/\langle g\rangle \to \tilde{C}/2S_4.$

The first map has the ramification points B_0, $B_1' = B_2'$, B_3, B_4 and no other one. $B_1' = B_2'$ is the singularity resolution of $B_1 = B_2$ (see 5.2.α). These four ramification points lie in the $2S_4$-orbit of P_3. The images of the other 12 points of this orbit are ramification points of branch order 3 of the second map of (5.2.6). But this map has degree 16. This is a contradiction because the branch order must divide the degree. Therefore $\tilde{C}$ is not isomorphic to the curve of type (III).

Next we study the branching situation of case (I):

$$\begin{array}{ccccc} \tilde{C} & \xrightarrow{(\mathbb{Z}/2\mathbb{Z})} & \tilde{C}/(\mathbb{Z}/2\mathbb{Z}) & \xrightarrow{D_3} & \tilde{C}/2D_3 \\ & & \wr\| & & \wr\| \\ & & \mathbb{P}^1 & & \mathbb{P}^1 \end{array}$$

D_3 is the symmetric group of the points $0, 1, \infty \in \mathbb{P}^1$. Let s be the generator of $\mathbb{Z}/2\mathbb{Z}$. The group D_3 contains an element of order 3. Therefore $2D_3$ contains an element g of order 3. Then $gs = sg$ has order 6. Hurwitz' genus formula applied to $2D_3$ yields successively

$$2 = 12\left(-2 + a\,\frac{1}{2} + b\,\frac{2}{3} + c\,\frac{3}{4} + d\,\frac{5}{6}\right), \qquad d \geqq 1,$$

$$26 = 6a + 8b + 9c + 10d, \qquad 2 \mid c, \qquad c = 0,$$

$$13 = 3a + 4b + 5d,$$

$$a = 1,\ b = 0,\ d = 2 \qquad \text{or} \qquad a = 0,\ b = 2,\ d = 1.$$

A preimage s_1 of an element of order 2 of D_3 in $2D_3$ has order 2 or 4. Since $c = 0$, it has order 2. Here we used 5.2.5. The element s_1 does not belong to $\langle sg\rangle \simeq \mathbb{Z}/6\mathbb{Z}$ because $s \neq s_1$. Therefore Fix s_1 and Fix sg are disjoint sets of points on C, hence the case $a = 0$ ($b = 2$, $c = 0$, $d = 1$) is not possible. So we have a $2D_3$-fixed point formation on C of the following type:

$$P_2, \qquad (2D_3)_{P_2} \simeq \mathbb{Z}/2\mathbb{Z}, \qquad |(2D_3)\cdot P_2| = 6,$$

$$P_6, \qquad (2D_3)_{P_6} \simeq \mathbb{Z}/6\mathbb{Z}, \qquad |(2D_3)\cdot P_6| = 2,$$

$$P_6', \qquad (2D_3)_{P_6'} \simeq \mathbb{Z}/6\mathbb{Z}, \qquad |(2D_3)\cdot P_6'| = 2.$$

Clearly $2D_3 \simeq D_6$ because this is the only non-commutative group of order 12 without elements of order 4. Now we can see that the fixed point constellation is not compatible with our 3-sheeted covering map $\tilde{C} \to \mathbb{P}^1$ with ramification points B_0, $B_1' = B_2'$, B_3, B_4. Let $\langle g\rangle \simeq \mathbb{Z}/3\mathbb{Z} \subset 2D_3$ be the corresponding Galois group. Then Fix g consists of the four points of the orbits of P_6 and P_6'. And $\tilde{C} \to \tilde{C}/\langle g\rangle$ is unramified outside of these points. Therefore the image of the orbit of P_2 on $\tilde{C}/\langle g\rangle$ consists of two points. Altogether we find six points on $\tilde{C}/\langle g\rangle \simeq \mathbb{P}^1$ with isotropy groups of order 2 under the action of $2D_3/\langle g\rangle \simeq K_4$. This is not possible because $\mathbb{P}^1 \to \mathbb{P}^1/K_4$ has exactly four ramification points. Proposition 5.2.4 is proved and moreover

5.2.7. COROLLARY. *The normalization of each hyperelliptic Picard curve is isomorphic to the genus-2-curve*

$$V^2 = U(U-1)(U-2)\left(U - \frac{1}{2}\right)(U+1).$$

5.2.8. COROLLARY. *Any two hyperelliptic Picard curves are isomorphic.*

Proof. We know that Aut $\tilde{C} \simeq 2D_6$ for each hyperelliptic Picard curve C. Now $\tilde{C}$ has exactly four points P with $3 \mid |(2D_6)_P|$ because the second morphism of the decomposition

$$\tilde{C} \to \tilde{C}/(\mathbb{Z}/3\mathbb{Z}) \to \tilde{C}/2D_6$$

is of degree 8. Hence it has no branch point of branch order $3m$, $m \geqq 1$. We use this fact in determining the fixed point constellation on $\tilde{C}$ by means of the Hurwitz genus formula applied to $\tilde{C} \to \tilde{C}/2D_6$:

$$2 = 24\left(-2 + a\,\frac{1}{2} + b\,\frac{2}{3} + c\,\frac{3}{4} + d\,\frac{5}{6} + e\,\frac{11}{12}\right),$$

$$25 = 6a + 8b + 9c + 10d + 11e, \qquad e \leqq 1;$$

and

$$4 = b\,\frac{24}{3} + d\,\frac{24}{6} + e\,\frac{24}{12} \text{ (our fact)},$$

$$4 = 8b + 4d + 2e, \qquad b = e = 0, \qquad d = 1,$$

$$25 = 6a + 9c + 10, \qquad 15 = 6a + 9c, \qquad a = c = 1.$$

Therefore the four fixed points $B_0, \tilde{B}_1 = \tilde{B}_2, B_3, B_4$ of g are the points of the orbit $(2D_6) \cdot B_0$ and $(2D_6)_{B_0} \simeq \mathbb{Z}/6\mathbb{Z}$. If C' is another hyperelliptic Picard curve, then there exists an isomorphism $\tilde{\alpha}: \tilde{C} \xrightarrow{\sim} \tilde{C}'$ by 5.2.4. Because of the transitive action of Aut $C' = \tilde{\alpha} \circ \text{Aut } C \circ \tilde{\alpha}^{-1}$ on $\{B_0', \tilde{B}_1' = \tilde{B}_2', B_3', B_4'\}$ we can without loss of generality assume that $\tilde{\alpha}(\tilde{B}_1) = \tilde{B}_1'$. Here $B_i', \tilde{B}_j'$ have the same meaning for C' as $B_i, \tilde{B}_j$ for C. The singularities $B_1 \in C$ and $B_1' \in C'$ are of the same type (see Table 5.2.α). Therefore there is an isomorphism $\alpha: C \to C'$ extending $\tilde{\alpha}|_{\tilde{C} \setminus \{\tilde{B}_1\}}$: $C \setminus \{B_1\} \xrightarrow{\sim} C' \setminus \{B_1'\}$.

§ 6. AUTOMORPHIC FORMS AND PICARD'S CURVE FAMILY

6.1. Arithmetic Classification of Picard Curves

As in 4.1 we denote by

$$\Phi_{[\Gamma',m]}: \mathfrak{B} \xrightarrow[(f_0:f_1:\cdots:f_M)]{} \mathbb{P}^{M-1}, \qquad M = \dim\,[\Gamma', m], \qquad m \geqq 1,$$

the holomorphic map induced by the basis $f_0, f_1, \ldots, f_M$ of $[\Gamma', m]$. The map $\Phi_{[\Gamma',m]}$ factorizes through $\mathfrak{B}/\Gamma' \xrightarrow{\varphi_m} \mathbb{P}^{M-1}$. After extension to the compactification $\widehat{\mathfrak{B}/\Gamma''}$ we get a commutative diagram

6.1.a.

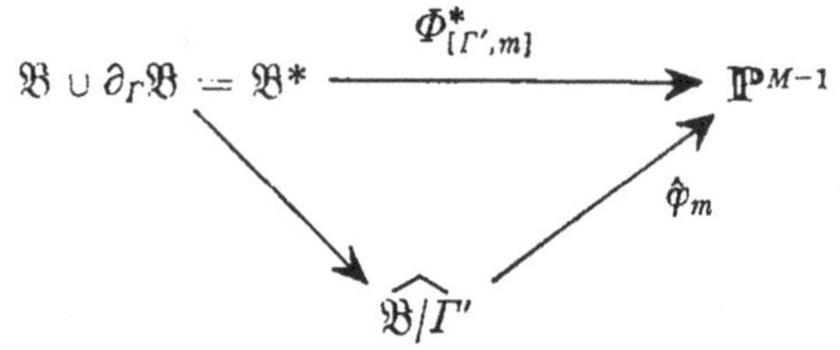

According to Proposition 4.3.1 φ_m and $\Phi_{m|\mathfrak{B}/\Gamma'}$, $\Phi_m = \Phi_{mK+m\mathfrak{T}}$, coincide, hence $\hat{\varphi}_m = \hat{\Phi}_m$. Taking into consideration Diagram 3.2.a we find in particular for $m = 1$ the commutative diagram

6.1.b.

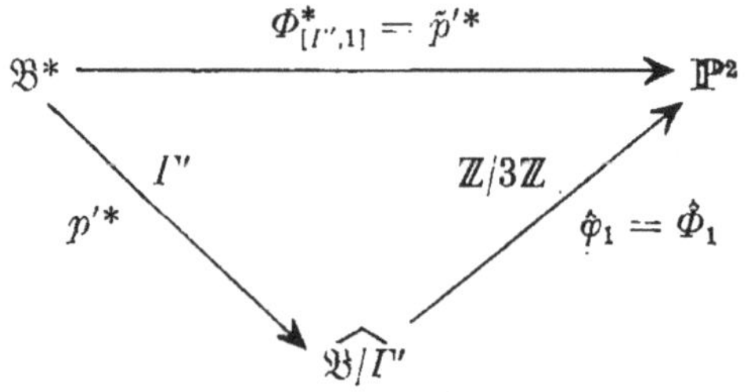

All these arrows represent quotient maps by Proposition 3.3.2. The image of $\Phi_{[\Gamma',1]}$ is $\mathbb{P}^2 \setminus \{P_1, P_2, P_3, P_4\}$ with the notations of 3.4.A (see also Diagram 3.3.b). In 4.4.1 we found the action of $\tilde{\Gamma}/\Gamma' = (\mathbb{Z}/3\mathbb{Z}) \times S_4$ on $\bigoplus_{m=0}^{\infty} [\Gamma'', m]$. Altogether we have

6.1.1. PROPOSITION. *There are four automorphic forms $\xi_1, \xi_2, \xi_3, \xi_4$ generating the $\mathbb{C}$-vector space $[\Gamma', 1] = [\tilde{\Gamma}', 1]_\chi \simeq \mathbb{C}^3$ such that $\xi_1 + \xi_2 + \xi_3 + \xi_4 = 0$ and $S_4 \simeq \Gamma/\Gamma' \simeq \tilde{\Gamma}/\tilde{\Gamma}'$ acts asymmetrically on them. The horizontal map in* 6.1.b *is*

$$\Phi^*_{[\Gamma',1]} = p'^*\colon \mathfrak{B}^* \xrightarrow[(\xi_1:\xi_2:\xi_3:\xi_4)]{\tilde{\Gamma}'} \mathbb{P}^2 = \mathbb{P}\left\{(x_1, x_2, x_3, x_4) \in \mathbb{C}^4;\ \sum_{i=1}^{4} x_i = 0\right\}$$

with S_4-coordinates x_1, x_2, x_3, x_4 on $\mathbb{P}^2$. Furthermore

$$\Phi^*_{[\Gamma',1]}(\partial_\Gamma \mathfrak{B}) = \{P_1, P_2, P_3, P_4\}, \qquad \Phi_{[\Gamma',1]}(\mathfrak{B}) = \mathbb{P}^2 \setminus \{P_1, P_2, P_3, P_4\}.$$

We want to obtain an ingenious connection between automorphic forms and the coefficients of Picard polynomials as in the theory of elliptic curves. This connection is that what we call the "arithmetic classification of Picard curves".

As in the Propositions 4.4.5 and 4.4.9 we denote by G_2, G_3, G_4 three generators of the ring of $\tilde{\Gamma}_\chi$-automorphic forms of weights 2, 3 or 4, respectively. Now we define the following map:

$$\mathfrak{Pic}\colon \mathfrak{B}^* \to \mathrm{PiC}^*,$$

$$b \mapsto \mathfrak{Pic}(b)\colon Y^3 = X^4 + G_2(b)\,X^2 + G_3(b)\,X + G_4(b).$$

$\mathfrak{Pic}$ has the following functorial property:

6.1.2. *If $b, b' \in \mathfrak{B}^*$ are $\tilde{\Gamma}$-equivalent, then the embedded Picard curves $\mathfrak{Pic}(b)$ and $\mathfrak{Pic}(b')$ are projectively equivalent.*

Proof. For $\gamma(b) = b'$, $\gamma \in \tilde{\Gamma}$, $j_\gamma^m(b)_\chi = \left(j_\gamma(b)_\chi\right)^m = \left(\chi(\gamma)\, j_\gamma(b)\right)^m$, we find the following transformations of the equation of $\mathfrak{Pic}(b)$:

$$Y^3 = X^4 + G_2(b)\,X^2 + G_3(b)\,X + G_4(b)$$

$$= X^4 + j_\gamma^2(b)_\chi\, G_2(\gamma b)\,X^2 + j_\gamma^3(b)_\chi\, G_3(\gamma b)\,X + j_\gamma^{-4}(b)_\chi\, G_4(\gamma b),$$

$$Y^3 = j_\gamma^4(b)\left(\left(\frac{X}{j_\gamma(b)_\chi}\right)^4 + G_2(b')\left(\frac{X}{j_\gamma(b)_\chi}\right)^2 + G_3(b')\,\frac{X}{j_\gamma(b)_\chi} + G_4(b')\right).$$

Obviously the last equation is equivalent to the equation of $\mathfrak{Pic}(b')$.

We denote the projective equivalence class of $\mathfrak{Pic}(b)$ by $\widehat{\mathfrak{Pic}}(b)$ and the isomorphism class by $\widehat{\widehat{\mathfrak{Pic}}}(b)$. One of the main results of this book is the following

6.1.3. THEOREM. *There are three algebraically independent generators G_2, G_3, G_4 of $\bigoplus_{m=0}^{\infty} [\tilde{\Gamma}, m]_\chi$ of weights 2, 3 or 4, respectively, such that the maps $\widehat{\mathfrak{Pic}}$ and $\widehat{\widehat{\mathfrak{Pic}}}$ factorize through $\mathfrak{B}^*/\tilde{\Gamma} = \widehat{\mathfrak{B}/\tilde{\Gamma}} \cong \mathbb{P}^2/S_4$. In the corresponding commutative Diagram 6.1.c the maps $\widehat{\mathfrak{Pic}}$ and $\widehat{\widehat{\mathfrak{Pic}}}$ are surjective and $\widehat{\mathfrak{pic}}$ is bijective. For $b, b' \in \mathfrak{B}^*$*

6.1. c

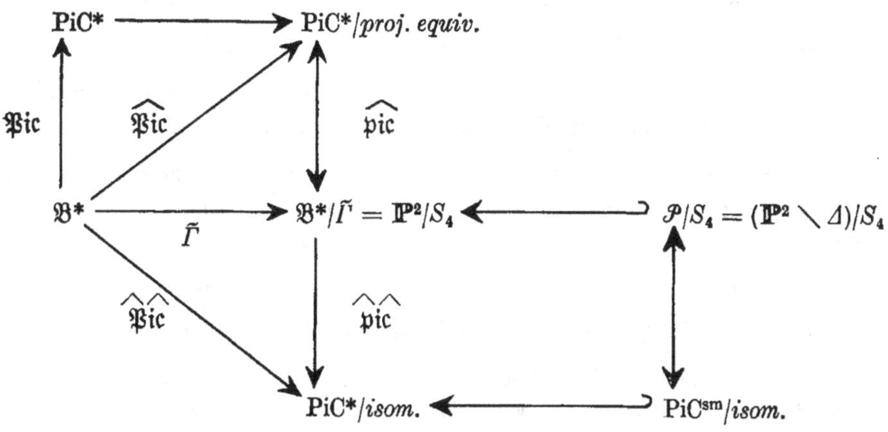

it holds that

$$\widehat{\mathfrak{Pic}}(b) = \widehat{\mathfrak{Pic}}(b') \quad \textit{iff } b \textit{ and } b' \textit{ are } \tilde{\Gamma}\textit{-equivalent.}$$

The map $\widehat{\widehat{\mathfrak{pic}}}$ *is injective on* $(\mathfrak{B}^*/\tilde{\Gamma}) \setminus (\tilde{\Gamma} \cdot \mathfrak{D}^*/\tilde{\Gamma}) = (\mathfrak{B}/\tilde{\Gamma}) \setminus (\tilde{\Gamma} \cdot \mathfrak{D}/\tilde{\Gamma})$ *with* (e.g.) $\mathfrak{D} = \mathfrak{D}_{24} = \{(0, z) \in \mathbb{C}^2; |z| < 1\}$. *The map* $\mathfrak{Pic}$ *maps* $\mathfrak{D}^* = \mathfrak{D} \cup \partial_{\Gamma_{\mathfrak{D}}}\mathfrak{D}$ *into the set* Pic^{sg*} *of Picard curves of genus smaller than* 3. *The preimage of* $\mathrm{PiC}^{sg}/(\textit{proj. equ.})$ *and of* $\mathrm{PiC}^{sg}/(\textit{isom.})$ *for* $\widehat{\mathfrak{Pic}}$ *and* $\widehat{\widehat{\mathfrak{Pic}}}$, *respectively, is* $\tilde{\Gamma}\mathfrak{D}^*$. *The image* $\widehat{\widehat{\mathfrak{pic}}}\,(\tilde{\Gamma}\mathfrak{D}^*/\tilde{\Gamma})$ *consists of three elements,* viz. $\mathrm{PiC}^h/\textit{isom.}$, $\mathrm{PiC}^e/\textit{isom.}$, $\mathrm{PiC}^r/\textit{isom.}$

Proof. Let $\xi_1, \xi_2, \xi_3, \xi_4$ be the four $\tilde{\Gamma}'_\chi$-automorphic forms of Proposition 6.1.1. Then we set (see 4.4)

$$G_2 = \sum_{1 \le i < j < 4} \xi_i \xi_j, \qquad G_3 = -\sum_{1 \le i < j < k \le 4} \xi_i \xi_j \xi_k, \qquad G_4 = \xi_1 \xi_2 \xi_3 \xi_4.$$

Then we can write

$$\mathfrak{Pic}(b)\colon Y^3 = \big(X - \xi_1(b)\big)\big(X - \xi_2(b)\big)\big(X - \xi_3(b)\big)\big(X - \xi_4(b)\big).$$

If $b, b' \in \mathfrak{B}^*$ are $\tilde{\Gamma}$-equivalent, then

$$\big(\xi_1(b) : \xi_2(b) : \xi_3(b) : \xi_4(b)\big) \in S_4 \cdot \big(\xi_1(b') : \xi_2(b') : \xi_3(b') : \xi_4(b')\big)$$

because $\tilde{\Gamma}/\tilde{\Gamma}'$ acts symmetrically on $\xi_1, \xi_2, \xi_3, \xi_4$. By Proposition 6.1.1 the map $(\xi_1 : \xi_2 : \xi_3 : \xi_4)\colon \mathfrak{B}^* \to \mathbb{P}^2$ is surjective. Composing this map with the quotient map $\mathbb{P}^2 \to \mathbb{P}^2/S_4$ and with the bijective map of Proposition 5.1.12 we see that our maps $\widehat{\widehat{\mathfrak{Pic}}}$ and $\widehat{\mathfrak{Pic}}$ are surjective. The map $\widehat{\mathfrak{pic}}$ is bijective by 5.1.12. The rectangle on the right-hand side of Diagram 6.1.c comes from Proposition 5.2.3. The equivalence assertion follows from the rectangle on the left-hand side of 6.1.c. The set PiC^{sg*} is identified with the set $\mathrm{PiC}^h \cup \mathrm{PiC}^e \cup \mathrm{PiC}^r$ of non-unstable singular Picard curves (see Table 5.2.α). Their projective equivalence classes are exactly classified by Δ. This follows from Propositions 5.1.12 and 5.2.3. The divisor Δ comes from $\sum_{i<j} \mathfrak{D}_{ij}$ on $\widehat{\mathfrak{B}/\Gamma'}$ (see 3.2.6) and this divisor is the image of $\partial_\Gamma \mathfrak{B} \cup \bigcup_{i<j} \mathfrak{D}_{ij}$ (see 1.3). These discs are $\tilde{\Gamma}$-equivalent. The last statement of the theorem comes from 5.2.8 and the row representing Δ in Table 5.2.α. The theorem is proved.

We conclude this section with an interesting interpretation of the $\tilde{\Gamma}$-fixed point formation on $\mathfrak{B}^*$. We use Proposition 3.4.4 together with the notations given there.

6.1.4. PROPOSITION. *$P \in \mathfrak{B}^*$ is a $\tilde{\Gamma}$-fixed point if and only if the normalized Picard curve* $\widetilde{\mathfrak{Pic}}(P)$ *has an automorphism group of order greater than* 3.

More precisely, up to projective equivalence, we give in Table 6.1.α the complete list for the special Picard curves corresponding to $\tilde{\Gamma}$-fixed points.

6.1.α.

(fixed point) class of P	representative equation	curve class	g	$\lvert\tilde{\Gamma}_P\rvert$	$\lvert\mathrm{Aut}\,\widetilde{\mathfrak{P}\mathrm{ic}}(P)\rvert$
$\mathfrak{B} \setminus (\tilde{\Gamma}\cdot\mathfrak{D} \cup \tilde{\Gamma}\cdot\mathfrak{E})$	$Y^3 = \prod_{i=1}^{4} (x - e_i)$, $\prod_{1\leq i<j\leq 4} (e_j - e_i) \neq 0$	PiC^{sm}	3	1	3
$\tilde{\Gamma}\cdot S$	$Y^3 = (X^3 - 1)\, X$			3	9
$\tilde{\Gamma}\cdot M$	$Y^3 = X^4 - 1$			4	48
$\mathfrak{E} \setminus (\tilde{\Gamma}\cdot O \cup \tilde{\Gamma}\cdot M)$	$Y^3 = (X^2 - 1)(X^2 - t)$, $t \neq 0, 1, -1$			2	6
$\mathfrak{D} \setminus (\tilde{\Gamma}\cdot O \cup \tilde{\Gamma}\cdot R)$	$Y^3 = (X + 1)^2 \big(X - (1 + c)\big) \times \big(X - (1 - c)\big)$, $c \neq 0, 1$	Pic^{h}	2	6	24
$\tilde{\Gamma}\cdot R$	$Y^3 = (X^2 - 1)\, X^2$ or $Y^3 = (X + 1)^2 (X + 2)\, X$			12	24
$\tilde{\Gamma}\cdot O$	$Y^3 = (X - 1)^2 (X + 1)^2$	PiC^{e}	1	48	∞
$\partial_{\Gamma}\mathfrak{B}$	$Y^3 = (X + 3)(X - 1)^3$	PiC^{r}	0	∞	∞

$\mathfrak{D} = \mathfrak{D}_{14}$, $\mathfrak{E} = \mathfrak{E}_{(12)(34)}$

The projection of $\tilde{\Gamma}\cdot\mathfrak{E}$ on $\mathbb{P}^2$ is $\Delta_{(12)(34)} + \Delta_{(13)(24)} + \Delta_{(14)(23)}$ in 3.4.A. The second and fourth columns are extensions of the corresponding rows in Table 5.2.α. The fifth column comes from Proposition 3.4.4. The automorphism groups of all (smooth) curves of genus 3 were calculated by Kuribayashi in [40]. The hyperelliptic normalized Picard curves form exactly one isomorphy class. The automorphism group is $2D_6$ (see 5.2.7 and type (II) in 5.2).

6.1.5. REMARK. The Proposition 6.1.4 indicates a deep connection between fixed points and curves with degenerate (higher) automorphism groups. In this respect the proof given is not quite satisfactory because it makes use of Kuribayashi's elementary calculations of automorphism groups. There should be a proof of a more general nature based on the fundamental group of the space of labelled smooth Picard curves $\pi_1(\mathscr{P})$ (see 6.3 below) and the additional action of the group S_4 on $\mathscr{P}$ via the monodromy action of $\pi_1(\mathscr{P})$ on the first cohomology groups of the curves together with the additional action of S_4. In this context S_4 is characterized as the group which is reponsible for the difference between the space of our labelled curves $\mathscr{P}$ and the space $\mathscr{P}/S_4$ of smooth Picard curves.

6.2. A Geometric Connection with Elliptic Curves and Modular Forms

6.2.1. DEFINITION. Two polynomials $p(X)$, $q(X) \in \mathbb{C}[X]$ are called *affinely equivalent*, if there exists an affine transformation $\alpha: \mathbb{A}^1(\mathbb{C}) \xrightarrow{\sim} \mathbb{A}^1(\mathbb{C})$ such that $q(X) = \alpha^* p(X)$. We say that a subset M of $\mathbb{C}^n$ *resolves algebraic equations of degree n*, if for each polynomial $q(X)$ of degree n there is an affinely equivalent polynomial $p(X)$ and an element $(x_1, \ldots, x_n) \in M$ such that

$$p(X) = (X - x_1)(X - x_2) \cdots (X - x_n).$$

We say that a set of n functions $f_1, \ldots, f_n$, $f_i: \mathfrak{M} \to \mathbb{C}$, resolves algebraic equations of degree n, if

$$M = \{(f_1(m), \ldots, f_n(m)); m \in \mathfrak{M}\} \cup \{(0, 0, \ldots, 0)\} \subseteq \mathbb{C}^n$$

resolves algebraic equations of degree n. A set F of functions $f: \mathfrak{M} \to \mathbb{C}$ resolves algebraic equations of degree n, if there exist n functions $f_1, \ldots, f_n \in \mathsf{F}$ resolving algebraic equations of degree n.

By Tschirnhaus transformations it is easy to see that

6.2.2. $\{f_1, \ldots, f_n\}$ *resolves algebraic equations of degree n, if*

$$\sum_{i=1}^{n} f_i = 0$$

and

$$(f_1 : \cdots : f_n): \mathfrak{M}^n \to \mathbb{P}^{n-2} = \left\{(x_1 : x_2 : \cdots : x_n) \in \mathbb{P}^{n-1}; \sum_i x_i = 0\right\}$$

is surjective.

6.2.3. PROPOSITION. *It holds that*

a) $[\tilde{\Gamma}', 1]_\chi$ *resolves algebraic equations of degree* 4.

b) $[\mathbb{S}\mathrm{l}_2(\mathbb{Z})(2), 1]$ *resolves algebraic equations of degree* 3.

Proof. a) is an immediate consequence of 6.1.1 and 6.2.2. b) is well-known. For the convenience of the reader we give a short proof, which is a mini-version of the proof of a). The group $\mathbb{S}\mathrm{l}_2(\mathbb{Z})(2)$ is the principal subgroup of $\mathbb{S}\mathrm{l}_2(\mathbb{Z})$ of level 2. It acts on $\mathfrak{H}^* = \mathfrak{H} \cup \partial_{\mathbb{S}\mathrm{l}_2(\mathbb{Z})}\mathfrak{H}$, $\mathfrak{H} = \{z \in \mathbb{C}; \operatorname{Im} z > 0\}$ the *Siegel upper half plane*, with the following properties (see [64], II, § 2):

$$\mu = [\mathbb{S}\mathrm{l}_2(\mathbb{Z}) : \mathbb{S}\mathrm{l}_2(\mathbb{Z})(2)] = 6,$$

$$\nu_i = 0 \quad \text{(number of elliptic points with stationary group of type } \mathbb{Z}/i\mathbb{Z}),$$

$$i_\infty = [\mathbb{S}\mathrm{l}_2(\mathbb{Z})_\infty : (\mathbb{S}\mathrm{l}_2(\mathbb{Z})(2))_\infty] = 2,$$

$$\nu_\infty = \frac{\mu}{i_\infty} = 3 \quad \text{(number of cusps modulo } \mathbb{S}\mathrm{l}_2(\mathbb{Z})(2)).$$

The proportionality formula for the genus g of lattice quotients of $\mathfrak{H}^*$ yields

$$g\big(\mathfrak{H}/\widehat{\mathbb{S}\mathrm{l}_2(\mathbb{Z})}\,(2)\big) = 1 + \frac{\mu}{12} - \frac{\nu_2}{4} - \frac{\nu_3}{3} - \frac{\nu_\infty}{2} = 0,$$

hence

$$\mathfrak{H}/\mathbb{S}\mathrm{l}_2(\mathbb{Z})\,(2) \cong \mathbb{P}^1 \setminus \{S_1, S_2, S_3\}.$$

From Mumford's proportionality formulas in [50] it follows that

$$\bigoplus_{m=0}^{\infty} [\mathbb{S}\mathrm{l}_2(\mathbb{Z})\,(2), m] \cong \bigoplus_{m=0}^{\infty} H^0\big(\mathbb{P}^1, \Omega^1_{\mathbb{P}^1}(\log(S_1 + S_2 + S_3))^m\big)$$

$$\cong \bigoplus_{m=0}^{\infty} H^0\big(\mathbb{P}^1, \mathcal{O}(-2P) \otimes \mathcal{O}(S_1 + S_2 + S_3)\big) \cong \bigoplus_{m=0}^{\infty} H^0\big(\mathbb{P}^1, \mathcal{O}(mP)\big)$$

$$\cong \mathbb{C}[Y_0, Y_1].$$

For each $m > 0$ we have a commutative diagram

$$\begin{array}{ccccc}
\mathfrak{H} & \xrightarrow{\mathbb{S}\mathrm{l}_2(\mathbb{Z})(2)} & \mathbb{P}^1 \setminus \{S_1, S_2, S_3\} & \hookrightarrow & \mathbb{P}^1 \\
\downarrow & & \downarrow \varphi_m & & \downarrow \Phi_m = \Phi_{mP} \\
\mathfrak{H}^* & \longrightarrow & \Phi_m^*(\mathfrak{H}^*) & \hookrightarrow & \mathbb{P}^{(mP)} \\
\end{array}$$

$$\Phi_m^* = \Phi^*_{[\mathbb{S}\mathrm{l}_2(\mathbb{Z})(2), m]}$$

Φ_1 is an isomorphism because $\mathcal{O}(P) \cong \mathcal{O}(1)$. Therefore φ_1 is an embedding and $\Phi_1^*(\mathfrak{H}^*) = \mathbb{P}^1$, $\Phi_1^*(\mathfrak{H}) = \mathbb{P}^1 \setminus \{S_1, S_2, S_3\}$. Therefore there are three $\mathbb{S}\mathrm{l}_2(\mathbb{Z})\,(2)$-automorphic forms $\varepsilon_1, \varepsilon_2, \varepsilon_3$ of weight 1, $\sum_{i=1}^{3} \varepsilon_i = 0$ such that the functions ε_i: $\mathfrak{H}^* \to \mathbb{C}$ satisfy the surjectivity condition of 6.3.2. This was to be proved.

6.2.4. REMARK. $\{\varepsilon_1, \varepsilon_2, \varepsilon_3\} \subset [\mathbb{S}\mathrm{l}_2(\mathbb{Z})\,(2), 1]$ can be chosen in such a way that $S_3 \cong \mathbb{S}\mathrm{l}_2(\mathbb{Z})/\mathbb{S}\mathrm{l}_2(\mathbb{Z})\,(2)$ acts symmetrically on this set.

We know that $[\mathbb{S}\mathrm{l}_2(\mathbb{Z}) : \mathbb{S}\mathrm{l}_2(\mathbb{Z})\,(2)] = 6$. Moreover, the quotient group has a faithful representation as the symmetric group of the set $(\mathfrak{H}^* \setminus \mathfrak{H})/\mathbb{S}\mathrm{l}_2(\mathbb{Z})\,(2)$, which is represented by the three cusps $\mathbb{P}\begin{pmatrix}1\\0\end{pmatrix}$, $\mathbb{P}\begin{pmatrix}0\\1\end{pmatrix}$, $\mathbb{P}\begin{pmatrix}1\\1\end{pmatrix}$. By factorization we see that

$$\mathbb{S}\mathrm{l}_2(\mathbb{Z})/\mathbb{S}\mathrm{l}_2(\mathbb{Z})\,(2) = \mathrm{Aut}\,(\mathbb{P}^1, \{S_1, S_2, S_3\}) = S_3.$$

Therefore the action on

$$\bigoplus_{m=0}^{\infty} [\mathbb{S}\mathrm{l}_2(\mathbb{Z})\,(2), m] \simeq \mathbb{C}[Y_0, Y_1] \simeq \mathbb{C}[X_1, X_2, X_3]/(X_1 + X_2 + X_3)$$

is induced by the symmetric action. Restriction to the homogeneous part of degree 1 yields 6.2.4.

6.2.5. DEFINITION. Using S_{n+2}-coordinates on $\mathbb{P}^n$, $n = 2, 1$, we call the rational map

$$l: \mathbb{P}^2 \dashrightarrow \mathbb{P}^1,$$

$$(x_1 : x_2 : x_3 : x_4) \mapsto (l_1 : l_2 : l_3)$$

with

$$l_1 = x_1x_2 + x_3x_4 + \frac{1}{6} \sum_{i=1}^{4} x_i^2,$$

$$l_2 = x_1x_3 + x_2x_4 + \frac{1}{6} \sum_{i=1}^{4} x_i^2,$$

$$l_3 = x_1x_4 + x_2x_3 + \frac{1}{6} \sum_{i=1}^{4} x_i^2$$

the *Lagrange map*.

This map mimics the reduction step, which transforms polynomials of degree 4 into polynomials of degree 3 (*Lagrange resolvent*) in the resolution procedure of algebraic equations of degree 4 up to affine equivalence:

$$\prod_{i=1}^{4} (X - x_i)$$

$$\mapsto \big(X - (x_1x_2 + x_3x_4)\big)\big(X - (x_1x_3 + x_2x_4)\big)\big(X - (x_1x_4 + x_2x_3)\big).$$

6.2.6. *The fundamental points of the Lagrange map l are the points*

$$P_1 = (-3 : 1 : 1 : 1),\ P_2 = (1 : -3 : 1 : 1),$$

$$P_3 = (1 : 1 : -3 : 1),\ P_4 = (1 : 1 : 1 : -3).$$

In particular l is regular on $\mathbb{P}^2 \setminus \{P_1, P_2, P_3, P_4\}$.

Proof. The fundamental points of l are the intersection points of the three quadrics $l_1 = 0$, $l_2 = 0$, $l_3 = 0$. Two of them have four intersection points by Bezout's theorem. The points P_1, P_2, P_3, P_4 are intersection points. Therefore these are all fundamental points.

The commutative Diagram 6.2.a shows that there is an interesting arithmetic connection between Picard curves and elliptic curves as well as between $\tilde{\Gamma}'_\chi$-automorphic forms and modular forms.

Here $\{\xi_1, \xi_2, \xi_3, \xi_4\}$ is a set of S_4-symmetric $\tilde{\Gamma}'_\chi$-automorphic forms of weight 1 (see 6.1.1), and $\{\varepsilon_1, \varepsilon_2, \varepsilon_3\}$ is a set of S_3-symmetric modular forms of level 2 and of weight 1 (see 6.2.4); $\sum_{i=1}^{4} \xi_i = \sum_{j=1}^{3} \varepsilon_j = 0$. Let $\sigma_3: \mathbb{P}^1 \to \mathbb{P}^1/S_3$ be the natural quotient morphism. By the construction of l it is easy to see that $\sigma_3 \circ l$ is stable under the

.a.

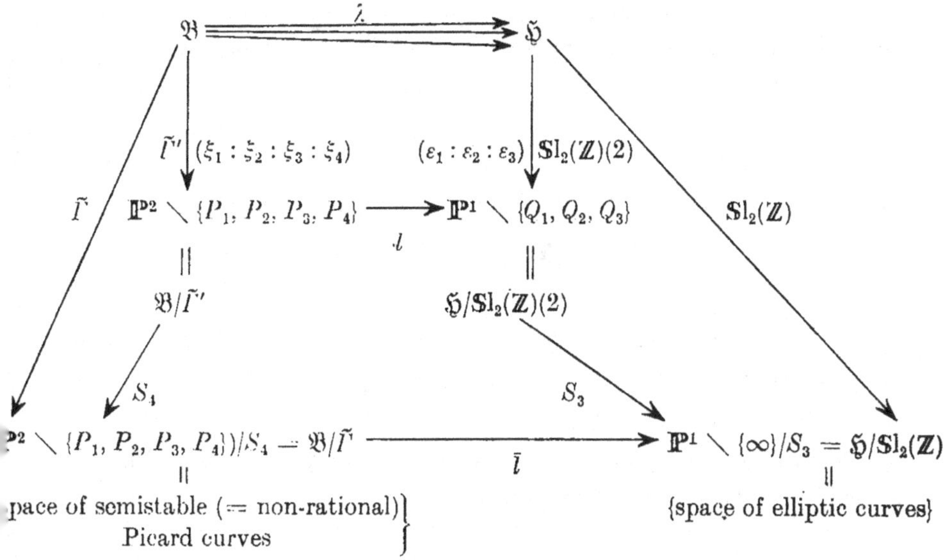

action of S_4. This defines the lower horizontal morphism $\bar{l}$ in 6.2.a; λ is the multi-valued lift of l which corresponds to $b \in B$ each $h \in \mathfrak{H}^*$ such that $l(b \bmod \tilde{\Gamma}') = h \bmod \mathbf{Sl}_2(\mathbb{Z})\,(2)$.

The Diagram 6.2.a shows that B parametrizes not only Picard curves but also all elliptic curves. Furthermore some $\tilde{\Gamma}' \cdot K_4$-automorphic forms can be expressed by modular forms of weight 1 and level 2 and certain correction factors coming from the Lagrange lift λ. For

$$V_1 = \xi_1\xi_2 + \xi_3\xi_4 + \frac{1}{6}\sum_{i=1}^{4} \xi_i^2, \qquad V_2 = \xi_1\xi_3 + \xi_2\xi_4 + \frac{1}{6}\sum_{i=1}^{4} \xi_i^2,$$

$$V_3 = \xi_1\xi_4 + \xi_2\xi_3 + \frac{1}{6}\sum_{i=1}^{4} \xi_i^2$$

(compare this with the $\tilde{\Gamma}' \cdot K_4$-part of Diagram 4.4.$\alpha$) we have

6.2.7. *There is a function $\Psi: \mathfrak{B} \times_{\mathbb{P}^1} \mathfrak{H}^* \to \mathbb{C}$ such that*

$$\mathfrak{v}(b) = \Psi(b, h) \cdot \mathfrak{n}(h), \qquad (b, h) \in \mathfrak{B} \times_{\mathbb{P}^1} \mathfrak{H}^* \quad (\text{i.e. } h \in \lambda(b))$$

where

$$\mathfrak{v}(b) = \big(V_1(b), V_2(b), V_3(b)\big), \qquad \mathfrak{n}(h) = \big(\varepsilon_1(h), \varepsilon_2(h), \varepsilon_3(h)\big).$$

The factor function Ψ has some interesting properties with respect to the action of $\tilde{\Gamma}' \times \mathbf{Sl}_2(\mathbb{Z})\,(2)$ on $\mathfrak{B} \times_{\mathbb{P}^1} \mathfrak{H}^*$. We shall not study Ψ in more detail because we shall be able to obtain another factor function ψ with only one argument together with a single-valued lift of l, if we restrict our considerations to $\mathbb{P}^2 \setminus \Delta$.

The preimage of a point $\mu = (\mu_1 : \mu_2 : \mu_3) \in \mathbb{P}^1$, $\sum \mu_i = 0$, for the Lagrange map $l: \mathbb{P}^2 \dashrightarrow \mathbb{P}^1$ (more precisely the closure in $\mathbb{P}^2$ of the preimage of $l|_{\mathbb{P}^2 \setminus \{P_1, P_2, P_3, P_4\}}$ is a quadric

6.2.8. $$\mathfrak{Q}_\mu : \mu_2 \left(X_1X_2 + X_3X_4 + \frac{1}{6} \sum_{i=1}^{4} X_i^2 \right) = \mu_1 \left(X_1X_3 + X_2X_4 + \frac{1}{6} \sum_{i=1}^{4} X_i^2 \right).$$

Each of these contains the points P_1, P_2, P_3, P_4. The preimages of

$$Q_1 = (-2:1:1), \qquad Q_2 = (1:-2:1), \qquad Q_3 = (1:1:-2)$$

are degenerate quadrics (see Figure 3.4.A)

$$\mathfrak{Q}_1 = \Delta_{14} + \Delta_{23}, \qquad \mathfrak{Q}_2 = \Delta_{24} + \Delta_{13}, \qquad \mathfrak{Q}_3 = \Delta_{34} + \Delta_{12}.$$

For example for $\mathfrak{Q}_3$ one finds immediately the equation

$$(X_1 - X_4)(X_2 - X_3) = 0.$$

We see that $\mathbb{P}^1$ parametrizes via l all quadrics on $\mathbb{P}^2$ going through the four points P_1, P_2, P_3, P_4 (see Figure 6.2.A).

6.2.A.

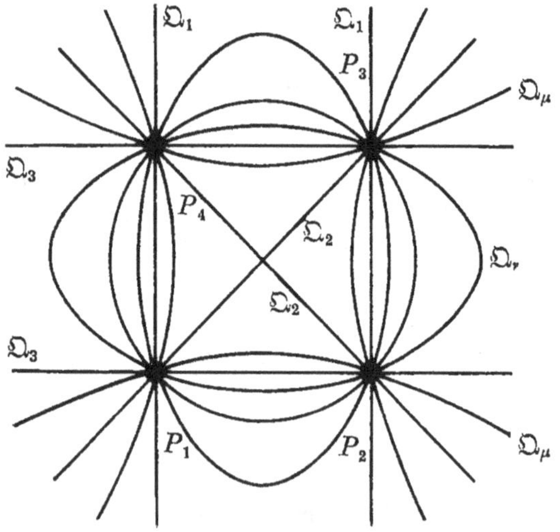

In order to get a fibration we have to remove the degenerate quadrics.

6.2.9. *The Lagrange map restricts to a fibration*

$$l_\Delta : \mathbb{P}^2 \setminus \Delta \to \mathbb{P}^1 \setminus \{Q_1, Q_2, Q_3\}$$

with fibres $\mathfrak{Q}_\mu \setminus \{P_1, P_2, P_3, P_4\}$ isomorphic to $\mathbb{P}^1 \setminus \{4$ points$\}$.

Proof. Obviously, $\mathfrak{Q}_\mu \cap \Delta = \{P_1, P_2, P_3, P_4\}$, hence $l_\Delta^{-1}(\mu) = \mathfrak{Q}_\mu \setminus \{P_1, P_2, P_3, P_4\}$ for any $\mu \in \mathbb{P}^1$. The only degenerate quadrics through P_1, P_2, P_3, P_4 are

$\mathfrak{Q}_1, \mathfrak{Q}_2, \mathfrak{Q}_3$. Therefore $\mathfrak{Q}_\mu$ is a smooth (irreducible) quadric for $\mu \neq Q_1, Q_2, Q_3$. These quadrics are isomorphic to $\mathbb{P}^1$.

Let $\mathfrak{U}$ be the universal cover of $\mathbb{P}^2 \setminus \Delta$. At the same time $\mathfrak{U}$ is the universal cover of $\mathfrak{B} \setminus \Gamma \cdot \mathfrak{D}$, $\mathfrak{D} = D_{24}$ (see Diagram 3.3.b). We denote the composition of $\mathfrak{B} \setminus \Gamma \cdot \mathfrak{D} \to \mathbb{P}^2 \setminus \Delta$ and l_Δ by $l_\mathfrak{D}$. Since $\mathfrak{U}$ is simply-connected $l_\mathfrak{D}$ lifts to a holomorphic map $\tilde{\lambda}: \mathfrak{U} \to \mathfrak{H}$. The lift $\tilde{\lambda}$ is uniquely determined if we impose that $\tilde{\lambda}(u_0) = h_0$ for a given pair $(u_0, h_0) \in \mathfrak{U} \times_{\mathbb{P}^1 \setminus \{Q_1, Q_2, Q_3\}} \mathfrak{H}$. Altogether we have the commutative Diagram 6.2.b.

6.2.b.

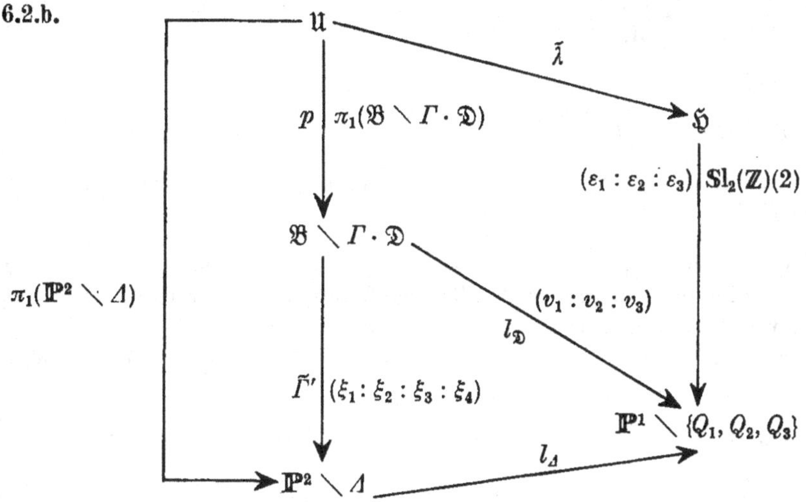

In terms of fundamental groups $\tilde{\lambda}$ corresponds to a representation of $\pi_1(\mathfrak{B} \setminus \Gamma \cdot \mathfrak{D})$ in $\mathrm{Sl}_2(\mathbb{Z})\,(2) = \pi_1(\mathbb{P}^1 \setminus \{Q_1, Q_2, Q_3\})$:

6.2.10. $l_{\mathfrak{D}*}: \pi_1(\mathfrak{B} \setminus \Gamma \cdot \mathfrak{D}) \to \mathrm{Sl}_2(\mathbb{Z})\,(2)$,

$$\tilde{\lambda}(\alpha u) = l_{\mathfrak{D}*}(\alpha)\, \tilde{\lambda}(u), \qquad u \in \mathfrak{U}, \qquad \alpha \in \pi_1(\mathfrak{B} \setminus \Gamma \cdot \mathfrak{D}).$$

The upper parallelogram in 6.2.b defines a function $\psi: \mathfrak{U} \to \mathbb{C}^*$ such that

6.2.11. $\mathfrak{v}(p(u)) = \psi(u) \cdot \mathfrak{n}(\tilde{\lambda}(u)), \qquad u \in \mathfrak{U}$,

with the notations of 6.2.7. For $\alpha \in \pi_1(\mathfrak{B} \setminus \Gamma \cdot \mathfrak{D})$ we have

$$\begin{aligned} \mathfrak{v}(p(u)) &= \mathfrak{n}(p(\alpha u)) = \psi(\alpha u) \cdot \mathfrak{n}(\tilde{\lambda}(\alpha u)) \\ &= \psi(\alpha u) \cdot \mathfrak{n}\big(l_{\mathfrak{D}*}(\alpha)\,(\tilde{\lambda}(u))\big) = \psi(\alpha u) \cdot j^{-1}_{l_{\mathfrak{D}*}(\alpha)}(\tilde{\lambda}(u)) \cdot \mathfrak{n}(\tilde{\lambda}(u)). \end{aligned}$$

Comparing this with (6.2.11) we find

6.2.12. $\psi(\alpha u) \cdot j^{-1}_{l_{\mathfrak{D}*}(\alpha)}(\tilde{\lambda}(u)) = \psi(u), \qquad u \in \mathfrak{U}, \qquad \alpha \in \pi_1(\mathfrak{B} \setminus \Gamma \cdot \mathfrak{D})$,

where $j_{l_{\mathfrak{D}*}(\alpha)}$ is the Jacobian of $l_{\mathfrak{D}*}(\alpha) \in \mathrm{Sl}_2(\mathbb{Z})\,(2)$.

The function ψ defines a $\pi_1(\mathfrak{B} \setminus \Gamma \cdot \mathfrak{D})$-action on the trivial line bundle $\mathfrak{U} \times \mathbb{C}$ over $\mathfrak{U}$, which extends the action on the base $\mathfrak{U}$:

$$\alpha\colon \mathfrak{U} \times \mathbb{C} \to \mathfrak{U} \times \mathbb{C},$$
$$(u, c) \mapsto \left(\alpha(u), j^{-1}_{l_{\mathfrak{D}*}(\alpha)}\left(\tilde{\lambda}(u)\right) \cdot c\right).$$

Indeed, by (6.2.10) we have

$$j^{-1}_{l_{\mathfrak{D}*}(\alpha\beta)}(\tilde{\lambda}u) = j^{-1}_{l_{\mathfrak{D}*}(\alpha)}\left(\tilde{\lambda}\left(\beta(u)\right)\right) \cdot j^{-1}_{l_{\mathfrak{D}*}(\alpha)}\left(\tilde{\lambda}(u)\right).$$

Formula (6.2.12) implies that ψ is a $\pi_1(\mathfrak{B} \setminus \Gamma \cdot \mathfrak{D})$-invariant nowhere vanishing section of $\mathfrak{U} \times \mathbb{C}$ over $\mathfrak{U}$.

6.2.13. DEFINITION. The quotient bundle

$$\mathcal{L} = \mathfrak{U} \times \mathbb{C}/\pi_1(\mathfrak{B} \setminus \Gamma \cdot \mathfrak{D})$$

with the action of $\pi_1(\mathfrak{B} \setminus \Gamma \cdot \mathfrak{D})$ defined above is called the *Lagrange bundle* over $\mathfrak{B} \setminus \Gamma \cdot \mathfrak{D}$.

Our factor function ψ is nothing else than the inverse image of a nowhere vanishing section of the Lagrange bundle. The function ψ is uniquely determined by the choice of our representation of $\pi_1(\mathfrak{B} \setminus \Gamma \cdot \mathfrak{D})$ corresponding to $\tilde{\lambda}$. Taking into account 6.2.11 we have proved the following lifting property for modular forms:

6.2.14. PROPOSITION. *There is a natural isomorphism of graded rings*

$$\Phi\colon \bigoplus_{m=0}^{\infty} [\mathrm{Sl}_2(\mathbb{Z})\,(2), m] \xrightarrow{\sim} \mathbb{C}[V_1, V_2] \subset \bigoplus_{m=0}^{\infty} [\tilde{\Gamma}' \cdot K_4, 2m]_\chi$$

which lets correspond to each modular form ε of level 2 and weight m a $\tilde{\Gamma}'_\chi \cdot K_4$-automorphic form V of weight $2m$ such that

$$V(b) = \psi(u)^m\, \varepsilon\left(\tilde{\lambda}(u)\right), \qquad b \in \mathfrak{B} \setminus \Gamma \cdot \mathfrak{D}, \qquad u \in p^{-1}(b),$$
$$\text{i.e. } p^*(V) = \psi^m \cdot \tilde{\lambda}^*(\varepsilon).$$

The representation of $\pi_1(\mathfrak{B} \setminus \Gamma \cdot \mathfrak{D})$ induced by $l_{\mathfrak{D}}$ is a subrepresentation of $\pi_1(\mathcal{P}) \to \mathrm{Sl}_2(\mathbb{Z})\,(2)$, $\mathcal{P} = \mathbb{P}^2 \setminus \Delta$, induced by l_Δ (see 6.2.b). The next proposition shows that the latter representation can be obtained via very familar morphisms.

6.2.15. PROPOSITION. *The representation* (6.2.10) *of* $\pi_1(\mathfrak{B} \setminus \Gamma \cdot \mathfrak{D})$ *is the subrepresentation of* $\pi_1(\mathcal{P})$ *induced by one of the following morphisms*:

(a)
$$\begin{array}{ccccc} \mathcal{P} & \hookrightarrow & \mathbb{P}^1 \times \mathbb{P}^1 & \longrightarrow & \mathbb{P}^1 \\ & \searrow^{p_1} & & & \cup \\ & & & & \mathbb{P}^1 \setminus \{Q_1, Q_2, Q_3\} \end{array}$$

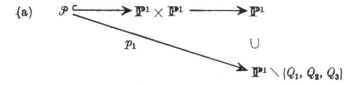

where the first horizontal morphism is the product compactification of $\mathcal{P}$ (see 3.5.a) *and the second is the projection onto the first factor;*

- *the quotient morphism* $\mathcal{P} \xrightarrow{q} \mathcal{P}/K_4$, *the universal cover of* $\mathcal{P}/K_4$ *being* $\mathfrak{H} \times \mathbb{C}$.

The maps l_Δ, p_1, q are connected with each other by means of the commutative agram 6.2.c.

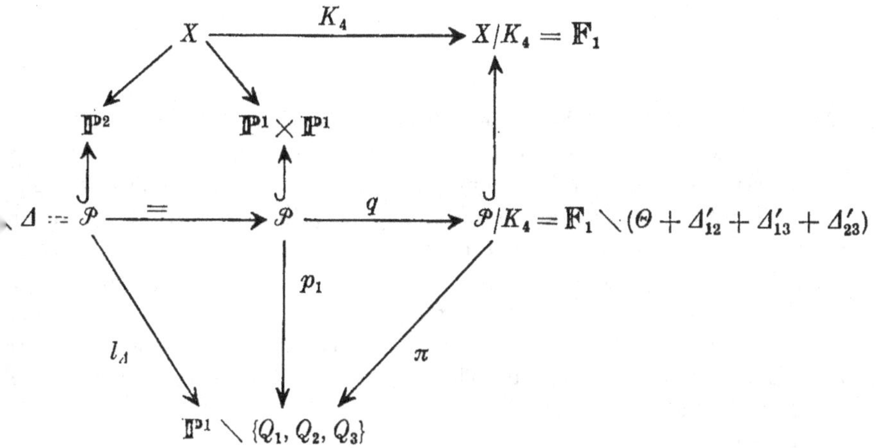

ere $X \to \mathbb{P}^2$ is the blowing up of the four points P_1, P_2, P_3, P_4 (see 3.5.A), is the section of the Hirzebruch surface $\mathbb{F}_1$ over $\mathbb{P}^1$ for $\pi\colon \mathbb{F}_1 \to \mathbb{P}^1$ with self-tersection number -1, and Δ'_{ij} is the image of Δ_{ij} on X (see 3.5.A) by $X \to X/K_4$, ich is a fibre of $\bar{\pi}$ (see 3.6.a, 3.6.3).

Proof. The surface $\mathcal{P}/K_4$ has a product comptactification $\mathbb{P}^1 \times \mathbb{P}^1$. Indeed, can blow up a point P of $\Delta'_{23} \setminus \Theta$, and then we blow down Δ'_{23} and obtain $\times \mathbb{P}^1$:

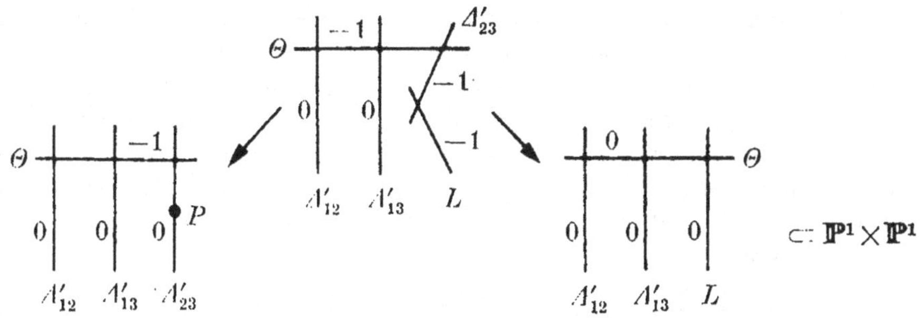

is the restriction of the projection $\mathbb{P}^1 \times \mathbb{P}^1 \to \mathbb{P}^1$ onto the first factor. There-re $\mathcal{P}/K_4 \cong (\mathbb{P}^1 \setminus \{Q_1, Q_2, Q_3\}) \times \mathbb{A}^1$, and the universal cover of $\mathcal{P}/K_4$ is $\mathfrak{H} \times \mathbb{C}$. Now it remains to show that l_Δ and p_1 are the same morphisms. The rational map. $l\colon \mathbb{P}^2 \dashrightarrow \mathbb{P}^1$ has P_1, P_2, P_3, P_4 as fundamental points by 6.2.6. After blowing up we get a commutative diagram

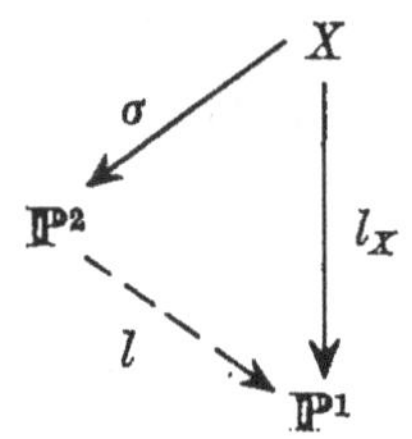

By 6.2.9 the fibres of l_X are the proper σ-transforms of the $\mathfrak{Q}_\mu$'s through P_1, P_2, P_3, P_4. These transforms are disjoint because any two of the $\mathfrak{Q}_\mu$'s have exactly four intersection points on $\mathbb{P}^2$ (Bezout), and these intersection points are P_1, P_2, P_3, P_4. Therefore they intersect transversally in P_i, that means that the proper transforms do not meet each other on the exceptional curves $\Theta_i = \sigma^{-1}(P_i)$. The constellation of the Θ_i's and the exceptional fibres of l_X is the same as in 3.5.A on X. As in that case we blow down $\Delta_{14} + \Delta_{24} + \Delta_{34}$ and obtain $\mathbb{P}^1 \times \mathbb{P}^1$. The proper transforms of the $\mathfrak{Q}_\mu$'s on $\mathbb{P}^1 \times \mathbb{P}^1$ have selfintersection number 0 and they intersect Θ_i transversally. This is only possible for the fibres of the projection of $\mathbb{P}^1 \times \mathbb{P}^1 = \Theta_1 \times \mathbb{P}^1$ onto the first factor. The restriction to $\mathscr{P} \subset \mathbb{P}^2$, X, $\mathbb{P}^1 \times \mathbb{P}^1$ yields $p_1 = l_\Delta$.

The essential point in Proposition 6.2.15 is the splitting of the universal cover of $\mathscr{P}/K_4$. Furthermore we are not far away from the cusp bundle $\mathbb{F}'_\varkappa$ of $\tilde{\Gamma}'' \cdot K_4$ (or $\Gamma'', \tilde{\Gamma}', \Gamma'' \cdot K_4$) in $\varkappa$. Indeed, $\overline{\mathbb{F}'_\varkappa}$ is the 3-sheeted cyclic cover of $X/K_4 = \mathbb{F}_1$ branched along $\Delta'_{12} + \Delta'_{13} + \Delta'_{23}$ (see 3.6.a, 3.6.b and 2.7.a). By 2.4.a and 2.4.B the preimage of $\mathscr{P}/K_4$ is the unramified 3-sheeted cover isomorphic to $T'_\varkappa \times \mathbb{P}^1 \setminus (T_4 + D_{12} + D_{13} + D_{23})$ in Figure 2.1.A of the graph T_4 of the Weierstrass $\wp'$-function (see 2.8.3). Altogether we have the commutative Diagram 6.2.d.

Around a cusp $\varkappa$ one has pretty series expressions for automorphic forms with Theta-functions of $T'_\varkappa$ as coefficients (see e.g. [28]). Using local coordinates around $T'_\varkappa$ on the cusp bundle $\mathbb{F}'_\varkappa$ one can find these series expressions. Indeed, an automorphic form can be understood as a special holomorphic function on an analytic neighbourhood of $T'_\varkappa$ on the cusp bundle. So Diagram 6.2.d gives a good geometric background for a deeper understanding of the connection between our automorphic forms and modular forms. In the same manner one can also descend to $\mathscr{P}/A_4$. Our surface fine classifications yield similar good fibration diagrams.

We close this section with a short look at the fibration $\overline{\mathfrak{B}/\tilde{\Gamma}} \to \mathbb{P}^1$ considered in 3.6.5. The fine stepwise fibre classification in 3.6, the interpretation of the fibration $X \to \mathbb{P}^1$ in terms of $(l_1 : l_2 : l_3)$ and its lift to $\mathfrak{B}$ by means of the $\tilde{\Gamma}'' \cdot K_4$-automorphic forms V_1, V_2, V_3 are summerized in the commutative Diagram 6.2.e.

.d.

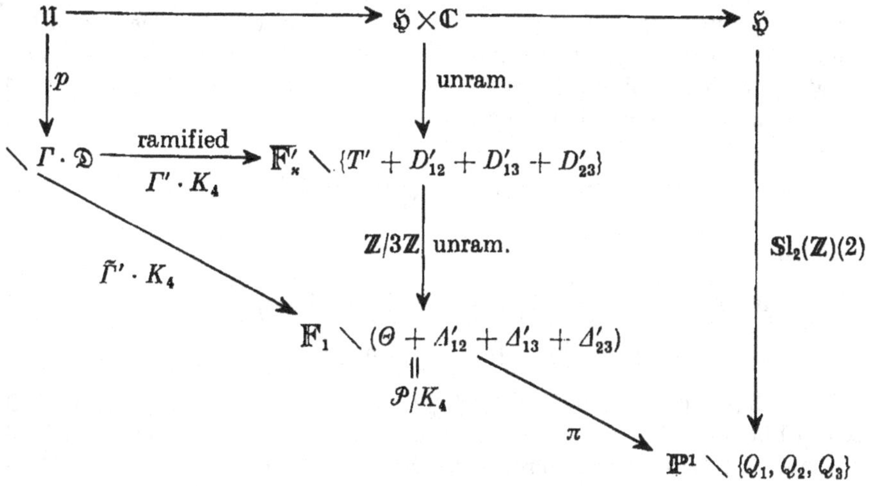
unram.
ramified
$\Gamma' \cdot K_4$
$\mathbb{Z}/3\mathbb{Z}$ unram.
$\mathbb{S}\mathrm{l}_2(\mathbb{Z})(2)$
$\tilde{\Gamma}' \cdot K_4$
$\mathbb{F}_1 \setminus (\Theta + \Delta'_{12} + \Delta'_{13} + \Delta'_{23})$
$\mathcal{P}/K_4$
π
$\mathbb{P}^1 \setminus \{Q_1, Q_2, Q_3\}$

2. 6

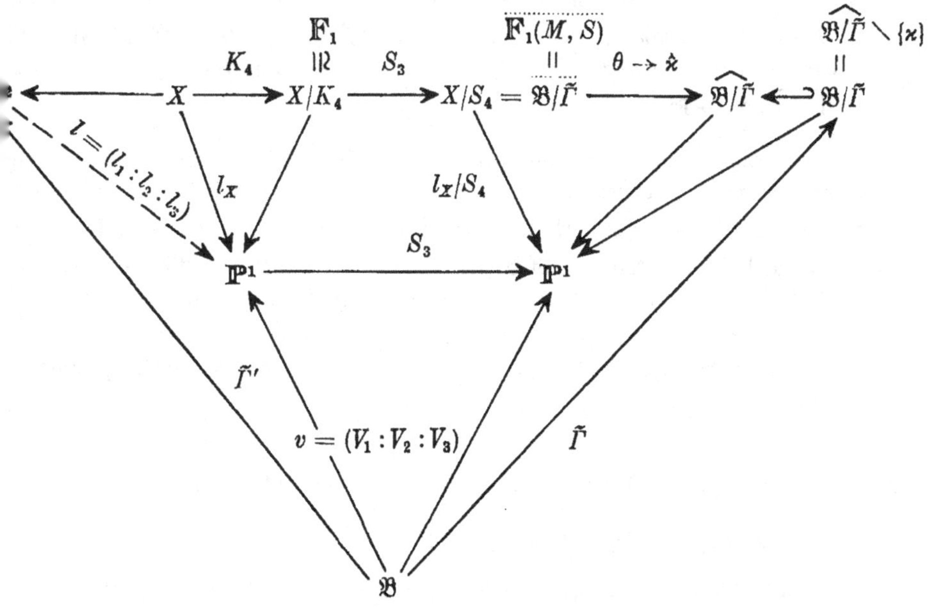
$\mathbb{F}_1$
K_4
S_3
$\theta \dashrightarrow \hat{\varkappa}$
X
X/K_4
X/S_4
$l = (l_1 : l_2 : l_3)$
l_X
l_X/S_4
S_3
$\mathbb{P}^1$
$\mathbb{P}^1$
$\tilde{\Gamma}'$
$v = (V_1 : V_2 : V_3)$
$\tilde{\Gamma}$

We define

$$g_4 = V_1V_2 + V_1V_3 + V_2V_3, \qquad g_6 = -V_1V_2V_3.$$

These are $\tilde{\Gamma}_\chi$-automorphic forms of weight 4 or 6, respectively. The ring of $\tilde{\Gamma}_\chi$-automorphic forms of even weights is generated by G_2, G_3^2, G_4 (see 4.4.5). Evaluation at the points $(1:0:0:-1)$, $(-2:1:1:0)$, $(1:1:1:-3) \in \mathbb{P}^2$ yields after an easy calculation the precise relations

$$g_4 = -\frac{1}{3}G_2^2 + 12G_4, \qquad g_6 = -\frac{2}{27}G_2^3 - G_3^2 + \frac{5}{9}G_2G_4.$$

Let $E(b)$ be the curve

$$E(b)\colon Y^2 = \big(X - V_1(b)\big)\big(X - V_2(b)\big)\big(X - V_3(b)\big)$$
$$= X^3 + g_4(b)\,X + g_6(b), \qquad b \in \mathfrak{B}.$$

$E(b)$ is an elliptic curve if and only if $b \notin \tilde{\Gamma}\cdot\mathfrak{D}$. The closure of the image of $\mathfrak{D}$ in X is a fibre of l_X. Therefore the closure of the image of $\tilde{\Gamma}\cdot\mathfrak{D}$ in X/S_4 is a fibre of the quotient l_X/S_4. These fibres parametrize the rational curves $E(b)$. More generally, the following result is an immediate consequence of all our previous considerations.

6.2.16. PROPOSITION. *The family* $\{E(b)\colon Y^2 = X^3 + g_4(b)\,X + g_6(b),\ b \in \mathfrak{B}\}$ *contains up to isomorphism all elliptic curves and the rational curve with a singularity of type* $Y^2 = X^2$ *(double point). The curves* $E(a)$ *and* $E(b)$ *are isomorphic if* $b \in \tilde{\Gamma}\cdot a$, *that means* $\bar{a} = a \bmod \tilde{\Gamma} = b \bmod \tilde{\Gamma} = \bar{b} \in \mathfrak{B}/\tilde{\Gamma} = \overline{\mathbb{F}_1(M,S)} \setminus \Theta$. *They are isomorphic if and only if* $\bar{a}$ *and* $\bar{b}$ *lie in the same fibre of the fibration* $l_X/S_4\colon \overline{\mathbb{F}_1(M,S)} \to \mathbb{P}^1$.

The projection l_X/S_4 has a remarkable section $\overline{M'S'}$ going through the surface singularities M', S' of types $\langle 2, 1\rangle$ and $\langle 3, 2\rangle$, respectively (see 3.6.A). The section $\overline{M'S'}$ is a horizontal section, that means that it does not intersect Θ. In order to find it recall the construction of $\overline{\mathbb{F}_1(M,S)}$ in the first part of Section 3.6 starting from $\mathbb{P}^2$ and three points $\hat{\Theta}, S, M \in \mathbb{P}^2$ in general position. The proper transform of the projective line $\overline{SM}$ through $S, M \in \mathbb{P}^2$ in $\overline{\mathbb{F}_1(M,S)}$ is really a section of l_X/S_4 going through S', M'. It would be well to understand the preimage of $\overline{M'S'}$ in $\mathfrak{B}$.

6.3. *Inversion of the Picard Integral Map by Means of Automorphic Forms*

Let $\mathfrak{C} \subset \mathscr{P} \times \mathbb{P}^2$ be the Picard curve family

$$WY^3 = X(X - W)(X - t_1W)(X - t_2W), \qquad (W : X : Y) \in \mathbb{P}^2,$$

$$t = (t_1, t_2) \in \mathscr{P}.$$

The fibre over t for the projection onto the first factor is the smooth Picard curve $\mathscr{C}_t$. We consider the holomorphic 1-forms

$$\eta = \eta_t = \frac{dx}{y} \in H^0(\mathscr{C}_t, \Omega^1_{\mathscr{C}_t}).$$

Each path π in $\mathscr{C}_t$ defines a holomorphic function germ $\int_\pi \eta_t \in \mathcal{O}_{\mathscr{P},t}$. The value in t of this function germ is denoted by $\oint\!\!\!\!\!\!\;\;\;_\pi \eta_t$, if π is not a loop and by $\oint_\pi \eta_t$, if π is a loop. We call $\oint\!\!\!\!\!\!\;\;\;_\pi \eta_t$ a *fractional period* (of η), if both the starting point g of π and the endpoint h lie in $\{0, 1, t_1, t_2, \infty\}$. This is justified by the following

6.3.1. LEMMA. *If* $\oint\!\!\!\!\!\!\;\;\;_\pi \eta_t$ *is a fractional period, then* $(\omega^2 - \omega) \oint\!\!\!\!\!\!\;\;\;_\pi \eta_t$ *is a period, that means that*

$$(\omega^2 - \omega) \oint\!\!\!\!\!\!\;\;\;_\pi \eta_t = \oint_\alpha \eta_t$$

for a suitable loop α *on* $\mathscr{C}_t$.

Proof. Without loss of generality we can assume that only the starting and end points, g and h, respectively, of π lie in the ramification locus $\operatorname{Ram} \mathscr{C}_t = \{0, 1, t_1, t_2, \infty\}$ of the cyclic covering map $p_t\colon \mathscr{C}_t \to \mathbb{P}^1$ $\big((w : x : y) \mapsto y\big)$. The projection $p_t(\pi)$ has exactly three lifts $\pi = \pi_1, \pi_2, \pi_3$ on $\mathscr{C}_t$. Since $y^3 = x(x-1) \times (x - t_1)(x - t_2)$ on the affine part $w = 1$ of $\mathscr{C}_t$ we find

$$\oint\!\!\!\!\!\!\;\;\;_\pi \eta = \omega^{k_i} \oint\!\!\!\!\!\!\;\;\;_{\pi_i} \eta, \qquad k_2 = 1,\ k_3 = 2 \qquad \text{(without loss of generality)};$$

hence

$$(\omega^2 - \omega) \oint\!\!\!\!\!\!\;\;\;_\pi \eta = \oint_{\pi_3 - \pi_2} \eta.$$

Consider the locally constant sheaf $\mathscr{H}_1(\mathfrak{C}, \mathbb{Z})$ of abelian groups over $\mathscr{P}$ with stalk $H_1(\mathscr{C}_t, \mathbb{Z})$ at $t \in \mathscr{P}$. For a given point $t \in \mathscr{P}$ we have an action of the fundamental group $\pi_1(\mathscr{P}) = \pi_1(\mathscr{P}, t)$ on $H_1(\mathscr{C}_t, \mathbb{Z})$, the monodromy action. This action induces an action on the abelian group of period integrals at t

$$\mathscr{H}^1(\mathscr{C}_t, \mathbb{Z}) = \left\{\int_\pi \eta ; \pi \in H_1(\mathscr{C}_t, \mathbb{Z})\right\}.$$

So each $\pi \in H_1(\mathscr{C}_t, \mathbb{Z}) = \pi_1(\mathscr{C}_t)/[\pi_1(\mathscr{C}_t), \pi_1(\mathscr{C}_t)]$ defines locally a multivalued holomorphic function. This multivalued function is globally defined on $\mathscr{P}$ because the monodromy corresponds to analytic extensions of our holomorphic functions along paths in $\mathscr{P}$ with starting point t. Using the same notations as for the germs of the multivalued period integral functions around t we write

$$\int_\pi \eta : \mathscr{P} \Rrightarrow \mathbb{C}.$$

Lemma 6.3.1 extends this definition to all paths π of $\mathscr{C}_t$ joining two points g, h of Ram $\mathscr{C}_t$. So for a fixed $t \in \mathscr{P}$ and a fixed path π in $\mathscr{C}_t$ of this type we have defined a multivalued holomorphic function

$$\int_g^h \eta = {}_\pi\!\int_g^h \eta : \mathscr{P} \Rrightarrow \mathbb{C}.$$

In order to understand the multivalence of these functions we consider simultaneously three period integrals $\int_{\alpha_i} \eta$, $i = 1, 2, 3$, $\alpha_i \in H_1(\mathscr{C}_t, \mathbb{Z})$. Picard (see [55]) found a canonical basis $(\alpha_1, \alpha_2, \alpha_4, \alpha_3, \alpha_5, \alpha_6)$ of $H_1(\mathscr{C}_t, \mathbb{Z})$ such that the period matrix with respect to the basis $\eta = \eta_1$, $\eta_2 = \dfrac{dx}{y^2}$, $\eta_3 = \dfrac{x\,dx}{y^2}$ of $H^0(\mathscr{C}_t, \Omega^1_{\mathscr{C}_t})$ has the following form:

$$\Pi = \left(\begin{array}{ccc|ccc} A_1 & A_2 & -\omega^2 A_1 & A_3 & \omega^2 A_2 & \omega A_3 \\ B_1 & B_2 & -\omega B_1 & B_3 & \omega B_2 & \omega^2 B_3 \\ C_1 & C_2 & -\omega C_1 & C_3 & \omega C_2 & \omega^2 C_3 \end{array}\right),$$

$$A_i = \oint_{\alpha_i} \eta_1, \qquad B_i = \oint_{\alpha_i} \eta_2, \qquad C_i = \oint_{\alpha_i} \eta_3, \qquad i = 1, 2, 3.$$

A triple $\alpha_1, \alpha_2, \alpha_3 \in H_1(\mathscr{C}_t, \mathbb{Z})$, which has an extension to a canonical basis of the above type will shortly be called an $\mathfrak{O}$*-canonical basis of* $H_1(\mathscr{C}_t, \mathbb{Z})$. Here $\mathfrak{O}$ denotes the ring $\mathfrak{O}_{\mathbb{Q}(\sqrt{-3})}$ of integers in $\mathbb{Q}(\sqrt{-3})$. Such a basis defines an $\mathfrak{O}$-structure on $H_1(\mathscr{C}_t, \mathbb{Z})$ such that $\alpha_1, \alpha_2, \alpha_3$ is an $\mathfrak{O}$-basis of this $\mathfrak{O}$-lattice.

6.3.A.

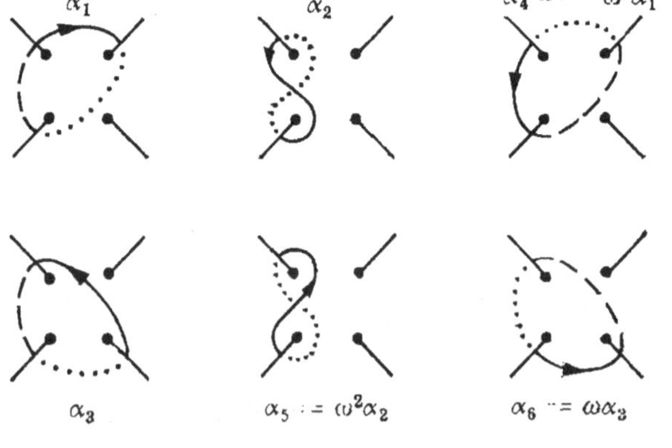

In order to obtain an idea of such an $\mathfrak{O}$-canonical basis together with its extension to the canonical basis of the above type we draw the Figure 6.3.A. Here $\mathscr{C}_t$ is understood in the classical way as the union of three copies of $\mathbb{P}^1$ with natural identifications along four cuts in $\mathbb{P}^1$ joining ∞ with 0, 1, t_1 and t_2, respectively. In order to distinguish the paths in the three copies of $\mathbb{P}^1$ we used solid lines, broken lines and dotted lines, respectively.

By means of elementary row operations corresponding to basis changes in $H^0(\mathscr{C}_t, \Omega^1_{\mathscr{C}_t})$ and the *Riemann first bilinear relations*

$$\Pi J \Pi^{\mathrm{T}} = 0, \qquad J = \left(\begin{array}{c|c} O & E \\ \hline -E & O \end{array}\right), \qquad E = E_3 = \begin{pmatrix} 1 & 0 & 0 \\ 0 & 1 & 0 \\ 0 & 0 & 1 \end{pmatrix}$$

(see e.g. [23], p. 231) it is easy to see that Π is completely determined by its first row (and the choice of the basis extension η_1, η_2, η_3 of η_1). The *second Riemann bilinear relation*

$$\mathrm{i} \cdot \Pi_1 J \overline{\Pi}_1^{\mathrm{T}} > 0, \qquad \mathrm{i} = \sqrt{-1},$$

for the first row Π_1 of Π comes down to the hermitian condition

$$(A_1, A_2, A_3)\, M \begin{pmatrix} 1 & 0 & 0 \\ 0 & 1 & 0 \\ 0 & 0 & -1 \end{pmatrix} \overline{M}^{\mathrm{T}} (\overline{A}_1, \overline{A}_2, \overline{A}_3)^{\mathrm{T}},$$

$$M = \begin{pmatrix} -\omega & 0 & \omega^2 \\ 0 & 1 & 0 \\ \omega & 0 & \omega \end{pmatrix} \in \mathfrak{Gl}_3(\mathfrak{O}).$$

So we have defined locally a multivalued holomorphic mapping

$$\Big(\int_{\alpha_1}\eta : \int_{\alpha_2}\eta : \int_{\alpha_3}\eta\Big) \colon \mathcal{U}' \Rrightarrow \mathfrak{B}', \qquad t \in \mathcal{U}' \subset \mathscr{P},$$

with $\mathfrak{B}'$ projectively equivalent in $\mathbb{P}^2$ to the complex unit ball $\mathfrak{B}$ (use M). In other words:

6.3.2. *There exists a $\mathfrak{Gl}_3(\mathfrak{O})$-linear combination I_1, I_2, I_3 of period integrals $\int_{\alpha_1}\eta$, $\int_{\alpha_2}\eta$, $\int_{\alpha_3}\eta$ such that the multivalued holomorphic mapping $(I_1 : I_2 : I_3) \colon \mathscr{P} \Rrightarrow \mathbb{P}^2$ defines locally around t a multivalued map into the ball $\mathfrak{B}$*

$$(I_1 : I_2 : I_3) \colon \mathcal{U} \Rrightarrow \mathfrak{B}, \qquad t \in \mathcal{U} \subset \mathscr{P}.$$

Picard's construction of the $\mathfrak{O}$-canonical basis $\alpha_1, \alpha_2, \alpha_3$ is very global. More generally the following statement is true:

6.3.3. PROPOSITION (Mostow/Deligne, [45], [14]). *Let Σ be the monodromy group of I_1, I_2, I_3; i.e. Σ is the image of $\pi_1(\mathscr{P}) = \pi_1(\mathscr{P}, t)$ of the monodromy representation $\pi_1(\mathscr{P}) \to \mathfrak{Gl}_3(\mathbb{C}I_1 \oplus \mathbb{C}I_2 \oplus \mathbb{C}I_3)$. There is a simply-connected dense* *subset $\mathscr{P}'$ of $\mathscr{P}$ such that the image of*

$$(I_1 : I_2 : I_3) \colon \mathscr{P}' \to \mathbb{P}^2$$

is a fundamental domain of $\mathfrak{B}$ with respect to the monodromy group Σ.

More information about Σ one can find again in two papers of Picard:

6.3.4. PROPOSITION (Picard, [55], [56]). *Σ is a subgroup of the congruence subgroup $\tilde{\Gamma}'$ of $\mathbb{U}((2, 1), \mathfrak{O})$.*

In fact Picard found explicitly five generators of Σ all lying in $\tilde{\Gamma}'$. We reproduce them at the end of this section together with an idea of how to construct them. For the proof of our main result we need

6.3.5. PROPOSITION (Mostow, [46], [45], [14]). *Σ is a lattice subgroup of $\mathbb{U}((2, 1), \mathbb{C})$.*

This means that the fundamental domain of Σ in $\mathfrak{B}$ has finite volume with respect to the Bergmann metric of $\mathfrak{B}$. Since the arithmetic group $\tilde{\Gamma}'$ has the same lattice property (see Borel [5]) and because $\Sigma \subseteq \tilde{\Gamma}'$ we have

6.3.6. COROLLARY. *Σ is an arithmetic ball lattice; it is a subgroup of $\tilde{\Gamma}'$ of finite index.*

Altogether the commutative Diagram 6.3.a of holomorphic mappings (one of them multivalued) describes the situation. The compactification X of $\mathscr{P}$ (see 3.5.a) has a compactification divisor with transversally intersecting irreducible components. By a theorem of Borel (see [6]) there exists an extension of $\mathscr{P} \to \mathfrak{B}/\tilde{\Gamma}'$ to a holomorphic mapping $X \to \widehat{\mathfrak{B}/\tilde{\Gamma}'} = \mathbb{P}^2$. By a theorem of Chow (see e.g. Mumford [51], IV, § 4B, (4.14)) this mapping has to be an algebraic morphism. The same conclusion holds for Σ instead of $\tilde{\Gamma}'$. Se we have a commutative diagram of (algebraic) morphisms

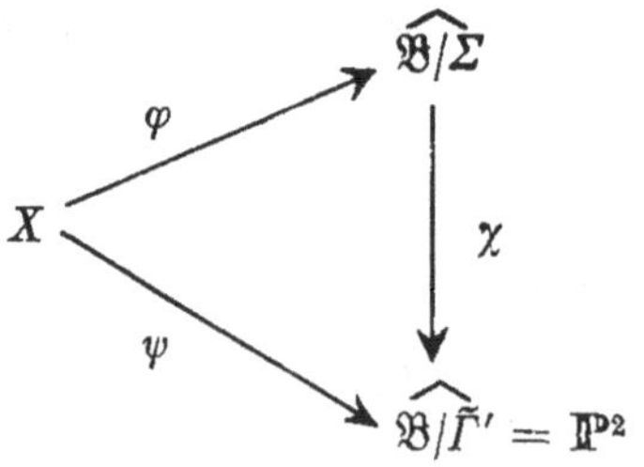

The upper morphism φ is bijective on certain dense subsets. Therefore it is a birational morphism. We know that $\varphi(\mathscr{P}) \subseteq \mathfrak{B}/\Sigma$. Therefore the cusp singularities are contractions of subcurves of the compactification curve $X \setminus \mathscr{P}$. The latter curve consists of smooth rational curves with selfintersection number -1. The cusp singularities are normal. By a result of Mumford (see [47]) each resolution of a normal surface singularity consists of a connected curve with negative definite intersection matrix of their components. For our cusp singularity re-

6.3.a.

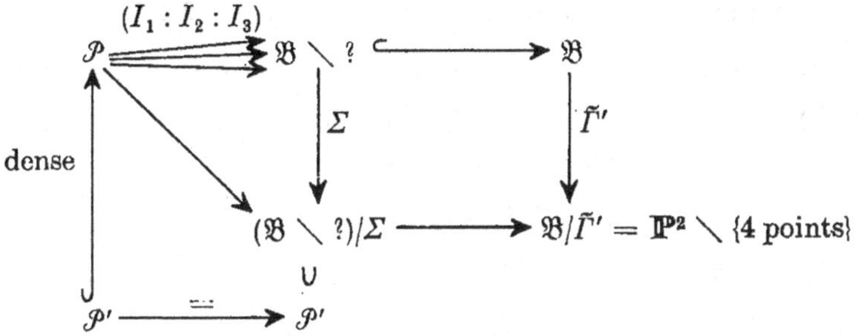

solution in X this is only possible for an irreducible component of $X \setminus \mathscr{P}$. But a point, which has a smooth rational curve with selfintersection number -1 as resolution curve is a regular point. So we will take note of

6.3.7. *The cusp singularities of $\widehat{\mathfrak{B}/\Sigma}$ are regular points.*

The same arguments work also for the possible singularities of $\mathfrak{B}/\Sigma$ because the curves of $X \setminus \mathscr{P}$ are the only ones on X with negative selfintersection numbers. Therefore:

6.3.8. *$\widehat{\mathfrak{B}/\Sigma}$ is smooth.*

Now χ is quasifinite, hence finite by the Zariski main theorem. So χ is a branched covering of smooth surfaces. The branch locus B is a divisor on $\mathbb{P}^2$ (purity of branch locus, see e.g. [23], V. 2). We will prove that

6.3.9. *$\chi\colon \widehat{\mathfrak{B}/\Sigma} \dashrightarrow \widehat{\mathfrak{B}/\tilde{\Gamma}'} = \mathbb{P}^2$ is the identity map.*

To this end note that $\mathbb{P}^2$ is simply-connected. Therefore $\mathbb{P}^2$ itself is the only unramified cover of $\mathbb{P}^2$. Therefore it suffices to prove that:

6.3.10. *The branch divisor of χ is trivial* (i. e. $B = 0$).

Consider the commutative diagram

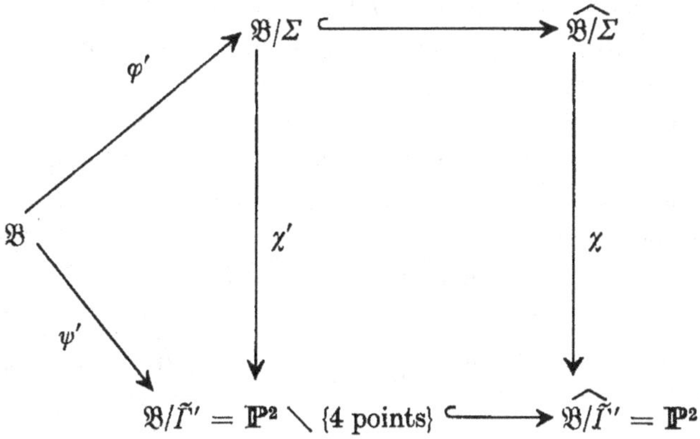

Now ψ' is unramified outside of $\Delta = \sum_{1 \leq i < j \leq 4} \Delta_{ij}$ (see 3.3.b, 3.4.A). Therefore χ' and χ are unramified outside of Δ. Each Δ_{ij} goes through two cusp "singularities" $P_i, P_j \in \widehat{\mathfrak{B}/\tilde{\Gamma}'}$, namely the cusp "singularities" are the triple points of Δ (see 3.3.b again). So our problem is reduced to:

6.3.11. *χ is not ramified around the cusp "singularities".*

We check 6.3.11 for a Γ-cusp $\varkappa$ locally around $\hat{\varkappa} \in \widehat{\mathfrak{B}/\Sigma}$. The covering $\widehat{\mathfrak{B}/\Sigma} \to \widehat{\mathfrak{B}/\tilde{\Gamma}'}$ is locally isomorphic around the corresponding images of $\varkappa$ to the covering of quotients of cusp bundles

$$\widehat{\mathbb{F}_\varkappa(\Sigma)}/G'_\varkappa \to \widehat{\mathbb{F}_\varkappa(\tilde{\Gamma}')}/G_\varkappa, \qquad G'_\varkappa = \Sigma_\varkappa/\Sigma_{\varkappa,u}, \qquad G_\varkappa = \tilde{\Gamma}'_\varkappa/\tilde{\Gamma}'_{\varkappa,u}$$

(see 1.2.c). This morphism enters into the commutative Diagram 6.3.b

6.3.b.
$$\begin{array}{ccccc}
\mathbb{F}'_\varkappa = \mathbb{F}_\varkappa(\Sigma) & \xrightarrow{G'_\varkappa} & \mathbb{F}_\varkappa(\Sigma)/G'_\varkappa & \xrightarrow{T'_\varkappa/G'_\varkappa} & \widehat{\mathbb{F}_\varkappa(\Sigma)}/G'_\varkappa \\
\downarrow & & \downarrow & & \downarrow \\
\mathbb{F}_\varkappa = \mathbb{F}_\varkappa(\tilde{\Gamma}') & \xrightarrow[G_\varkappa]{} & \mathbb{F}_\varkappa(\tilde{\Gamma}')/G_\varkappa & \xrightarrow[T_\varkappa/G_\varkappa]{} & \widehat{\mathbb{F}_\varkappa(\tilde{\Gamma}')}/G_\varkappa
\end{array}$$

where the horizontal arrows on the right-hand side represent contractions.

The cusp singularities are in both cases regular points because $\widehat{\mathfrak{B}/\Gamma'} = \mathbb{P}^2$ and because of 6.3.7. In particular they cannot be contractions of elliptic curves. Therefore $G_\varkappa, G'_\varkappa \neq 1$, $T_\varkappa/G_\varkappa \cong T'_\varkappa/G'_\varkappa \cong \mathbb{P}^1$. We know that $[\tilde{\Gamma}' : \Gamma'] = 3$ (see 3.2.3) and $\Gamma'_\varkappa = \Gamma'_{\varkappa,u}$ (see 1.2). Therefore

$$1 < [\Sigma_\varkappa : \Sigma_{\varkappa,u}] \leq [\tilde{\Gamma}'_\varkappa : \tilde{\Gamma}'_{\varkappa,u}] \leq [\tilde{\Gamma}' : \Gamma'] = 3,$$

hence $G_\varkappa \cong G'_\varkappa \cong \mathbb{Z}/3\mathbb{Z}$. By the Hurwitz genus formula applied to $T_\varkappa^{(\prime)} \xrightarrow{G_\varkappa^{(\prime)}} \mathbb{P}^1$ these Galois coverings have exactly three branch points. These are cyclic surface "singularities" in $\mathbb{F}'_\varkappa/G_\varkappa$ of type $\langle 3, e\rangle$, $e \in \{0, 1, 2\}$. The curve on the singularity resolution $\overline{\mathbb{F}_\varkappa/G_\varkappa}$ or $\overline{\mathbb{F}'_\varkappa/G'_\varkappa}$, respectively, consisting of $\mathbb{P}^1$ and the resolution curves can be contracted to a regular point because this contraction is the cusp "singularity" (see 1.2), which is a regular point in both cases Σ and $\tilde{\Gamma}'$. Knowing the resolution curves $-3/$ and -2×-2 of the cyclic singularities $\langle 3, 1\rangle$ or $\langle 3, 2\rangle$, respectively, we can see that there are only two possibilities of such "resolutions" of our regular cusp "singularities" on $\mathbb{F}'_\varkappa/G'_\varkappa$:

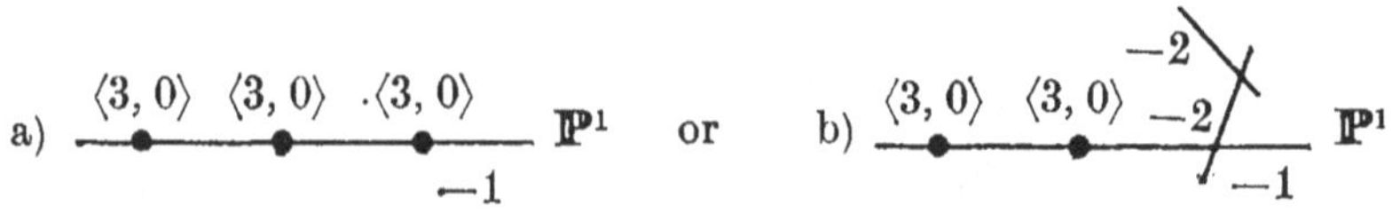

Formula (3.6.2) for selfintersection numbers of quotient curves on quotient surfaces applied to $\mathbb{F}'_\varkappa \to \mathbb{F}'_\varkappa/G'_\varkappa$ yields the relations

$$-1 = \frac{1}{3}\,(T'_\varkappa)^2 - \begin{cases} 0, & \text{case a)},\\ 2/3, & \text{case b)},\end{cases}$$

hence

$$(T'^2_\varkappa) = \begin{cases} -3, & \text{case a)},\\ -1, & \text{case b)}.\end{cases}$$

In 1.2 we proved that $\left(T_\varkappa(\Gamma'')^2\right) = -3$. On the other hand we know that $\mathbb{F}_\varkappa(\tilde{\Gamma}') = \mathbb{F}(\tilde{\Gamma}'_{\varkappa,u}) = \mathbb{F}(\Gamma'_\varkappa)$, hence $(T^2_\varkappa) = -3$. So we have case a) in the cusps of $\tilde{\Gamma}'$. Now we will exclude case b) also for Σ. For this purpose we decompose the left vertical morphism into two abelian Galois coverings:

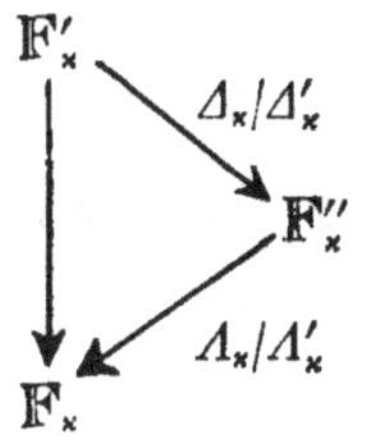

Here $\mathbb{F}''_\varkappa$ is the cusp bundle of the auxiliary cusp lattice $\Lambda_\varkappa \cdot \Sigma_{\varkappa,u}$, where $\Delta_\varkappa$ $\Delta_\varkappa(\tilde{\Gamma}')$, $\Delta'_\varkappa = \Delta_\varkappa(\Sigma_\varkappa)$, $\Lambda_\varkappa = \Lambda_\varkappa(\tilde{\Gamma}') = \tilde{\Gamma}'_{\varkappa,u}/\Lambda_\varkappa$, $\Lambda'_\varkappa = \Lambda_\varkappa(\Sigma)$ with the notations of 1.2, $\mathbb{F}''_\varkappa \to \mathbb{F}_\varkappa$ is unramified and $\mathbb{F}'_\varkappa \to \mathbb{F}''_\varkappa$ is totally ramified along $T'_\varkappa \Rightarrow T''_\varkappa$, $T''_\varkappa = T_\varkappa(\Delta_\varkappa \cdot \Sigma_\varkappa)$. Now let $\sigma \in \Sigma_\varkappa \setminus \Sigma_{\varkappa,u} \subseteq \tilde{\Gamma}'_\varkappa \setminus \tilde{\Gamma}'_{\varkappa,u}$. Then σ acts on $\mathbb{F}'_\varkappa$, $\mathbb{F}''_\varkappa$, $\mathbb{F}_\varkappa$ and generates the Galois groups $G'_\varkappa \cong \mathbb{Z}/3\mathbb{Z}$ of $\mathbb{F}'_\varkappa \to \mathbb{F}'_\varkappa/G'_\varkappa$. The map σ has exactly three fixed points on $T'_\varkappa$. Let R' be one of them and R'', R the images of R' in $\mathbb{F}''_\varkappa$ and $\mathbb{F}_\varkappa$, respectively. Since $\mathbb{F}''_\varkappa \to \mathbb{F}_\varkappa$ is unramified the isotropy groups $\langle\sigma\rangle_{R''}$ and $\langle\sigma\rangle_R$ on $\mathbb{F}''_\varkappa$ and $\mathbb{F}_\varkappa$, respectively, are of the same type. This type is $\langle 3, 0\rangle$ as we already know from situation a) for $\mathbb{F}_\varkappa \to \mathbb{F}_\varkappa/G_\varkappa$. The group element σ normalizes $\Delta'_\varkappa$, hence also $\Delta_\varkappa$ by an easy calculation with unipotent matrices as described in 1.2. Therefore $\Sigma_{\varkappa,u}$ is a normal subgroup of $\Gamma''_\varkappa = \langle \Delta_\varkappa \cdot \Sigma_{\varkappa,u}, \sigma\rangle$. The isotropy group $(\Gamma''_\varkappa/\Sigma_{\varkappa,u})_{R'}$ is abelian. It is generated by σ and $\Gamma''_{\varkappa,u}/\Sigma_{\varkappa,u} = \Delta_\varkappa \cdot \Sigma_{\varkappa,u}/\Sigma_{\varkappa,u} \cong \Delta_\varkappa/\Lambda'_\varkappa$. It is not difficult to see that $(\Gamma''_\varkappa/\Sigma_{\varkappa,u})_{R'}$ contains a vertical reflection, i.e. a reflection with reflection curve through R' transversally to $T'_\varkappa$, if and only if $\langle\sigma\rangle_{R''}$ contains a reflection. The latter condition is satisfied because $\langle\sigma\rangle_{R''}$ is of type $\langle 3, 0\rangle$. But a vertical reflection of $(\Gamma''_\varkappa/\Sigma_{\varkappa,u})_{R'}$ $\cong \langle\sigma\rangle \times (\Delta_\varkappa/\Lambda'_\varkappa)$ must be a power of σ because the elements of $\Lambda_\varkappa/\Delta'_\varkappa$ are horizontal reflections, that means that they fix pointwise $T'_\varkappa$. So σ itself is a (vertical) reflection in R'. Therefore $\langle\sigma\rangle_{R'}$ is of type $\langle 3, 0\rangle$. This is true for all three fixed points of σ on $T'_\varkappa$. So we proved that we have situation a) for $\mathbb{F}'_\varkappa \to \mathbb{F}'_\varkappa/G'_\varkappa$.

Now we go back to Diagram 6.3.b. All surfaces occuring in the diagram are smooth. The horizontal morphisms in the left square are branched along three vertical curves with branch order 3. The cover $\mathbb{F}'_\varkappa \to \mathbb{F}_\varkappa$ is unramified outside $T'_\varkappa, T_\varkappa$. Therefore $\mathbb{F}'_\varkappa/G'_\varkappa \to \mathbb{F}_\varkappa/G_\varkappa$ is unramified outside $T'_\varkappa/G'_\varkappa, T_\varkappa/G_\varkappa$. After the contraction of these curves we see that the right vertical morphism in Diagram 6.3.b is unramified. This morphism is locally isomorphic to χ around the cusp singularities belonging to $\varkappa$. This proves 6.3.11, 6.3.10 and 6.3.9.

6.3.12. THEOREM.

(i) *The monodromy group Σ of $(I_1 : I_2 : I_3): \mathcal{P} \Rrightarrow \mathfrak{B}$ is equal to $\tilde{\Gamma}'$.*

(ii) *There are three $\tilde{\Gamma}'_\chi$-automorphic forms ξ_1, ξ_2, ξ_3 of weight 1 generating the whole ring $\bigoplus_{m=0}^{\infty} [\tilde{\Gamma}', m]_\chi$ of $\tilde{\Gamma}'_\chi$-automorphic forms such that*

$$(\xi_1 : \xi_2 : \xi_3): \mathfrak{B} \to \mathbb{P}^2 \setminus \{P_1, P_2, P_3, P_4\}$$

is inverse to $(I_1 : I_2 : I_3)$. More precisely we have a commutative diagram

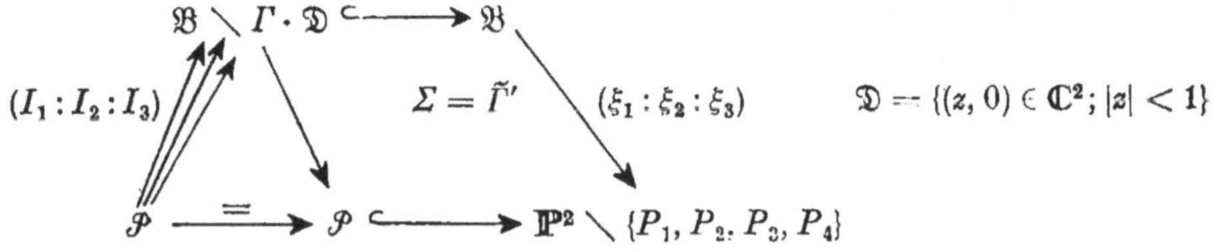

Proof. (i) follows immediately from 6.3.8.

(ii) We know that $[\Gamma', 1] = [\tilde{\Gamma}', 1]_\chi \simeq \mathbb{C}^3$ and $\Phi_{[\Gamma',1]} = (\xi_1 : \xi_2 : \xi_3)$ for a basis ξ_1, ξ_2, ξ_3 of $[\Gamma', 1]$ maps $\mathfrak{B}$ onto $\mathbb{P}^2 \setminus \{4 \text{ points}\}$ (see 6.1). Without loss of generality we can assume that these four points are the given ones P_1, P_2, P_3, P_4 because $\mathrm{Gl}_3(\mathbb{C})$ acts transitively on $\mathbb{P}^2$. At the same time $\Phi_{[\Gamma',1]}$ is the quotient map of $\tilde{\Gamma}' = \Sigma$ (see 6.1.b). Further $\mathfrak{B} \to \mathfrak{B}/\Sigma$ is by definition of Σ and the fundamental domain property described in Proposition 6.3.3 the inverse of the integral map. This proves (ii).

At the end of this section we will say something about the five generators of $\Sigma = \tilde{\Gamma}'$ found by Picard. It is easy to see that $\pi_1(\mathcal{P})$ has really five generators. Indeed, the projection of $\mathbb{C} \times \mathbb{C}$ onto the first factor restricts to a fibration $\mathcal{P} \to \mathbb{C} \setminus \{0, 1\}$ with fibres isomorphic to $\mathbb{C} \setminus \{3 \text{ points}\}$. The exact homotopy sequence yields the short exact sequence

$$1 \to F_3 = \pi_1(\mathbb{C} \setminus \{3 \text{ points}\} \to \pi_1(\mathcal{P}) \to F_2 = \pi_1(\mathbb{C} \setminus \{0, 1\}) \to 1$$

where F_k denotes the free group generated by k elements. Since F_2 is free the sequence splits, that means that $\pi_1(\mathcal{P})$ is a semidirect product of F_2 and F_3. Therefore $\pi_1(\mathcal{P})$ has five generators. Via the representation of $\pi_1(\mathcal{P}) = \pi_1(\mathcal{P}, t)$ in the space of period integrals $\mathcal{H}^1(\mathcal{C}_t, \mathbb{Z})$ we see that $\Sigma = \tilde{\Gamma}'$ has five generators.

Picard's method to construct explicitly five generators can in fact be understood in terms of *braid groups*. It suffices to describe the action of $\pi_1(\mathcal{P}, t)$ on

$H_1(\mathscr{C}_t, \mathbb{Z})$. An element of $\pi_1(\mathscr{P}, t)$ is represented by a loop l in $\mathscr{P}$ beginning and ending in t:

$$l: I = [0, 1] \to \mathscr{P},$$
$$s \mapsto t(s) = \big(t_1(s), t_2(s)\big).$$

Let $\mathbb{C}_s^* = \mathbb{C} \setminus \{t_1(s), t_2(s), 0, 1\}$, $s \in I$. Then the complement $S = S(l)$ of $\bigcup_{s\in I} s \times \mathbb{C}_s^*$ in $I \times \mathbb{C}$ is a braid with four strings

$$S_1 = \{s \times t_1(s);\, s \in I\}, \qquad S_2 = \{s \times t_2(s);\, s \in I\}, \qquad S_3 = I \times 0, \qquad S_4 = I \times 1.$$

So one has a natural representation of $\pi_1(\mathscr{P}, t)$ as a "restricted" braid group $B = B(t)$ of braids beginning and ending in $(t_1, t_2, 0, 1) = \big(t_1(0) = t_1(1), t_2(0) = t_2(1), 0, 1\big)$ with four strings of which two are constant. The structure of braid groups is well understood (see E. Artin [3]). In particular one easily finds five simple generators as described in the following figures:

It is not difficult to find the action of these generators on the $\mathfrak{D}$-canonical basis of $H_1(\mathscr{C}_t, \mathbb{Z})$ and on the corresponding basis of the lattice of period integrals. In [56] Picard wrote down explicit matrix forms of these five generators. Translating them into $\mathfrak{B}$-coordinates by means of conjugation with $\begin{pmatrix} 1 & 0 & 0 \\ 0 & \omega^2 & -1 \\ 0 & \omega & 1 \end{pmatrix}$ we get

6.3.13. PROPOSITION. *The monodromy group $\tilde{\Gamma}'$ is generated by the five elements*

$$\begin{pmatrix} \omega & 0 & 1-\omega^2 \\ \omega^2-\omega & 1 & \omega-1 \\ 0 & 0 & 1 \end{pmatrix}, \quad \begin{pmatrix} \omega & 0 & 1-\omega \\ 1-\omega & 1 & \omega-1 \\ 0 & 0 & 1 \end{pmatrix}, \quad \begin{pmatrix} \omega^2 & 0 & 0 \\ 0 & 1 & 0 \\ 0 & 0 & 1 \end{pmatrix},$$

$$\begin{pmatrix} 1 & 0 & 0 \\ 0 & \omega^2 & \omega-1 \\ 0 & \omega-1 & -2\omega^2 \end{pmatrix}, \quad \begin{pmatrix} \omega & \omega^2-\omega & 0 \\ 0 & 1 & 0 \\ 1-\omega^2 & \omega-1 & 1 \end{pmatrix}.$$

Remark. We have seen in Chapter I that the $\tilde{\Gamma}'_\chi$-automorphic forms $\xi_1, \xi_2, \xi_3, \xi_4$, $\sum \xi_i = 0$, of Nebentypus play a fundamental role. They do not belong to the usual ring of automorphic forms. In order to determine this ring one must additionnally know that $\varrho(\zeta) = \zeta$, $\varrho = \operatorname{diag}(\omega, 1, 1)$ (see 4.4.1, $z = \zeta$). This was first proved by Feustel in [i] (also $\varrho(\xi_i) = \omega\xi_i$). He used results of Shiga [vii]. A second proof can be found in [v], based on the general dimension formula for cusp forms of arithmetic ball lattices (see [iii]).

Chapter II

The Gauss-Manin Connection of Cycloelliptic Curve Families

§ 1. EULER-PICARD SYSTEMS OF PARTIAL DIFFERENTIAL EQUATIONS

1.1. *Euler-Picard Modules*

Let S be a principal open affine subscheme of

$$\mathbb{A}^r = \mathbb{A}^r_{\mathbb{C}} = \operatorname{Spec} \mathbb{C}[t], \qquad \mathbb{C}[t] = \mathbb{C}[t_1, \ldots, t_r],$$

say

$$S = \operatorname{Spec} \mathbb{C}[S], \qquad \mathbb{C}[S] = \mathbb{C}[t]_\Pi = \mathbb{C}[t_1, \ldots, t_r, \Pi^{-1}], \qquad \Pi \in \mathbb{C}[t].$$

The derivations $D_i = \dfrac{\partial}{\partial t_i} : \mathbb{C}(t) \to \mathbb{C}(t)$ restrict to derivations on $\mathbb{C}[S]$. Let $\mathbb{C}[S, D]$ be the non-commutative ring $\mathbb{C}[t_1, \ldots, t_r, \Pi^{-1}, D_1, \ldots, D_r]$. This ring is to be understood as a subring of the endomorphism ring $\operatorname{End}_{\mathbb{C}} \mathbb{C}[S]$. We call it the *ring of (rational) differential operators over S*.

A *simple quadratic differential operator over S* is an element of the form

1.1.1. $$D_i D_j - \sum_{k=0}^{r} a^{(k)} D_k \in \mathbb{C}[S, D], \qquad a^{(k)} \in \mathbb{C}[S], \qquad D_0 = 1.$$

We consider the free $\mathbb{C}[S, D]$-module $\mathbb{C}[S, D]^{r^2}$ and denote the canonical basis elements by 1_{ij}. A *simple quadratic system of differential operators over S* is a $\mathbb{C}[S, D]$-homomorphism

$$\varepsilon : \mathbb{C}[S, D]^{r^2} \to \mathbb{C}[S, D], \qquad 1_{ij} \mapsto \varepsilon_{ij}$$

such that ε_{ij} is simply quadratic of type (1.1.1) and $\varepsilon_{ij} = \varepsilon_{ji}$ for all i, j. We call also the set $\{\varepsilon_{ij}\}$ a *simple quadratic system of differential operators over S*.

A pair (M, m) consisting of a $\mathbb{C}[S, D]$-module M and an element $m \in M$ is called a *principal simple quadratic module* (of type ε), if $M = \mathbb{C}[S, D]\, m$ and the short sequence

1.1.2: $$\mathbb{C}[S, D]^{r^2} \xrightarrow{\varepsilon} \mathbb{C}[S, D] \xrightarrow{\mu} M, \qquad 1 \longmapsto m$$

with the simple quadratic system ε is a complex of $\mathbb{C}[S, \mathrm{D}]$-modules; i.e. $\mu \circ \varepsilon = 0$. Equivalently we can say that m satisfies the *system of simple quadratic differential equations*

1.1.3. $\varepsilon_{ij}\xi = 0, \qquad i, j = 1, \ldots, r.$

The principal simple quadratic module (M, m) of type ε is said to be *free*, if for any other principle simple quadratic module (N, n) of the same type the correspondence $m \mapsto n$ extends (uniquely) to a homomorphism of $\mathbb{C}[S, \mathrm{D}]$-modules.

1.1.4. LEMMA-DEFINITION. *For a principal simple quadratic module* (M, m) *of type* ε *the following conditions* (i)—(iv) *are equivalent*:

(i) (M, m) *is free*;

(ii) $M \simeq \mathbb{C}[S, \mathrm{D}]/(\varepsilon_{ij}]$, $m \mapsto 1 \bmod (\varepsilon_{ij}]$, *where* $(\varepsilon_{ij}]$ *denotes the left ideal in* $\mathbb{C}[S, \mathrm{D}]$, *generated by the elements* ε_{ij}, $i, j = 1, \ldots, r$;

(iii) *the complex* (1.1.2) *is an exact sequence*;

(iv) *if* $\delta\xi = 0$, $\delta \in \mathbb{C}[S, \mathrm{D}]$, *is a differential equation satisfied by* m, *then* δ *belongs to* $(\varepsilon_{ij}]$.

The conditions (i)—(iv) *are consequences of the following two equivalent conditions*:

(v) $M = \mathbb{C}[S]\, m \oplus \mathbb{C}[S]\, \mathrm{D}_1 m \oplus \cdots \oplus \mathbb{C}[S]\, \mathrm{D}_r m \qquad$ ($\mathbb{C}([S]$-*direct sum*);

(vi) *there is no linear differential operator* $\lambda = \sum_{k=0}^{r} a_k D_k$, $a_k \in \mathbb{C}[S]$, *such that* $\lambda m = 0$.

If the last two (and consequently all other) conditions are satisfied, then we call (M, m) a *primitive principal* (simple quadratic) $\mathbb{C}[S, \mathrm{D}]$-module. A direct sum of primitive principal (simple quadratic) $\mathbb{C}[S, \mathrm{D}]$-modules is called a *primitive* (simple quadratic) module.

Proof. The equivalence assertions are obviously true. We prove the implication (vi) $\Rightarrow$ (iv). Assume that m satisfies $\delta\xi = 0$. Then we can find an operator of order 1 $\delta' = \delta \bmod (\varepsilon_{ij}]$ satisfying $\delta' m = 0$ because of the simple quadratic structure of the ε_{ij}'s. From (vi) it follows that $\delta' = 0$, hence $\delta \in (\varepsilon_{ij}]$.

Generalizing a definition given by Darboux (see [9]) we call a simple quadratic differential operator over S of the form

1.1.5. $$\varepsilon_{ij} = \mathrm{D}_i \mathrm{D}_j + \frac{1}{t_j - t_i} (a\mathrm{D}_j - b\mathrm{D}_i), \qquad i \neq j, \qquad a, b \in \mathbb{C},$$

an *Euler* (*partial differential*) *operator* (over S). Here we assume that $(t_j - t_i)^{-1}$ is an element of $\mathbb{C}[S]$. The corresponding differential equation (1.1.3) is called an *Euler partial differential equation*. An *Euler system of differential operators* is a set $\{\varepsilon_{ij}\}$ of $r^2 - r$ differential operators of type (1.1.5), where $\varepsilon_{ij} = \varepsilon_{ji}$ for all pairs i, j. The corresponding system of differential equations is called an *Euler system of partial differential equations*.

From now on we fix the notations

1.1.6. $\mathbb{C}[S] = \mathbb{C}[t_1, \ldots, t_r, \Pi_t^{-1}], \qquad \Pi_t = \prod_{-1 \leq i < j \leq r} (t_j - t_i),$

$$t_{-1} = 1, \qquad t_0 = 0.$$

Let $\mathfrak{b} = (b_{-1}, b_0, b_1, \ldots, b_r) \in \mathbb{N}^{r+2}$ be a tuple of natural numbers.

1.1.7. DEFINITION. For integers l, n ($n \neq 0$) we call

$$\varepsilon_{ij}^{l,n,\mathfrak{b}} = \mathrm{D}_i \mathrm{D}_j + \frac{l}{n(t_j - t_i)} (b_j \mathrm{D}_i - b_i \mathrm{D}_j), \qquad i \neq j,$$

the (i, j)th *Euler operator of type* $(l, n, \mathfrak{b})$. The mth *Picard-operator of type* $(k, l, n, \mathfrak{b})$ is the simple quadratic operator

$$\varepsilon_{mm}^{k,l,n,\mathfrak{b}} = \mathrm{D}_m^2 + \frac{l}{n(t_m - 1)\, t_m} \left\{ b_m \sum_{0<i\neq m}' \frac{t_i^2 - t_i}{t_i - t_m} \mathrm{D}_i - \left[\sum_{0<i\neq m} \frac{b_i(t_m - 1)\, t_i}{t_i - t_m} \right.\right.$$

$$\left.\left. + \frac{nB}{l}(t_m - 1) - b_m t_m - b_{-1} \right] \mathrm{D}_m - (B + 1)\, b_m \right\},$$

$$m \in \{1, \ldots, r\}, \qquad B = k - \frac{l}{n} b, \qquad b = \sum_{i=-1}^{r} b_i, \qquad k \in \mathbb{Z}, \qquad k \geqq 0.$$

The (k, l)th *Euler-Picard system* of differential operators of type $(n, \mathfrak{b})$ is the system $\{\varepsilon_{ij}^{k,l,n,\mathfrak{b}}\}_{i,j \in \{1,\ldots,r\}}$, where we set $\varepsilon_{ij}^{k,l,n,\mathfrak{b}} = \varepsilon_{ij}^{l,n,\mathfrak{b}}$ for $i \neq j$. The corresponding system of partial differential equations we call the *Euler-Picard system* of the corresponding type.

A principal simple quadratic module of type $\varepsilon^{k,l,n,\mathfrak{b}}$ is called a *principal Euler-Picard module of type* $(k, l, n, \mathfrak{b})$. An *Euler-Picard module of type* $(n, \mathfrak{b})$ is a $\mathbb{C}[S, \mathrm{D}]$-module M, which admits a finite decomposition (not necessarily direct)

$$M = \sum_{i=1}^{N} M_i,$$

(M_i, m_i) a principal Euler-Picard module of type $(k_i, l_i, n, \mathfrak{b})$.

If the sum is direct and the summands (M_i, m_i) are primitive principal Euler-Picard modules, then we call M a *primitive Euler-Picard module* of type $(n, \mathfrak{b})$.

1.1.8. REMARK. In the case $r = 1$ there are no Euler operators and an Euler-Picard system consists then of one Picard operator only. For historical reasons we call it a *Picard-Fuchs operator*. The corresponding ordinary differential equation is a *Picard-Fuchs equation* (see Katz [36]). Instead of Euler-Picard modules one can use the notation of *Picard-Fuchs modules*.

1.2. Affine Cycloelliptic Curve Families

A closed point s of $S = \operatorname{Spec} \mathbb{C}[t, \Pi_t^{-1}]$ can be identified with an r-tuple $(s_1, \ldots, s_r)$ of complex numbers such that $s_i \neq s_j$ for $i \neq j$ and $s_i \neq 1, 0$ for all $i = 1, \ldots, r$. We set $s_{-1} = 1$, $s_0 = 0$. We let correspond to each s an affine plane curve $\mathfrak{Y}_s \subset \mathbf{A}^2$

defined by the equation

$$\ddot{\mathfrak{Y}}_s \colon Y^n = \prod_{i=-1}^{r} (X - s_i)^{b_i}$$

for fixed natural numbers $b_{-1}, b_0, b_1, \ldots, b_r, n, n > 1$. The curve $\ddot{\mathfrak{Y}}_s$ is called the *affine cycloelliptic curve* over s (of type $(n, \mathfrak{b})$). By considering the projective closure $\bar{\mathfrak{Y}}_s$ of $\ddot{\mathfrak{Y}}_s$ we can see that it is not essential to restrict ourselfs to the fixed values $s_{-1} = 1$ and $s_0 = 0$. The curves are arranged in a curve family

$$\{\ddot{\mathfrak{Y}}_s\}_{s \in S}, \qquad S \text{ parameter space}.$$

More precisely we define

$$\ddot{\mathfrak{Y}} = \operatorname{Spec} \mathbb{C}[S]\,[X, Y] / \left(Y^n - \prod_{i=-1}^{r} (X - t_i)^{b_i} \right) = \operatorname{Spec} \mathbb{C}[S]\,[x, y].$$

The inclusion $\mathbb{C}[S] \subset \mathbb{C}[S]\,[x, y]$ corresponds ot the projection $\ddot{\pi} \colon \ddot{\mathfrak{Y}} \to S$ onto the parameter space S. Altogether we have for each $s \in S$ a commutative Diagram 1.2.a.

1.2.a.

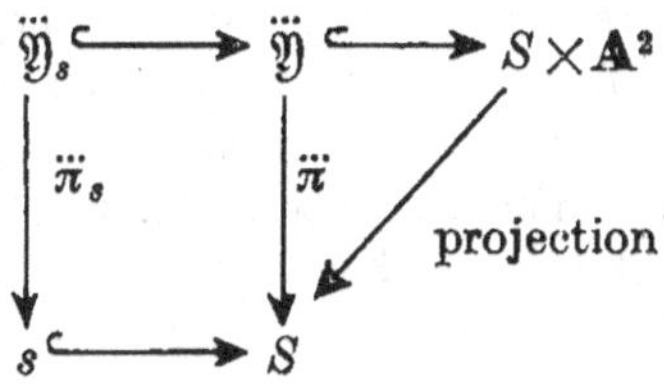

1.2.1. DEFINITION. The family $\ddot{\mathfrak{Y}}/S$, that means $\ddot{\pi}$, is called the *affine cycloelliptic curve family* over S of type $(n, \mathfrak{b})$, $\mathfrak{b} = (b_{-1}, b_0, \ldots, b_r)$.

Let $\mathbf{A}_S^1 = \mathbf{A}_S^1[x] = \operatorname{Spec} \mathbb{C}[S]\,[x]$. Then $\ddot{\pi}$ factors through $\mathbf{A}_S^1$. The morphism $\ddot{\mathfrak{Y}} \to \mathbf{A}_S^1$ is quasifinite. We throw out the ramification of this morphism by defining

$$\mathfrak{Y} = \operatorname{Spec} \mathbb{C}[S]\,[x, y, \Pi_x^{-1}] = \operatorname{Spec} \mathbb{C}[S]\,[x, y, y^{-1}], \quad \Pi_x = \prod_{i=-1}^{r} (x - t_i).$$

For each $s \in S$ we now have a commutative Diagram 1.2.b.

1.2. b

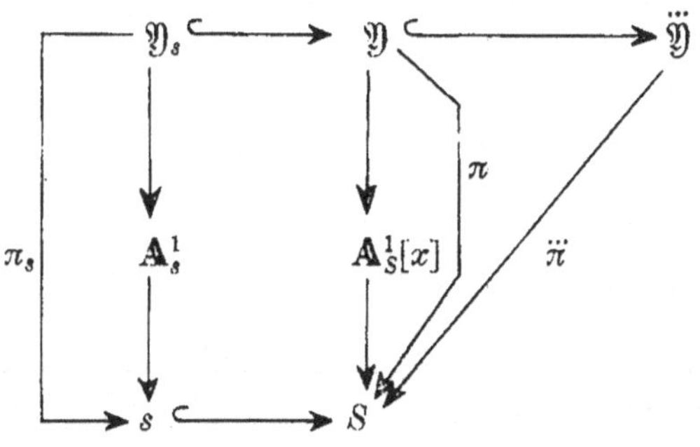

1.2.2. DEFINITION. The family $\mathfrak{Y}/S$ is called the *étale cycloelliptic curve family* over S of type $(n, \mathfrak{b})$.

The étale morphisms define cyclic étale coverings over the image varieties (see Diagram 1.2.c).

1.2.c.

$$\begin{array}{ccccc} & & \mathfrak{Y}_s & \hookrightarrow & \mathfrak{Y} \\ & & \downarrow & & \downarrow \text{ cyclic étale} \\ \operatorname{Spec} \mathbb{C}[x, \Pi_x(s)^{-1}] & = & \mathbf{A}'_s & \hookrightarrow & \mathbf{A}'_S = \operatorname{Spec} \mathbb{C}[S]\,[x, \Pi_x^{-1}] \\ & & \cap & & \cap \\ & & \mathbf{A}^1_s & \hookrightarrow & \mathbf{A}^1_S[x] \end{array}$$

The Galois group G is isomorphic to $\mathbb{Z}/n\mathbb{Z}$.

For later use we define some important special families.

1.2.3. DEFINITION. Cycloelliptic families of type $\big(n, (1, 1, \ldots, 1)\big)$ are called *general cycloelliptic families* (of type (n, r)). A cycloelliptic family is called *primitive*, if $\text{g.c.d.}(n, b) = \text{g.c.d.}(n, b_i) = 1$ for all $i = -1, 0, \ldots, r$.

Examples of primitive general families are the general families of type $(2, r)$, r odd. These are *hyperelliptic families*. We will see that from the point of view of differential equations the primitive general families of type $(n, n - 1)$ are the best generalizations of the elliptic curve family, which has type $(2, 1)$. We call these families *generalized Picard curve families* of type n. For $n = 3$ we obtain the *Picard curve family* discussed in Chapter I. This discussion suggests to consider non-general cycloelliptic families as limits of general cycloelliptic families.

1.3. The Relative de Rham Cohomology Group $H^1_{\mathrm{DR}}(\mathfrak{Y}/S)$

Let $\mathfrak{Y}/S$ be an étale cycloelliptic curve family (see 1.2.2). The morphism $\pi\colon \mathfrak{Y} \to S$ defines an exact sequence

1.3.1. $\quad 0 \to \pi^*\Omega^1_S \to \Omega^1_{\mathfrak{Y}} \to \Omega^1_{\mathfrak{Y}/S} \to 0$

of locally free sheaves of $\mathcal{O}_Y$-modules. The *sheaf* $\Omega^1_{\mathfrak{Y}/S}$ *of relative differential forms* has rank equal to 1. The sequence (1.3.1) is exact because $\mathfrak{Y}/S$ is a smooth family (see [27], III, § 10). The *de Rham complex* of $\mathfrak{Y}/S$ is the sequence of $\mathcal{O}_S$-modules

1.3.2. $\quad 0 \to \mathcal{O}_{\mathfrak{Y}} \xrightarrow[\mathrm{d}_{\mathfrak{Y}/S}]{} \Omega^1_{\mathfrak{Y}/S} \to 0,$

where $\mathrm{d}_{\mathfrak{Y}/S}$ is the composition

$$\mathcal{O}_{\mathfrak{Y}} \xrightarrow[\mathrm{d}_{\mathfrak{Y}}]{} \Omega^1_{\mathfrak{Y}} \to \Omega^1_{\mathfrak{Y}/S},$$
$$f \mapsto \mathrm{d}f.$$

By this definition we can see that $\mathrm{d}_{\mathfrak{Y}/S}$ is trivial on $\pi^*\Omega^1_S$. Kernel and cokernel of $\mathrm{d}_{\mathfrak{Y}/S}$ in (1.3.2) yield an exact sequence

1.3.3. $0 \to d\mathcal{O}_{\mathfrak{Y}} \to \Omega^1_{\mathfrak{Y}/S} \to \mathcal{H}^1_{DR}(\mathfrak{Y}/S) \to 0$

of sheaves of $\mathcal{O}_S$-modules. Altogether we have a commutative Diagram 1.3.a

1.3.a.

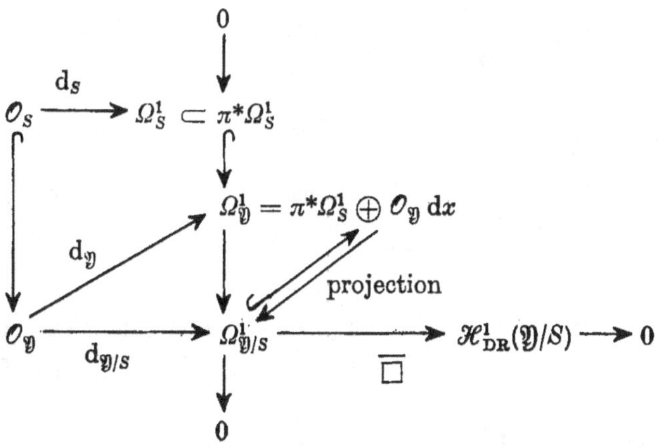

with an exact column and an exact row. The diagram is to be understood in the category of $\mathcal{O}_S$-modules. We fix the notation of the $\mathbb{C}[S]$-module of global sections of $\mathcal{H}^1_{DR}(\mathfrak{Y}/S)$:

$$H^1_{DR}(\mathfrak{Y}/S) = \Gamma_S(\mathcal{H}^1_{DR}(\mathfrak{Y}/S)) = \mathbb{C}[\mathfrak{Y}]\, dx/\{df;\ f \in \mathbb{C}[\mathfrak{Y}]\}.$$

We will endow $H^1_{DR}(\mathfrak{Y}/S)$ with the structure of an $\mathbb{C}[S, D]$-module (see 1.1).

Let B be a commutative algebra over the ring A with unit element 1 and M an object of the category of B-modules **mod** B. As usual the B-module of A-derivations of B in M is defined by

$$\mathrm{Der}_A(B, M) = \{\delta \in \mathrm{Hom}_A(B, M);\ \delta(bb') = b\delta(b') + b'\delta(b) \text{ for } b, b' \in B\}.$$

We set $\mathrm{Der}_A(B) = \mathrm{Der}_A(B, B)$. The *module of relative differential forms* of B over A is denoted by $\Omega_{B/A}$. If $A = \mathbb{C}$, then we write Ω_B instead of $\Omega_{B/\mathbb{C}}$. We have available „universal derivations" $d_{B/A}\colon B \to \Omega_{B/A}$.

1.3.4. LEMMA. *Any derivation* $\delta' \in \mathrm{Der}_{\mathbb{C}}\, \mathbb{C}[\mathbf{A}'_S]$ *has a unique extension* $\delta \in \mathrm{Der}_{\mathbb{C}}\, \mathbb{C}[\mathfrak{Y}]$.

Proof. If δ is an extension of δ', then

1.3.5. $$\delta(y) = \delta(y^n)/ny^{n-1} = \delta\left(\prod_{-1}^{r}(x - t_i)^{b_i}\right)\Big/ny^{n-1} = \delta'\left(\prod_{-1}^{r}(x - t_i)^{b_i}\right)\Big/ny^{n-1}.$$

So δ is uniquely determined because y generates $\mathbb{C}[\mathfrak{Y}]$ over $\mathbb{C}[\mathbf{A}'_S]$. Conversely, it is easy to see that δ defined by (1.3.5) and by the derivation rules is an extension of δ'.

Using the same notations we define for each $i \in \{1, \ldots, r\}$ an extension of $\mathrm{D}_i \in \mathrm{Der}_{\mathbb{C}}\, \mathbb{C}[S]$ to $\mathrm{Der}_{\mathbb{C}}\, \mathbb{C}[\mathfrak{Y}]$ in two steps:

1.3.b.
$$\begin{array}{ccccc} \mathbb{C}[S] & \hookrightarrow & \mathbb{C}[\mathbf{A}'_S] & \hookrightarrow & \mathbb{C}[\mathfrak{Y}] \\ \downarrow \mathrm{D}_i & & \downarrow \mathrm{D}_i & & \downarrow \mathrm{D}_i \\ \mathbb{C}[S] & \hookrightarrow & \mathbb{C}[\mathbf{A}'_S] & \hookrightarrow & \mathbb{C}[\mathfrak{Y}] \end{array}$$

By the lemma it suffices to define the extension D_i on $\mathbb{C}[\mathbf{A}'_S] = \mathbb{C}[S]\,[x, \Pi_x^{-1}]$. This is done by extension of

$$\mathrm{D}_i : t_j \mapsto \delta_{ij} \text{ (Kronecker symbol)}, \qquad x \mapsto 0$$

to a derivation of $\mathbb{C}(t)\,(x)$ by means of the derivation rules. This derivation restricts to $\mathbb{C}[\mathbf{A}'_S]$. In order to define an action of $\mathbb{C}[S, \mathrm{D}]$ on $\Omega_{\mathbb{C}[\mathfrak{Y}]/\mathbb{C}[S]}$ we remark that

$$\Omega^1_{\mathfrak{Y}} = \pi^* \Omega^1_S \oplus \mathcal{O}_{\mathfrak{Y}}\, \mathrm{d}x, \qquad \Omega_{\mathbb{C}[\mathfrak{Y}]/\mathbb{C}[S]} \cong \mathbb{C}[\mathfrak{Y}]\, \mathrm{d}x \text{ (identification)}.$$

Therefore

$$f\, \mathrm{d}x \mapsto \mathrm{D}_i(f)\, \mathrm{d}x, \qquad f \in \mathbb{C}[\mathfrak{Y}],$$

is a well-defined $\mathbb{C}$-linear endomorphism of $\Omega_{\mathbb{C}[\mathfrak{Y}]/\mathbb{C}[S]}$, which we also denote by D_i. The derivation $\mathrm{D}_i : \Omega^1_{\mathfrak{Y}/S} \to \Omega^1_{\mathfrak{Y}/S}$ has the following connection property:

1.3.7. $$\mathrm{D}_i(f\omega) = \mathrm{D}_i(f)\, \omega + f \cdot \mathrm{D}_i(\omega), \qquad \omega \in \Omega_{\mathbb{C}[\mathfrak{Y}]}, \qquad f \in \mathbb{C}[\mathfrak{Y}].$$

Indeed, for $\omega = g\, \mathrm{d}x$ we find that

$$\begin{aligned} \mathrm{D}_i(f\omega) &= \mathrm{D}_i(f \cdot g\, \mathrm{d}x) = \mathrm{D}_i(fg) \cdot \mathrm{d}x = g\mathrm{D}_i(f) \cdot \mathrm{d}x + f\mathrm{D}_i(g) \cdot \mathrm{d}x \\ &= \mathrm{D}_i(f) g\, \mathrm{d}x + f\mathrm{D}_i(g\, \mathrm{d}x) = \mathrm{D}_i(f)\, \omega + f\mathrm{D}_i(\omega). \end{aligned}$$

1.3.8. LEMMA. $\mathrm{d}_{\mathfrak{Y}/S}$ *commutes with* D_i; i.e. *the diagram*

1.3.c.
$$\begin{array}{ccc} \mathcal{O}_{\mathfrak{Y}} & \xrightarrow{\mathrm{d}_{\mathfrak{Y}/S}} & \Omega^1_{\mathfrak{Y}/S} \\ \mathrm{D}_i \downarrow & & \downarrow \mathrm{D}_i \\ \mathcal{O}_{\mathfrak{Y}} & \xrightarrow[\mathrm{d}_{\mathfrak{Y}/S}]{} & \Omega^1_{\mathfrak{Y}/S} \end{array}$$

is commutative.

Proof. First we note that

1.3.9. $$\mathrm{d}(y^k) = \frac{k}{n}\, y^k \sum_{j=-1}^{r} b_j \frac{d(x - t_j)}{x - t_j} \qquad \text{for} \qquad \mathrm{d} = \mathrm{d}_{\mathfrak{Y}} : \mathbb{C}[\mathfrak{Y}] \to \Omega_{\mathbb{C}[\mathfrak{Y}]},$$

1.3.10. $$\mathrm{D}_i(y^k) = -\frac{k}{n}\, y^k \frac{b_i}{x - t_i}.$$

This follows from the product rules

1.3.11. $\mathrm{d}G = G \sum_{j=1}^{m} \frac{\mathrm{d}G_j}{G_j}$,

1.3.12. $\mathrm{D}_i G = G \sum_{j=1}^{m} \frac{\mathrm{D}_i G_j}{G_j}$

for $G = \prod_{j=1}^{m} G_j \in \mathbb{C}[\mathfrak{Y}]$

and from

$$y^n = \prod_{-1}^{r} (x - t_j)^{b_j}, \qquad \mathrm{d}(y^k) = ky^{k-1}\,\mathrm{d}y, \qquad \mathrm{D}_i(y^k) = ky^{k-1}\mathrm{D}_i(y).$$

For example, we obtain by means of these rules

$$\mathrm{D}_i(y^k) = ky^{k-1}\mathrm{D}_i y = \frac{ky^{k-1}}{ny^{n-1}}\,\mathrm{D}_i(y^n) = \frac{ky^{k-1}}{ny^{n-1}}\,y^n \sum_{j=-1}^{r} \frac{b_j \mathrm{D}_i(x - t_j)}{x - t_j}$$
$$= -\frac{kb_i y^k}{n(x - t_i)}.$$

Now we prove that

1.3.13. $\mathrm{D}_i \mathrm{d}_{\mathfrak{Y}/S}(y) = \mathrm{d}_{\mathfrak{Y}/S}\mathrm{D}_i(y)$.

We find

$$n\mathrm{D}_i(\mathrm{d}_{\mathfrak{Y}/S} y) = n\mathrm{D}_i(\mathrm{d}y \bmod \pi^*\Omega_S^1) = n\mathrm{D}_i(\mathrm{d}y) \bmod \pi^*\Omega_S^1$$
$$= \mathrm{D}_i\left(y \sum_{-1}^{r} \frac{b_j \mathrm{d}x}{x - t_j}\right) \bmod \pi^*\Omega_S^1$$
$$= \left[\mathrm{D}_i(y) \cdot \sum_j \frac{b_j}{x - t_j} + y\mathrm{D}_i\left(\sum_j \frac{b_j}{x - t_j}\right)\right] \mathrm{d}x$$
$$= \left[-\frac{b_i y}{n(x - t_i)} \cdot \sum_j \frac{b_j}{x - t_j} + \frac{b_i y}{(x - t_i)^2}\right] \mathrm{d}x$$
$$= \frac{-b_i}{x - t_i}\,\mathrm{d}y - b_i y \cdot \mathrm{d}(x - t_i)^{-1}$$
$$= \mathrm{d}\left(y \cdot \frac{-b_i}{x - t_i}\right) = n\,\mathrm{d}\mathrm{D}_i(y) \quad \text{(everything mod } \pi^*\Omega_S^1\text{)}.$$

Formula (1.3.13) extends to the powers of y. Namely,

$$\mathrm{D}_i(\mathrm{d}_{\mathfrak{Y}/S} y^k) = \mathrm{D}_i(\mathrm{d}y^k) \bmod \pi^*\Omega_S^1 = \mathrm{D}_i(ky^{k-1}\,\mathrm{d}y)$$
$$= k(k-1)\,y^{k-2}\mathrm{D}_i(y) \cdot \mathrm{d}y + ky^{k-1}\mathrm{D}_i(\mathrm{d}y)$$
$$= k(k-1)\,y^{k-2}\mathrm{D}_i(y) \cdot \mathrm{d}y + ky^{k-1}\,\mathrm{d}\mathrm{D}_i(y)$$
$$= k\,\mathrm{d}\big(y^{k-1}\mathrm{D}_i(y)\big) = \mathrm{d}D_i(y^k) = \mathrm{d}_{\mathfrak{Y}/S} D_i(y^k)$$

(everything mod $\pi^*\Omega_S^1$).

For the extension of the commutativity rule to $\mathbb{C}[\mathfrak{Y}] = \sum_{k \geq 0} \mathbb{C}[\mathbb{A}'_S]\, y^k$ we remark that

$$\mathrm{D}_i \mathrm{d}_{\mathfrak{Y}/S} f = \mathrm{d}_{\mathfrak{Y}/S} \mathrm{D}_i f = 0 \quad \text{for} \quad f \in \mathbb{C}[\mathbb{A}^1_S] \subset \mathbb{C}(t)\,(x).$$

This is easily checked for $f = t_j$, x by taking into account that $\mathrm{D}_i(x) = 0$ and $\mathrm{d}_{\mathfrak{Y}/S}(t_j) = 0$ for $j = 1, \ldots, r$. So we get for $f \in \mathbb{C}[\mathbb{A}^1_S] \subset \mathbb{C}(t)\,(x)$

$$\begin{aligned}
&\mathrm{D}_i \mathrm{d}_{\mathfrak{Y}/S}(y^k \cdot f) = \mathrm{D}_i(y^k \mathrm{d}_{\mathfrak{Y}/S}(f) + f \mathrm{d}_{\mathfrak{Y}/S} y^k) \\
&= y^k \cdot \mathrm{D}_i \mathrm{d}_{\mathfrak{Y}/S} f + \mathrm{D}_i(y^k) \cdot \mathrm{d}_{\mathfrak{Y}/S}(f) + \mathrm{D}_i(f) \cdot \mathrm{d}_{\mathfrak{Y}/S} y^k + f \cdot \mathrm{D}_i \mathrm{d}_{\mathfrak{Y}/S} y^k \\
&= \mathrm{D}_i(y^k) \cdot \mathrm{d}_{\mathfrak{Y}/S} f + f \cdot \mathrm{d}_{\mathfrak{Y}/S}(\mathrm{D}_i(y^k)) + y^k \mathrm{d}_{\mathfrak{Y}/S}(\mathrm{D}_i f) + \mathrm{D}_i(f) \cdot \mathrm{d}_{\mathfrak{Y}/S} y^k \\
&= \mathrm{d}_{\mathfrak{Y}/S}(f \cdot \mathrm{D}_i(y^k)) + \mathrm{d}_{\mathfrak{Y}/S}(y^k \mathrm{D}_i(f)) = \mathrm{d}_{\mathfrak{Y}/S} \mathrm{D}_i(y^k \cdot f),
\end{aligned}$$

q.e.d.

The Diagram 1.3.c in the lemma defines a $\mathbb{C}$-linear action of D_i on $H^1_{\mathrm{DR}}(\mathfrak{Y}/S)$. Indeed, by (1.3.3) or 1.3.a we can extend it to the following commutative diagram of $\mathbb{C}$-module sheaves over S:

1.3.d.

$$\begin{array}{ccccccc}
\mathcal{O}_{\mathfrak{Y}} & \xrightarrow{\mathrm{d}_{\mathfrak{Y}/S}} & \Omega^1_{\mathfrak{Y}/S} & \longrightarrow & \mathcal{H}^1_{\mathrm{DR}}(\mathfrak{Y}/S) & \longrightarrow & 0 \\
\downarrow{\scriptstyle \mathrm{D}_i} & & \downarrow{\scriptstyle \mathrm{D}_i} & & \downarrow{\scriptstyle \mathrm{D}_i} & & \\
\mathcal{O}_{\mathfrak{Y}} & \xrightarrow[\mathrm{d}_{\mathfrak{Y}/S}]{} & \Omega^1_{\mathfrak{Y}/S} & \longrightarrow & \mathcal{H}^1_{\mathrm{DR}}(\mathfrak{Y}/S) & \longrightarrow & 0
\end{array}$$

The rows are exact. So we see that $H^1_{\mathrm{DR}}(\mathfrak{Y}/S)$ has the structure of a $\mathbb{C}[S, \mathrm{D}]$-module.

Finally we note some rules for the action of D_i on $H^1_{\mathrm{DR}}(\mathfrak{Y}/S)$, which we will frequently use in the next sections.

1.3.14. For $f \in \mathbb{C}[\mathfrak{Y}]$, $h \in \mathbb{C}[S]$, $\overline{\omega} \in H^1_{\mathrm{DR}}(\mathfrak{Y}/S)$, $\overline{\omega} = \omega \bmod \mathrm{d}_{\mathfrak{Y}/S}\mathbb{C}[\mathfrak{Y}]$, $\omega \in \Gamma_S(\Omega^1_{\mathfrak{Y}/S})$, one has that

$$\mathrm{D}_i(\overline{f \cdot \omega}) = \overline{\mathrm{D}_i(f) \cdot \omega} + \overline{f \cdot \mathrm{D}_i(\omega)}, \qquad \mathrm{D}_i(h\overline{\omega}) = \mathrm{D}_i(h) \cdot \overline{\omega} + h \cdot \mathrm{D}_i(\overline{\omega}),$$

$$\mathrm{D}_i(\overline{f\, \mathrm{d}x}) = \overline{\mathrm{D}_i(f) \cdot \mathrm{d}x}, \qquad \mathrm{D}_i(h\, \overline{\mathrm{d}x}) = \mathrm{D}_i(h) \cdot \overline{\mathrm{d}x}.$$

G-action

We will now see that $H^1_{\mathrm{DR}}(\mathfrak{Y}/S)$ has also the structure of a G-module, where $G \simeq \mathbb{Z}/n\mathbb{Z}$ is the Galois group of the étale covering $\mathfrak{Y} \to \mathbb{A}'_S$ (see 1.2.c). Let g be the generator of G defined by the following action:

$$y \longmapsto \zeta^{-1} y, \qquad \zeta = \mathrm{e}^{2\pi i/n}$$

$$\begin{array}{ccc}
\mathbb{C}[\mathfrak{Y}] & \xrightarrow[g]{\sim} & \mathbb{C}[\mathfrak{Y}] \\
& \nwarrow \quad \nearrow & \\
& \mathbb{C}[\mathbb{A}'_S] &
\end{array}$$

We have the following isotypical decomposition in **mod** $\mathbb{C}[\mathbf{A}'_S]$:

1.3.15. $\mathbb{C}[\mathfrak{Y}] = \mathbb{C}[\mathfrak{Y}]_0 \oplus \mathbb{C}[\mathfrak{Y}]_1 \oplus \cdots \oplus \mathbb{C}[\mathfrak{Y}]_{n-1}$,

$$\mathbb{C}[\mathfrak{Y}]_l = \frac{1}{y^l}\, \mathbb{C}[\mathbf{A}'_S] = \{f \in \mathbb{C}[\mathfrak{Y}];\, g(f) = \zeta^l \cdot f\}.$$

1.3.16. LEMMA. *Let $\delta \in \mathrm{Der}_{\mathbb{C}}\, \mathbb{C}[\mathfrak{Y}]$ be the extension of a $\mathbb{C}$-derivation of $\mathbb{C}[\mathbf{A}'_S]$ as described in Lemma* 1.3.4. *Then $g\delta = \delta g$. In particular $\mathrm{D}_i g = g\mathrm{D}_i$ for $i = 1, \ldots, r$.*

Proof. It is easy to see that $\delta' = g^{-1}\delta g$ is an element of $\mathrm{Der}_{\mathbb{C}}(\mathbb{C}[\mathfrak{Y}])$. The restrictions of δ and δ' on $\mathbb{C}[\mathbf{A}'_S]$ coincide. Any derivation of $\mathbb{C}[\mathbf{A}'_S]$ has a unique extension to $\mathbb{C}[\mathfrak{Y}]$ by Lemma 1.3.4. Therefore

$$g^{-1}\delta g = \delta' = \delta, \qquad \delta g = g\delta \quad \text{on} \quad \mathbb{C}[\mathfrak{Y}].$$

Via inverse images we have a G-action on $\Omega^1_{\mathfrak{Y}}$, which is compatible with the multiplication by the structure sheaf $\mathcal{O}_{\mathfrak{Y}}$. Explicitly the action of g is given by

1.3.17. $f \cdot \omega \xmapsto[g\otimes 1]{} g(f) \cdot \omega$,

$$f \cdot \omega \in \Omega_{\mathbb{C}[\mathfrak{Y}]} = \mathbb{C}[\mathfrak{Y}] \otimes_{\mathbb{C}[\mathbf{A}'_S]} \Omega_{\mathbb{C}[\mathbf{A}'_S]} = \mathbb{C}[\mathfrak{Y}] \cdot \Omega_{\mathbb{C}[\mathbf{A}'_S]}$$
$$= \mathbb{C}[\mathfrak{Y}]\, \mathrm{d}t_1 \oplus \cdots \oplus \mathbb{C}[\mathfrak{Y}]\, \mathrm{d}t_r \oplus \mathbb{C}[\mathfrak{Y}]\, \mathrm{d}x.$$

So we have a commutative diagram

1.3.e.
$$\begin{array}{ccc}
\mathbb{C}[\mathfrak{Y}] & \xrightarrow{\ \mathrm{d}_{\mathfrak{Y}}\ } & \Omega_{\mathbb{C}[\mathfrak{Y}]} \\
\downarrow g & & \downarrow g \\
\mathbb{C}[\mathfrak{Y}] & \xrightarrow{\ \mathrm{d}_{\mathfrak{Y}}\ } & \Omega_{\mathbb{C}[\mathfrak{Y}]}
\end{array}$$

Indeed, it is easy to check that $g^{-1} \circ \mathrm{d}_{\mathfrak{Y}} \circ g \colon \mathbb{C}[\mathfrak{Y}] \to \Omega_{\mathbb{C}[\mathfrak{Y}]}$ is a $\mathbb{C}$-derivation. Hence, it suffices to prove that this derivation coincides with $\mathrm{d}_{\mathfrak{Y}}$ applied to elements $f \in \mathbb{C}[\mathbf{A}'_S]$ and to y. It holds that $g^{-1}\mathrm{d}_{\mathfrak{Y}}g(f) = g^{-1}\mathrm{d}_{\mathfrak{Y}}(f) = \mathrm{d}_{\mathfrak{Y}}(f)$ because $\mathrm{d}_{\mathfrak{Y}}(f) = \mathrm{d}_{\mathbf{A}'_S}(f) \in \Omega_{\mathbb{C}[\mathbf{A}'_S]}$. From (1.3.9) we derive

$$g^{-1}\mathrm{d}_{\mathfrak{Y}}g(y) = \zeta^{-1}g^{-1}\mathrm{d}_{\mathfrak{Y}}y = \zeta^{-1}g^{-1}\left(\frac{1}{n}\, y \sum_{-1}^{r} b_j\, \frac{d(x - t_j)}{x - t_j}\right)$$
$$= \frac{1}{n}\, y \sum_{-1}^{r} b_j\, \frac{\mathrm{d}(x - t_j)}{x - t_j} = \mathrm{d}_{\mathfrak{Y}}y.$$

The sheaf $\Omega^1_{\mathfrak{Y}}$ has an isotypical decomposition in **mod** $\mathcal{O}_{\mathbf{A}'_S}$:

$$\Omega^1_{\mathfrak{Y}} = (\Omega^1_{\mathfrak{Y}})_0 \oplus (\Omega^1_{\mathfrak{Y}})_1 \oplus \cdots \oplus (\Omega^1_{\mathfrak{Y}})_{n-1},$$

$$(\Omega^1_{\mathfrak{Y}})_l = \frac{1}{y^l}\, \Omega^1_{\mathbf{A}'_S}, \qquad (\Omega_{\mathbb{C}[\mathfrak{Y}]})_l = \{\omega \in \Omega_{\mathbb{C}[\mathfrak{Y}]};\, g\omega = \zeta^l \cdot \omega\}.$$

By Diagram 1.3.e d is compatible with the isotypical decomposition (1.3.15) and the above one. So 1.3.e splits into the direct sum of n diagrams of $\mathcal{O}_{\mathbf{A}'_S}$-modules

1.3.e.
$$\begin{array}{ccc} (\mathcal{O}_{\mathfrak{Y}})_l & \xrightarrow{\mathrm{d}} & (\Omega^1_{\mathfrak{Y}})_l \\ {\scriptstyle g}\downarrow{\scriptstyle \zeta^l} & & {\scriptstyle g}\downarrow{\scriptstyle \zeta^l} \\ (\mathcal{O}_{\mathfrak{Y}})_l & \xrightarrow[\mathrm{d}]{} & (\Omega^1_{\mathfrak{Y}})_l \end{array}$$

By (1.3.17), G acts on $\pi^*\Omega^1_S$. The exact sequence (1.3.1) yields an action of G on $\Omega^1_{\mathfrak{Y}/S}$. From the Diagrams 1.3.a and 1.3.e we deduce an exact commutative diagram

$$\begin{array}{ccccccc} \mathcal{O}_{\mathfrak{Y}} & \xrightarrow{\mathrm{d}_{\mathfrak{Y}/S}} & \Omega^1_{\mathfrak{Y}/S} & \longrightarrow & \mathcal{H}^1_{\mathrm{DR}}(\mathfrak{Y}/S) & \longrightarrow & 0 \\ {\scriptstyle g}\downarrow & & {\scriptstyle g}\downarrow & & {\scriptstyle g}\downarrow & & \\ \mathcal{O}_{\mathfrak{Y}} & \xrightarrow[\mathrm{d}_{\mathfrak{Y}/S}]{} & \Omega^1_{\mathfrak{Y}/S} & \longrightarrow & \mathcal{H}^1_{\mathrm{DR}}(\mathfrak{Y}/S) & \longrightarrow & 0 \end{array}$$

So $H^1_{\mathrm{DR}}(\mathfrak{Y}/S)$ has the structure of a G-module. From Lemma 1.3.16 and Diagram 1.3.d it follows easily that g commutes with D_i on $\Omega^1_{\mathfrak{Y}}$ and on $H^1_{\mathrm{DR}}(\mathfrak{Y}/S)$. So $H^1_{\mathrm{DR}}(\mathfrak{Y}/S)$ is a $\mathbb{C}[S, \mathrm{D}, g] = \mathbb{C}[S]\,[\mathrm{D}_1, \ldots, \mathrm{D}_r, g]$-module,

$$\mathbb{C}[S, \mathrm{D}, g] = \mathbb{C}[S, \mathrm{D}] \oplus \mathbb{C}[S, \mathrm{D}]\, g \oplus \cdots \oplus \mathbb{C}[S, \mathrm{D}]\, g^{n-1},$$
$$g^n = 1, \qquad \varphi \cdot g = g \cdot \varphi \quad \text{for} \quad \varphi \in \mathbb{C}[S, \mathrm{D}].$$

With respect to the G-action on $H^1_{\mathrm{DR}}(\mathfrak{Y}/S)$ we have an isotypical decomposition of $\mathbb{C}[S, \mathrm{D}]$-modules:

1.3.18.
$$\begin{aligned} H^1_{\mathrm{DR}}(\mathfrak{Y}/S) &= H^1_{\mathrm{DR}}(\mathfrak{Y}/S)_0 \oplus \cdots \oplus H^1_{\mathrm{DR}}(\mathfrak{Y}/S)_{n-1}, \\ H^1_{\mathrm{DR}}(\mathfrak{Y}/S)_l &= (\Omega_{\mathbb{C}[\mathfrak{Y}]/\mathbb{C}[S]})_l / \mathrm{d}_{\mathfrak{Y}/S}(\mathbb{C}[\mathfrak{Y}])_l \\ &= \mathbb{C}[\mathbf{A}'_S] \cdot \frac{\mathrm{d}x}{y^l} \Big/ \mathrm{d}_{\mathfrak{Y}/S}\left(\frac{1}{y^l}\, \mathbb{C}[\mathbf{A}'_S]\right), \qquad l = 0, 1, \ldots, n-1. \end{aligned}$$

1.4. Euler-Picard Structure of $H^1_{\mathrm{DR}}(\mathfrak{Y}/S)$

We want to investigate the $\mathbb{C}[S, \mathrm{D}]$-structure of $H^1_{\mathrm{DR}}(\mathfrak{Y}/S)_l$. For this purpose we investigate the principal modules generated by elements $\overline{\frac{x^k}{y^l}\,\mathrm{d}x} = $ class of $\frac{x^k}{y^l}\,\mathrm{d}x$ in $H^1_{\mathrm{DR}}(\mathfrak{Y}/S)_l$, $k, l \in \mathbb{Z}$, $l \neq 0$.

1.4.1. PROPOSITION. $\left(\mathbb{C}[S, \mathrm{D}] \cdot \overline{\frac{x^k}{y^l}\,\mathrm{d}x}, \overline{\frac{x^k}{y^l}\,\mathrm{d}x}\right)$ *is a principal Euler-Picard module of type* $(k, l, n, \mathfrak{b})$, $\mathfrak{b} = (b_{-1}, b_0, b_1, \ldots, b_r)$.

For the complicated proof we need some fundamental and some auxiliary formulas.

1.4.2. $$\mathrm{D}_i\left(\frac{x^k}{y^l}\,\mathrm{d}x\right) = \frac{b_i l}{n(x-t_i)}\cdot\frac{x^k}{y^l}\,\mathrm{d}x \quad \text{in} \quad \Omega^1_{\mathfrak{Y}/S}.$$

For the proof we use (1.3.6) and (1.3.10):

$$\mathrm{D}_i\left(\frac{x^k}{y^l}\,\mathrm{d}x\right) = \left[\mathrm{D}_i\left(\frac{x^k}{y^l}\right)\right]\mathrm{d}x = \frac{lx^k b_i}{ny^l(x-t_i)}\,\mathrm{d}x = \frac{b_i l}{n(x-t_i)}\cdot\frac{x^k}{y^l}\,\mathrm{d}x.$$

1.4.3. $$\mathrm{D}_i^2\left(\frac{x^k}{y^l}\,\mathrm{d}x\right) = \left(\frac{b_i^2 l^2}{n^2}+\frac{b_i l}{n}\right)\frac{1}{(x-t_i)^2}\cdot\frac{x^k}{y^l}\,\mathrm{d}x \quad \text{in} \quad \Omega^1_{\mathfrak{Y}/S}.$$

Indeed,

$$\begin{aligned}
\mathrm{D}_i^2\left(\frac{x^k}{y^l}\,\mathrm{d}x\right) &= \mathrm{D}_i\left(\frac{b_i l}{n(x-t_i)}\cdot\frac{x^k}{y^l}\,\mathrm{d}x\right)\\
&= \frac{b_i l}{n}\left[\mathrm{D}_i\left(\frac{1}{x-t_i}\right)\frac{x^k}{y^l}\,\mathrm{d}x + \frac{1}{x-t_i}\,\mathrm{D}_i\left(\frac{x^k}{y^l}\,\mathrm{d}x\right)\right]\\
&= \frac{b_i l}{n}\left[\frac{1}{(x-t_i)^2}\cdot\frac{x^k}{y^l}\,\mathrm{d}x + \frac{b_i l}{n(x-t_i)^2}\cdot\frac{x^k}{y^l}\,\mathrm{d}x\right]\\
&= \left(\frac{b_i^2 l^2}{n^2}+\frac{b_i l}{n}\right)\frac{1}{(x-t_i)^2}\cdot\frac{x^k}{y^l}\,\mathrm{d}x.
\end{aligned}$$

Analogously we calculate for $i \neq j$

$$\begin{aligned}
\mathrm{D}_j\mathrm{D}_i\frac{x^k}{y^l}\,\mathrm{d}x &= \mathrm{D}_j\left(\frac{b_i l}{n(x-t_i)}\cdot\frac{x^k}{y^l}\,\mathrm{d}x\right) = 0 + \frac{b_i l}{n(x-t_i)}\,\mathrm{D}_j\left(\frac{x^k}{y^l}\,\mathrm{d}x\right)\\
&= \frac{b_i b_j l^2}{n^2(x-t_i)(x-t_j)}\cdot\frac{x^k}{y^l}\,\mathrm{d}x,
\end{aligned}$$

hence

1.4.4. $$\mathrm{D}_j\mathrm{D}_i\frac{x^k}{y^l}\,\mathrm{d}x = \frac{b_i b_j l^2}{n^2(x-t_i)(x-t_j)}\cdot\frac{x^k}{y^l}\,\mathrm{d}x \quad \text{in} \quad \Omega^1_{\mathfrak{Y}/S}, \quad i \neq j.$$

Now we look for relations between the elements $\overline{\mathrm{D}_i\mathrm{D}_j\frac{x^k}{y^l}\,\mathrm{d}x}$, $\overline{\mathrm{D}_i\frac{x^k}{y^l}\,\mathrm{d}x}$ and $\overline{\frac{x^k}{y^l}\,\mathrm{d}x}$ in $H^1_{\mathrm{DR}}(\mathfrak{Y}/S)$. For this purpose we set

$$G_{kl}^{(m)} = \frac{x^k}{y^l}\cdot\frac{x(x-1)}{x-t_m} = \frac{x^{k+1}(x-1)}{y^l(x-t_m)}$$

and calculate the exact differential forms $\mathrm{d}G_{kl}^{(m)}$ in terms of the right-hand sides of (1.4.2), (1.4.3) and (1.4.4). This will lead us to

1.4.5. LEMMA. *It holds that*

$$\Delta^{(m)} \cdot \overline{\frac{x^k}{y^l}\,\mathrm{d}x} = 0 \quad \textit{for} \quad 1 \leqq m \leqq r,$$

$$\Delta^{(m)} = \Delta_{kl}^{(m)} = \frac{n}{l}\,(t_m - 1) \sum_{i=1}^{r} t_i \mathrm{D}_i \mathrm{D}_m + b_m \sum_{0<i\neq m} t_i \mathrm{D}_i$$

$$+ \left[-B(t_m - 1)\frac{n}{l} + b_m t_m + b_{-1}\right] \mathrm{D}_m - b_m(B+1) \in \mathbb{C}[S, \mathrm{D}],$$

with b, B *as in* 1.1.7.

Proof. To abbreviate we write $G = G_{kl}^{(m)}$, $\mathrm{d} = \mathrm{d}_{\mathfrak{Y}}$. Using the product rule (1.3.11) and (1.3.9) we find

$$\mathrm{d}G = G\left[-\frac{l}{y}\,\mathrm{d}y + \frac{k+1}{x}\,\mathrm{d}x + \frac{1}{x-1}\,\mathrm{d}x - \frac{\mathrm{d}x}{x-t_m}\right]$$

$$= \left[\frac{(n-b_{-1}l)/n}{x-1} + \frac{\big((k+1)n - b_0 l\big)/n}{x} + \frac{(-n-b_m l)/n}{x-t_m}\right.$$

$$\left. + \sum_{0<i\neq m} \frac{-lb_i/n}{x-t_i}\right] \frac{x(x-1)}{x-t_m} \cdot \frac{x^k}{y^l}\,\mathrm{d}x.$$

We put

$$c_{-1} = (n - b_{-1}l)/n, \qquad c_0 = \big((k+1)\,n - b_0 l\big), \qquad c_m = (-n - b_m l)/n,$$

$$c_i = -lb_i/n \quad \text{for} \quad 1 \leqq i \leqq r, \qquad i \neq m.$$

Then

$$\mathrm{d}G = \left[\sum_{-1}^{r} \frac{c_i}{x-t_i}\right] \frac{x(x-1)}{x-t_m} \cdot \frac{x^k}{y^l}\,\mathrm{d}x$$

$$= \left[\frac{c_{-1}x + c_0(x-1)}{x-t_m} + \frac{c_m x(x-1)}{(x-t_m)^2} + \sum_{0<i\neq m} \frac{c_i x(x-1)}{(x-t_m)\,(x-t_i)}\right] \frac{x^k}{y^l}\,\mathrm{d}x.$$

In the numerators we would like to have only elements of $\mathbb{C}[S]$. For this purpose we write

$$x = (x - t_m) + t_m, \qquad (x-1) = (x-t_m) + (t_m - 1)$$

in the first or second summand, respectively, and

$$x = (x - t_i) + t_i, \qquad (x-1) = (x-t_m) + (t_m - 1)$$

in the other summands. Then we find

$$\mathrm{d}G = \left\{(c_{-1} + c_0) + \frac{c_{-1}t_m + c_0(t_m - 1)}{x - t_m} + c_m + \frac{c_m t_m(t_m - 1)}{(x-t_m)^2} + \sum_{0<i\neq m} c_i\right.$$

$$+ \sum_{0<i\neq m} \frac{c_i(t_m - 1)}{x - t_m} + \sum_{0<i\neq m} \frac{c_i t_i}{x - t_i} + \sum_{0<i\neq m} \frac{c_i t_i (t_m - 1)}{(x - t_i)(x - t_m)}\Bigg\} \cdot \frac{x^k}{y^l}\,\mathrm{d}x$$

$$= \Bigg[\sum_{-1}^{r} c_i + \frac{(c_{-1} + c_m) t_m + \sum_{0\leq i} c_i(t_m - 1)}{x - t_m} + \sum_{0<i\neq m} \frac{c_i t_i}{x - t_i}$$

$$+ \sum_{0<i} \frac{c_i t_i (t_m - 1)}{(x - t_i)(x - t_m)}\Bigg] \cdot \frac{x^k}{y^l}\,\mathrm{d}x.$$

Now from $\Omega^1_{\mathfrak{Y}}$ pass to $\Omega^1_{\mathfrak{Y}/S}$ and use the substitutions (1.4.2), (1.4.3), (1.4.4), with b, B as in 1.1.7, $\sum_{-1}^{r} c_i = (k+1) - \frac{l}{n} b = B + 1$ and

$$(c_{-1} + c_m) t_m + \sum_{0\leq i} c_i(t_m - 1)$$

$$= c_m t_m + c_{-1} + \sum_{-1}^{r} c_i(t_m - 1) = c_m t_m + c_{-1}(B+1)(t_m - 1)$$

$$= \left(-1 - \frac{b_m l}{n}\right) t_m + \left(1 - \frac{b_{-1} l}{n}\right) + B(t_m - 1) + (t_m - 1)$$

$$= B(t_m - 1) - \frac{b_m l}{n} t_m - \frac{b_{-1} l}{n}.$$

Then we find

$$\mathrm{d}_{\mathfrak{Y}/S} G = \Bigg\{(B+1) + \frac{n}{b_m l}\left[B(t_m - 1) - \frac{b_m l}{n} t - \frac{b_{-1} l}{n}\right] \mathrm{D}_m - \sum_{0<i\neq m} t_i \mathrm{D}_i'$$

$$- \frac{n}{b_m l} \sum_{0<i\neq m} (t_m - 1)\, t_i \mathrm{D}_i \mathrm{D}_m - \frac{n}{b_m l} (t_m - 1)\, t_m \mathrm{D}_m\Bigg\} \cdot \frac{x^k}{y^l}\,\mathrm{d}x.$$

Now take images in $H^1_{\mathrm{DR}}(\mathfrak{Y}/S)$; then the left-hand side is 0. Using the last rule of 1.3.14 Lemma 1.4.5 follows immediately.

After these preparations we come to the

Proof of Proposition 1.4.1. Lemma 1.4.5 gives to us a system of r differential equations satisfied by $\overline{x^k\,\mathrm{d}x/y^l}$. We will enlarge this system to a system of r^2 differential equations. This will be done by means of a rigorous classical trick (Picard, Appell) which involves differential operators of order 3. For $p \neq m$ we calculate

$$\mathrm{D}_p \cdot \Delta^{(m)}$$

$$= \frac{n}{l}(t_m - 1) \sum_{0<i} t_i \mathrm{D}_i \mathrm{D}_m \mathrm{D}_p + \frac{n}{l}(t_m - 1)\, \mathrm{D}_m \mathrm{D}_p + b_m \sum_{0<i\neq m} t_i \mathrm{D}_i \mathrm{D}_p + b_m \mathrm{D}_p$$

$$+ \left[-\frac{Bn}{l}(t_m - 1) + b_m t_m + b_{-1}\right] \mathrm{D}_m \mathrm{D}_p - b_m(B+1)\, \mathrm{D}_p$$

$$= \frac{n}{l} (t_m - 1) \sum_{0<i} t_i D_i D_m D_p + b_m \sum_{0<i\neq m} t_i D_i D_p$$
$$+ \left[\frac{n}{l} (1 - B) (t_m - 1) + b_m t_m + b_{-1}\right] D_m D_p - b_m B D_p .$$

Interchanging m and p we obtain

$$D_m \cdot \Delta^{(p)} = \frac{n}{l} (t_p - 1) \sum_{0<i} t_i D_i D_m D_p + b_p \sum_{0<i\neq p} t_i D_i D_m$$
$$+ \left[\frac{n}{l} (1 - B) (t_p - 1) + b_p t_p + b_{-1}\right] D_m D_p - b_p B D_m .$$

We put

$$\Delta^{(m,p)} = (t_p - 1)\, D_p \Delta^{(m)} - (t_m - 1)\, D_m \Delta^{(p)} .$$

Then

$$\Delta^{(m,p)} = b_m(t_p - 1) \sum_{0<i\neq m} t_i D_i D_p - b_p(t_m - 1) \sum_{0<i\neq p} t_i D_i D_m$$
$$+ [(b_m t_m + b_{-1}) (t_p - 1) - (b_p t_p + b_{-1}) (t_m - 1)]\, D_m D_p$$
$$+ B b_p (t_m - 1)\, D_m - B b_m (t_p - 1)\, D_p .$$

These are $r^2 - r$ differential operators of second order ($m \neq p$). In order to remove almost all second order summands $D_u D_v$ we calculate

$$\Delta^{(m,p)} + \frac{l}{n} b_p \Delta^{(m)} - \frac{l}{n} b_m \Delta^{(p)}$$
$$= b_{-1}(t_p - t_m)\, D_m D_p + b_{-1} \frac{l}{n} (b_p D_m - b_m D_p)$$
$$= b_{-1}(t_p - t_m)\, \varepsilon_{mp} , \qquad \varepsilon_{mp} = \varepsilon_{mp}^{l,n,b}$$

with the notations of Definition 1.1.7.

Obviously $D_p \cdot \Delta^{(m)}$, $D_m \cdot \Delta^{(p)}$, $\Delta^{(m,p)}$ and finally ε_{mp} lie in the left ideal of $\mathbb{C}[S, D]$ generated by $\Delta^{(1)}, \ldots, \Delta^{(r)}$. Because of Lemma 1.4.5 $\overline{x^k\, dx/y^l}$ satisfies the $r(r-1)$ Euler differential equations

1.4.6. $\varepsilon_{mp} \overline{x^k\, dx/y^l} = 0, \qquad \varepsilon_{mp} = \varepsilon_{mp}^{l,n,b}, \qquad 1 \leqq m, p \leqq r, \qquad m \neq p .$

Now we can remove by substitution all mixed summands $D_u D_v$, $u \neq v$, in $\Delta^{(m)}$. Write (1.4.6) as

$$D_i D_m \overline{x^k\, dx/y^l} = - \frac{l}{n(t_i - t_m)} (-b_i D_m + b_m D_i)\, \overline{x^k\, dx/y^l} .$$

Then we obtain

$$Q^{(m)} \overline{x^k\, dx/y^l} = 0$$

with

$$\begin{aligned}
Q^{(m)} &= \frac{n}{l}(t_m-1)\,t_m\mathrm{D}_m^2 + \sum_{0<i\neq m}\frac{(t_m-1)\,t_i}{t_i-t_m}(-b_i\mathrm{D}_m+b_m\mathrm{D}_i)\\
&\quad + b_m\sum_{0<i\neq m}t_i\mathrm{D}_i + \left[-\frac{n}{l}B(t_m-1)+b_mt_m+b_{-1}\right]\mathrm{D}_m - b_m(B+1)\\
&= \frac{n}{l}(t_m-1)\,t_m\mathrm{D}_m^2 + \sum_{0<i\neq m}\frac{b_m(t_m-1)\,t_i}{t_i-t_m}\mathrm{D}_i - \left(\sum_{0<i\neq m}\frac{b_i(t_m-1)\,t_i}{t_i-t_m}\right)\mathrm{D}_m\\
&\quad + \sum_{0<i\neq m}b_mt_i\mathrm{D}_i + \left[-\frac{nB}{l}(t_m-1)+b_mt_m+b_{-1}\right]\mathrm{D}_m - (B+1)\,b_m\\
&= \frac{n}{l}(t_m-1)\,t_m\mathrm{D}_m^2 + b_m\sum_{0<i\neq m}\frac{t_i^2-t_i}{t_i-t_m}\mathrm{D}_i\\
&\quad - \left[\sum_{0<i\neq m}\frac{b_i(t_m-1)\,t_i}{t_i-t_m}+\frac{nB}{l}(t_m-1)-b_mt_m-b_{-1}\right]\mathrm{D}_m - (B+1)\,b_m\,.
\end{aligned}$$

Let

$$\varepsilon_{mm} = \frac{l}{n(t_m-1)\,t_m}\,Q^{(m)}\,.$$

Then

1.4.7. $\varepsilon_{mm}\overline{x^k\,\mathrm{d}x/y^l} = 0\,, \qquad \varepsilon_{mm} = \varepsilon_{mm}^{k,l,n,\mathfrak{b}} \qquad$ (see 1.1.7).

Taking into account the definition of a principal Euler-Picard module (see 1.1), (1.4.6) and (1.4.7) we see that the proof of Proposition 1.4.1 is finished.

1.4.8. PROPOSITION. *For* $0 \leqq l \leqq n-1$ *and* $l' \equiv l \bmod n$, $l' \gg 0$ $(H^1_{\mathrm{DR}}(\mathfrak{Y}/S))_l$, $\overline{\mathrm{d}x/y^{l'}}$) *is a principal Euler-Picard module of type* $(0, l', n, \mathfrak{b})$.

For the proof we again need some preparations. We denote the free principal Euler-Picard module of type $(0, l', n, \mathfrak{b})$ by $\mathscr{E}_{l'}$. By definition (see 1.1.7) $\mathscr{E}_{l'}$ can be written as

$$\mathscr{E}_{l'} = \mathbb{C}[S]\,e_0 + \mathbb{C}[S]\,e_1 + \cdots + \mathbb{C}[S]\,e_r$$

with $\mathbb{C}[S]$-generators $e_0, e_1, \ldots, e_r$,

$$e_i = \mathrm{D}_ie_0 \quad \text{for} \quad i = 1, \ldots, r\,.$$

These satisfy the linear differential equations

1.4.9. $\mathrm{D}_ie_j = \sum_{k=0}^{r} a_{ij}^{(k)}e_k\,, \qquad a_{ij}^{(k)} \in \mathbb{C}[S]\,, \qquad i, j = 1, \ldots, r\,,$

coming from the Euler-Picard operators of type $(0, l', n, \mathfrak{b})$:

$$\varepsilon_{ij} = \mathrm{D}_i\mathrm{D}_j - \sum_{k=0}^{r} a_{ij}^{(k)}\mathrm{D}_k\,, \qquad \varepsilon_{ij} = \varepsilon_{ij}^{0,l',n,\mathfrak{b}}\,.$$

On the other hand we have also the $\mathbb{C}[S]$-decomposition

$$\mathbb{C}[S, \mathrm{D}]\,\overline{\mathrm{d}x/y^{l'}} = \mathbb{C}[S]\,\overline{\mathrm{d}x/y^{l'}} + \mathbb{C}[S]\cdot \mathrm{D}_1\,\overline{\mathrm{d}x/y^{l'}} + \cdots + \mathbb{C}[S]\cdot \mathrm{D}_r\,\overline{\mathrm{d}x/y^{l'}}$$

and a $\mathbb{C}[S, \mathrm{D}]$-homomorphism

$$\varphi = \varphi_{l'}: \mathscr{E}_{l'} \to \mathbb{C}[S, \mathrm{D}]\,\overline{\mathrm{d}x/y^{l'}} \hookrightarrow H^1_{\mathrm{DR}}(\mathfrak{Y}/S)_l$$

determined by the $\mathbb{C}[S]$-linear extension of the correspondence

$$e_0 \mapsto \overline{\mathrm{d}x/y^{l'}}, \quad e_1 \mapsto \mathrm{D}_1\,\overline{\mathrm{d}x/y^{l'}}, \ldots, \quad e_r \mapsto \mathrm{D}_r\,\overline{\mathrm{d}x/y^{l'}}.$$

We have to prove that

1.4.10. $\varphi_{l'}: \mathscr{E}_{l'} \to H^1_{\mathrm{DR}}(\mathfrak{Y}/S)_{l'}$ is surjective.

Let $\mathfrak{M}$ be a subset of $\Omega_{\mathbb{C}[\mathfrak{Y}]/\mathbb{C}[S]}$ Then we denote the image of $\mathfrak{M}$ in $H^1_{\mathrm{DR}}(\mathfrak{Y}/S)$ by $\overline{\mathfrak{M}}$. We consider in $\Omega_{\mathbb{C}[\mathfrak{Y}]/\mathbb{C}[S]} = \mathbb{C}[\mathfrak{Y}]\cdot \mathrm{d}x$ the infinite ascending chain of $\mathbb{C}[S]$-submodules of $(\Omega_{\mathbb{C}[\mathfrak{Y}]/\mathbb{C}[S]})_l$:

1.4.11. $$\cdots \subseteqq \mathbb{C}[S]\,[x]\,\frac{\mathrm{d}x}{y^l} \subseteqq \mathbb{C}[S]\,[x]\,\frac{\mathrm{d}x}{y^{l+n}} \subseteqq \cdots \subseteqq \mathbb{C}[S]\,[x]\,\frac{\mathrm{d}x}{y^{l+an}} \subseteqq \cdots, \quad a \to \infty.$$

Obviously,

$$(\Omega_{\mathbb{C}[Y]/\mathbb{C}[S]})_l = \bigcup_{a=0}^{\infty} \mathbb{C}[S]\,[x]\,\frac{\mathrm{d}x}{y^{l+an}} = \bigcup_{a>0}^{\infty} \mathbb{C}[S]\,[x]\,\frac{\mathrm{d}x}{y^{l+an}}.$$

Taking images in $H^1_{\mathrm{DR}}(\mathfrak{Y}/S)$ it follows that

1.4.12. $$H^1_{\mathrm{DR}}(\mathfrak{Y}/S)_l = \bigcup_{a=0}^{\infty} \overline{\mathbb{C}[S]\,[x]\,\frac{\mathrm{d}x}{y^{l+an}}} = \bigcup_{\substack{l' \equiv l \bmod n \\ l' \geqq 0}} \overline{\mathbb{C}[S]\,[x]\,\frac{\mathrm{d}x}{y^{l'}}}.$$

First we will prove that

1.4.13. $\overline{x^k \mathrm{d}x/y^{l'}} \in \operatorname{Im} \varphi_{l'}$ *for* $k \geqq 0$ *and* $l' \neq 2n/b$.

It is convenient to give this the following formulation:

1.4.14. $\overline{x^k \mathrm{D}_j(\mathrm{d}x/y^{l'})} = \overline{\mathrm{D}_j(x^k\,\mathrm{d}x/y^{l'})} = \mathrm{D}_j(\overline{x^k\,\mathrm{d}x/y^{l'}}) \in \operatorname{Im} \varphi_{l'}$

for $j = 0, \ldots, r\ (\mathrm{D}_0 = 1), \quad k \geqq 0, \quad l' \equiv l \bmod n, \quad l' \neq 2n/b.$

We carry out an induction procedure over j and k. The starting point of the induction will be

1.4.15. $\overline{\mathrm{d}x/y^{l'}}, \overline{x\,\mathrm{d}x/y^{l'}} \in \operatorname{Im} \varphi_{l'}$ *for* $l' \neq 2n/b$.

The first assertion is trivial because $\overline{\mathrm{d}x/y^{l'}} = \varphi_{l'}(e_0) \in \operatorname{Im} \varphi_{l'}$. For the second element we calculate in $\Omega_{\mathbb{C}[\mathfrak{Y}]}$

$$\mathrm{d}\left(\frac{x(x-1)}{y^{l'}}\right) = x(x-1)\left[\frac{1 - b_0 l'/n}{x} + \frac{1 - b_{-1} l'/n}{x-1} - \frac{l'}{n}\sum_{i=1}^{r} \frac{b_i}{x - t_i}\right]\frac{\mathrm{d}x}{y^{l'}}.$$

After the substitution

$$x(x-1) = [(x-t_i)+t_i]\cdot[(x-t_i)+(t_i-1)]$$

we find

$$\mathrm{d}\left(\frac{x(x-1)}{y^{l'}}\right) = \Bigg[(x-1)\,(1-b_0 l'/n) + x(1-b_{-1}l'/n)$$
$$-\frac{l'}{n}\sum_1^r \left(b_i(x-t_i) + b_i(2t_i-1) + \frac{b_i t_i(t_i-1)}{x-t_i}\right)\Bigg]\frac{dx}{y^{l'}}.$$

Passing to $\Omega_{\mathbb{C}[\mathfrak{Y}]/\mathbb{C}[S]}$ we can replace $\dfrac{l'b_i}{n(x-t_i)}\cdot\dfrac{\mathrm{d}x}{y^l}$ by $\mathrm{D}_i(\mathrm{d}x/y^{l'})$ (see (1.4.2)). Then in $H^1_{\mathrm{DR}}(\mathfrak{Y}/S)$ we have that

$$0 = \left(2-\frac{l'}{n}\sum_{-1}^r b_i\right)\overline{\frac{x\,\mathrm{d}x}{y^{l'}}} + \left[-1+\sum_{-1}^r\frac{b_i l'}{n} - \frac{l'}{n}\sum_{-1}^r b_i t_i\right]\overline{\frac{\mathrm{d}x}{y^{l'}}}$$
$$-\sum_1^r t_i(t_i-1)\,\mathrm{D}_i\,\overline{\frac{\mathrm{d}x}{y^{l'}}}.$$

Now 1.4.15 follows from $\overline{\mathrm{d}x/y^{l'}}$, $\mathrm{D}_i\,\overline{\mathrm{d}x/y^{l'}} \in \operatorname{Im}\varphi_{l'}$ and $l' \neq 2b/n$.

For the induction procedure we arrange the elements $\mathrm{D}_j\,\overline{\dfrac{x^k}{y^{l'}}\,\mathrm{d}x}$ in a table (see 1.4.α).

1.4.α

$\overline{\mathrm{d}x/y^{l'}}$,	$\overline{x\,\mathrm{d}x/y^{l'}}$,	…,	$\overline{x^{m-1}\,\mathrm{d}x/y^{l'}}$	$\overline{x^m\,\mathrm{d}x/y^{l'}}$,	$\overline{x^{m+1}\,\mathrm{d}x/y^{l'}}$, …
$\mathrm{D}_1\,\overline{\mathrm{d}x/y^{l'}}$,	$\mathrm{D}_1\,\overline{x\mathrm{d}x/y^{l'}}$,	…	…,	$\mathrm{D}_1\overline{x^m\,\mathrm{d}x/y^{l'}}$,	$\mathrm{D}_1\overline{x^{m+1}\,\mathrm{d}x/y^{l'}}$, …
⋮				⋮	⋮
$\mathrm{D}_j\,\overline{\mathrm{d}x/y^{l'}}$,	$\mathrm{D}_j\,\overline{x\mathrm{d}x/y'}$,	…	…,	$\mathrm{D}_j\overline{x^m\mathrm{d}x/y^{l'}}$,	$\mathrm{D}_j\overline{x^{m+1}\,\mathrm{d}x/y^{l'}}$, …
⋮				⋮	⋮
$\mathrm{D}_r\,\overline{\mathrm{d}x/y^{l'}}$,	$\mathrm{D}_r\overline{x\,\mathrm{d}x/y^{l'}}$,	…	…,	$\mathrm{D}_r\overline{x^m\,\mathrm{d}x/y^{l'}}$,	$\mathrm{D}_r\overline{x^{m+1}\,\mathrm{d}x/y^{l'}}$, …

Then we carry out two induction steps described in the following scheme

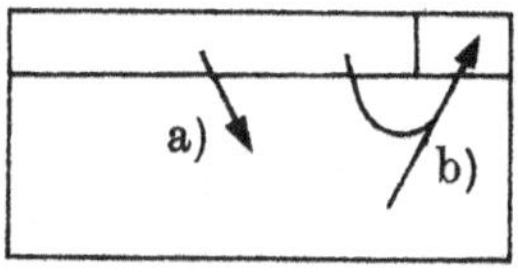

More precisely we prove

1.4.16. LEMMA. *Assume that* $m \geqq 2$, $l' \neq 2n/b$ *and*

$$\overline{\mathrm{d}x/y^{l'}},\ \overline{x\,\mathrm{d}x/y^{l'}},\ \ldots,\ \overline{x^{m-1}\mathrm{d}x/y^{l'}} \in \operatorname{Im} \varphi_{l'}.$$

Then

a) $\mathrm{D}_j\overline{x^k\,\mathrm{d}x/y^{l'}} \in \operatorname{Im}\varphi_{l'}$ *for* $j = 1, \ldots, r$, $k = 0, \ldots, m$;

b) $\overline{x^m\,\mathrm{d}x/y^{l'}} \in \operatorname{Im}\varphi_{l'}$.

Proof. a) For $k = 0$ there is nothing to prove (see 1.4.15). Assume that $k > 0$. In $\Omega_{\mathbb{C}[\mathfrak{Y}]/\mathbb{C}[S]}$ we have

$$\mathrm{D}_j\,\frac{x^k\,\mathrm{d}x}{y^{l'}} = x^k\mathrm{D}_j\,\frac{\mathrm{d}x}{y^{l'}} = x^k\,\frac{b_jl'/n}{x - t_j}\cdot\frac{\mathrm{d}x}{y^{l'}} = \frac{\big((x - t_j) + t_j\big)^k}{x - t_j}\cdot\frac{b_jl'}{n}\,\frac{\mathrm{d}x}{y^{l'}}$$

$$= \big((x - t_j)^{k-1} + \cdots + kt_j^{k-1}\big)\,\frac{b_jl'}{n}\cdot\frac{\mathrm{d}x}{y^{l'}} + t_j^k\mathrm{D}_j\,\frac{\mathrm{d}x}{y^{l'}}.$$

The class in $H^1_{\mathrm{DR}}(\mathfrak{Y}/S)$ of the element on the right-hand side is an element of $\operatorname{Im}\varphi_{l'}$ by our assumption and 1.4.15.

b) Using the substitutions $x - t_i = (x - t_j) + (t_j - t_i)$ we calculate in $\Omega_{\mathbb{C}[\mathfrak{Y}]}$

$$\mathrm{d}\left(\frac{x^{m-1}(x - 1)\,(x - t_i)}{y^{l'}}\right) = x^{m-1}(x - 1)\,(x - t_i)\left[\frac{1 - b_{-1}l'/n}{x - 1} + \frac{1 - b_0l'/n}{x}\right.$$

$$\left.- \sum_{j=1}^{r}\frac{b_jl'/n}{x - t_j}\right]\frac{\mathrm{d}x}{y^{l'}} = \Big[x^{m-1}(x - t_i)\,(1 - b_{-1}l'/n) + x^{m-2}(x - 1)\,(x - t_i)\,(1 - b_0l'n)$$

$$\left. - \frac{l'}{n}\sum_{j=1}^{r} b_jx^{m-1}(x - 1) - \frac{l'}{n}\sum_{j=1}^{r}\frac{b_jx^{m-1}(x - 1)\,(t_j - t_i)}{x - t_j}\right]\frac{\mathrm{d}x}{y^{l'}}$$

$$= \left[\left(2 - \frac{l'}{n}\sum_{-1}^{r} b_j\right) - \frac{l'}{n}\sum_{j=1}^{r}\frac{b_j(t_j - t_i)}{x - t_j}\right]\frac{x^m\,\mathrm{d}x}{y^{l'}}$$

$$+ \left[\frac{l'}{n}\sum_{j=1}^{r}\frac{b_j(t_j - t_i)}{x - t_j}\right]\frac{x^{m-1}\,\mathrm{d}x}{y^{l'}} + P(x)\,\frac{\mathrm{d}x}{y^{l'}},$$

$P(x) \in \mathbb{C}[S]\,[x]$, $\quad \deg_x P \leqq m - 1$.

Taking images in $H^1_{\mathrm{DR}}(\mathfrak{Y}/S)$ we obtain the relation

$$0 = \left(2 - \frac{l'b}{n}\right)\overline{x^m\,\mathrm{d}x/y^{l'}} - \sum_{j=1}^{r}(t_j - t_i)\,\mathrm{D}_j\overline{x^m\,\mathrm{d}x/y^{l'}}$$

$$+ \sum_{j=1}^{r}(t_j - t_i)\,\mathrm{D}_j\overline{x^{m-1}\,\mathrm{d}x/y^{l'}} + \overline{P(x)\,\mathrm{d}x/y^{l'}}.$$

By a) and the assumption of the lemma it follows that $\overline{x^m\,\mathrm{d}x/y^{l'}} \in \operatorname{Im}\varphi_{l'}$. This proves our lemma together with 1.4.14 and 1.4.13.

1.4.17. $\operatorname{Im}\varphi_{l'} = \mathbb{C}[S, \mathrm{D}]\cdot\overline{\mathbb{C}[S]\,[x]\cdot\mathrm{d}x/y^{l'}}$ *for* $l' \neq 2n/b$.

Proof. Im $\varphi_{l'}$ is the $\mathbb{C}[S, \mathrm{D}]$-submodule of $H^1_{\mathrm{DR}}(\mathfrak{Y}/S)$ generated by $\overline{\mathrm{d}x/y^{l'}}$. So the image of $\varphi_{l'}$ is contained in the $\mathbb{C}[S, \mathrm{D}]$-module on the right-hand side of 1.4.17. The latter module is generated by the elements $\overline{x^k\,\mathrm{d}x/y^{l'}}$. They all lie in Im $\varphi_{l'}$ by 1.4.13. This yields the inverse inclusion.

Proof of Proposition 1.4.8. By (1.4.11) and 1.4.17 we have an infinite ascending chain of $\mathbb{C}[S]$-modules of finite type

1.4.18. $\operatorname{Im}\varphi_l \subseteqq \operatorname{Im}\varphi_{l+n} \subseteqq \cdots \subseteqq \operatorname{Im}\varphi_{l+an} \subseteqq \cdots, \qquad a \to \infty$

(exclude Im $\varphi_{2n/b}$ from the chain if $\mathbb{Z} \ni 2n/b \equiv l \bmod n$) with the property (see (1.4.12))

$$H^1_{\mathrm{DR}}(\mathfrak{Y}/S)_l = \bigcup_{\substack{l' \gg 0\\ l' \equiv l \bmod n}} \operatorname{Im}\varphi_{l'}.$$

It suffices to show that the chain (1.4.18) stabilizes. For any closed point $s \in S$ we consider the local chain

1.4.18.(s)
$$\begin{array}{ccccc} (\operatorname{Im}\varphi_l)(s) & \subseteqq \cdots \subseteqq & (\operatorname{Im}\varphi_{l+an})(s) & \subseteqq \cdots, \\ \| & & \| & \\ \operatorname{Im}\varphi_l(s) & & \operatorname{Im}\varphi_{l+an}(s) & \end{array}$$

where

$$\varphi_{l'}(s) : (\mathscr{E}_{l'})(s) \to H\big({}^1_{\mathrm{DR}}(\mathfrak{Y}/S)\big)(s)$$

denotes the corresponding localisation of $\varphi_{l'}$, more precisely, the restriction to the closed point. Since $\mathscr{E}_{l'}$ is a $\mathbb{C}[S]$-module with $r+1$ generators, Im $\varphi_{l'}(s)$ is a $\mathbb{C}$-vector space of dimension not greater than $r+1$. Therefore the chain (1.4.18) (s) stabilizes. By the Nakayama Lemma* (see e.g. [52], III, § 2) there is a Zariski-open neighbourhood $U(s)$ of s in Spec $\mathbb{C}[S]$ and a suitable number $L(s)$ such that Im $\varphi_{l'} = $ Im $\varphi_{l''}$ on $U(s)$ for $l'' \geqq l' \geqq L(s)$. The open covering $\{U(s)\}_{s \in S}$ of S contains a finite covering $\{U_i\}_{i \in I}$ because S is noetherian. The corresponding constants $L(s)$ are denoted by L_i. Let L be the maximum of all L_i, $i \in I$. Then

$$\operatorname{Im}\varphi_{l'} = \operatorname{Im}\varphi_{l''} \quad \text{for} \quad l'' \geqq l' \geqq L, \qquad l', l'' \equiv l \bmod n,$$

on each U_i, hence on S. This concludes the proof of Proposition 1.4.8

1.4.19. COROLLARY. *$H^1_{\mathrm{DR}}(\mathfrak{Y}/S)$ is an Euler-Picard module over S. It is the $\mathbb{C}[S, \mathrm{D}]$-direct sum of n principal Euler-Picard modules:*

$$H^1_{\mathrm{DR}}(\mathfrak{Y}/S) = \bigoplus_{l=0}^{n-1} H^1_{\mathrm{DR}}(\mathfrak{Y}/S)_l,$$

1.4.20. $H^1_{\mathrm{DR}}(\mathfrak{Y}/S)_l = \overline{\mathbb{C}[S]\,[x]\,\mathrm{d}x/y^{l'}}$ *for* $l' \equiv l \bmod n, \qquad l' \gg 0.$

---

* Nakayama Lemma: Let X be a noetherian scheme, $\mathscr{F}$ a coherent sheaf of $\mathcal{O}_X$-modules, $x \in X$, $\mathscr{F}(x) = 0$, then $\mathscr{F}|_U = 0$ for a suitable open neighbourhood U of x.

Proof. We know that

$$H^1_{\mathrm{DR}}(\mathfrak{Y}/S)_l = \mathrm{Im}\, \varphi_{l'} = \mathbb{C}[S, \mathrm{D}] \cdot \overline{\mathbb{C}[S]\,[x]\,\mathrm{d}x/y^{l'}} = \mathrm{Im}\, \varphi_{l'+n}$$

(see 1.4.17). Therefore it suffices to remark that

$$\mathrm{D}_i \overline{\frac{x^k\,\mathrm{d}x}{y^{l'}}} = \overline{\frac{b_i l'}{n(x - t_i)} \frac{x^k \mathrm{d}x}{y^{l'}}} \in \overline{\mathbb{C}[S]\,[x] \frac{\mathrm{d}x}{y^{l'+n}}} \subseteqq \mathrm{Im}\, \varphi_{l'+n} = \mathrm{Im}\, \varphi_{l'}.$$

We want to decide in which cases $H^1_{\mathrm{DR}}(\mathfrak{Y}/S)_l$ is a primitive principal Euler-Picard module. For this purpose it is convenient to replace the $\mathbb{C}[S]$-generators $\overline{\mathrm{d}x/y^{l'}}$, $\mathrm{D}_1\,\overline{\mathrm{d}x/y^{l'}}, \ldots, \mathrm{D}_r\,\overline{\mathrm{d}x/y^{l'}}$ of $H^1_{\mathrm{DR}}(\mathfrak{Y}/S)_l = \mathrm{Im}\, \varphi_{l'}$ by generators of type $\overline{x^k\,\mathrm{d}x/y^{l'}}$, $k \geqq 0$.

1.4.21. LEMMA. *If $n \nmid bl'$, then*

$$\mathbb{C}[S]\,[x] \cdot \frac{\mathrm{d}x}{y^{l'}} \subseteqq \mathbb{C}[S] \frac{\mathrm{d}x}{y^{l'}} + \cdots + \mathbb{C}[S] \frac{x^r\,\mathrm{d}x}{y^{l'}} + \mathrm{d}\, \frac{\mathbb{C}[S]\,[x]}{y^{l'}} \quad \text{in } \Omega_{\mathbb{C}[\mathfrak{Y}]}.$$

Proof. It suffices to prove that, for $m > 0$, $x^{m+r}\,\mathrm{d}x/y^{l'}$ is an element of the space on the right-hand side of the inclusion in 1.4.21. This will be done by induction. For this purpose we calculate

$$\mathrm{d} \left(\frac{x^{m-1} \prod_{-1}^{r} (x - t_i)}{y^{l'}} \right) = \frac{x^{m-1} \cdot \Pi_x}{y^{l'}} \left(-\frac{l'}{n} \sum_{-1}^{r} \frac{b_i}{x - t_i} + \sum_{-1}^{r} \frac{1}{x - t_i} + \frac{m-1}{x} \right) \mathrm{d}x$$

$$= x^{m-1} \Pi_x \left(\frac{m-1}{x} + \sum_{-1}^{r} \frac{(1 - b_i l'/n)}{x - t_i} \right) \frac{\mathrm{d}x}{y^{l'}}$$

$$= \left\{ (m-1)\, x^{m-2}\, \Pi_x + \sum_{i=-1}^{r} \left[(1 - b_i l'/n)\, x^{m-1} \prod_{\substack{j=-1 \\ j \neq i}}^{r} (x - t_j) \right] \right\} \frac{\mathrm{d}x}{y^{l'}}$$

$$= x^{m+r} \left[(m-1) - \sum_{i=-1}^{r} b_i l'/n + r + 2 \right] \frac{\mathrm{d}x}{y^{l'}} + P(x) \frac{\mathrm{d}x}{y^{l'}}$$

$$= [(m + r + 1) - bl'/n] \frac{x^{m+r}}{y^{l'}}\, \mathrm{d}x + P(x) \frac{\mathrm{d}x}{y^{l'}},$$

where $P(x) \in \mathbb{C}[S]\,[x]$, $\deg_x P < m + r$. Now our statement follows from the induction hypothesis and from $bl'/n \notin \mathbb{Z}$; q.e.d.

1.4.22. COROLLARY. *If $n \nmid lb$, then for $l' \equiv l \bmod n$, $l' \gg 0$,*

$$H^1_{\mathrm{DR}}(\mathfrak{Y}/S)_l = \mathbb{C}[S] \cdot \overline{\mathrm{d}x/y^{l'}} + \mathbb{C}[S] \cdot \overline{x\,\mathrm{d}x/y^{l'}} + \cdots + \mathbb{C}[S] \cdot \overline{x^r \mathrm{d}x/y^{l'}}.$$

This follows immediately from 1.4.20 and 1.4.21.

1.4.23. THEOREM. *If $n \nmid lb$ and $n \nmid lb_i$ for $i = -1, 0, 1, \ldots, r$, then*

(i) *$H^1_{\mathrm{DR}}(\mathfrak{Y}/S)_l$ is a primitive principal Euler-Picard module of type $(0, l, n, \mathfrak{b})$ with generator $\overline{\mathrm{d}x/y^l}$.*

(ii) *$H^1_{\mathrm{DR}}(\mathfrak{Y}/S)_l = \mathbb{C}[S] \cdot \overline{\mathrm{d}x/y^l} \oplus \mathbb{C}[S] \cdot \overline{x\,\mathrm{d}x/y^l} \oplus \cdots \oplus \mathbb{C}[S] \cdot \overline{x^r\,\mathrm{d}x/y^l}$ as $\mathbb{C}[S]$-module.*

For the proof we use the localisations of $\mathscr{H}^1_{\mathrm{DR}}(\mathfrak{Y}/S)$ in the general point $\operatorname{Spec} \mathbb{C}(S) = \operatorname{Spec} \mathbb{C}(t)$ and in the closed points s of S (see Diagramm 1.2.c). The central fact is the following

1.4.24. LEMMA (see Katz [36]). *Let K be a field of characteristic 0, $\dddot{\mathfrak{Y}}_K$ the affine plane curve*

$$y^n = \sum_{i=-1}^{r} (x - t_i)^{b_i}, \qquad t_i \in K,$$

and $\mathfrak{Y}_K = \dddot{\mathfrak{Y}}_K \setminus \{t_{-1}, t_0, \ldots, t_r\}$ the étale part over $\mathbb{A}^1_K$. If $n \nmid lb$ and $n \nmid lb_i$ for $i = -1, 0, \ldots, r$, then the classes $\overline{\mathrm{d}x/y^l}, \overline{x\,\mathrm{d}x/y^l}, \ldots, \overline{x^r\,\mathrm{d}x/y^l}$ in the K-vector space $H^1_{\mathrm{DR}}(\mathfrak{Y}_K/K) = \Omega_{K[\mathfrak{Y}_K]/K}/\mathrm{d}K[\mathfrak{Y}_K]$ of $\mathrm{d}x/y^l, x\,\mathrm{d}x/y^l, \ldots, x^r\,\mathrm{d}x/y^l$, respectively, are linearly independent.

Proof. For the convenience of the reader we draw the diagram

$$\begin{array}{ccc} \mathfrak{Y}_K & \hookrightarrow & \dddot{\mathfrak{Y}}_K \\ \text{étale} \downarrow & & \downarrow G \cong \mathbb{Z}/n\mathbb{Z} \\ \mathbb{A}'_K & \hookrightarrow & \mathbb{A}^1_K \end{array}$$

Now assume that $\overline{\mathrm{d}x/y^l}, \overline{x\,\mathrm{d}x/y^l}, \ldots, \overline{x^r\,\mathrm{d}x/y^l}$ are linearly dependent. Taking into account the G-action we find an element

$$\textbf{1.4.25.} \quad R(x) \in K[\mathbb{A}'_K] = K\left[x, \prod_{-1}^{r} (x - t_i)^{-1}\right], \qquad R(x) \neq 0,$$

such that

$$\mathrm{d}\left(\frac{R(x)}{y^l}\right) = \sum_{0}^{r} \alpha_i \frac{x^i\,\mathrm{d}x}{y^l} \quad \text{in } \Omega_{K[\mathfrak{Y}_K]/K}, \qquad \alpha_i \in K.$$

The quotient rule for d yields

$$\frac{\mathrm{d}R(x)}{R(x)} - l\,\frac{\mathrm{d}y}{y} = \frac{1}{R(x)} \sum_{0}^{r} \alpha_i x^i\,\mathrm{d}x,$$

and by (1.3.9) we get the identity

$$\textbf{1.4.26.} \quad \frac{\mathrm{d}R(x)}{R(x)} - \frac{l}{n} \sum_{-1}^{r} \frac{b_i}{x - t_i}\,\mathrm{d}x = \left(\sum_{0}^{r} \alpha_i x^i\,\mathrm{d}x\right) \Big/ R(x).$$

1.4.27. $R(x) \in K(\mathbb{A}'_K) = K(x)$ has zeros in t_i for $i = -1, 0, \ldots, r$.

For the proof we calculate residues of differential forms in $\Omega_{K(x)/K}$ in the points $P_i \in \mathbb{P}^1_K$ for $i = -1, 0, \ldots, r, \infty$ with local parameters $x - t_i$ or $1/x$, respectively (see Chevalley [8]). Let v_i denote the valuation with respect to the corresponding parameter. We find

1.4.28. $$\operatorname{res}_{P_i} \frac{\mathrm{d}R(x)}{R(x)} = v_i(R(x)) \in \mathbb{Z}, \qquad i = -1, 0, \ldots, r, \infty,$$

and

$$\operatorname{res}_{P_i} \left(\frac{l}{n} \sum_{-1}^{r} \frac{b_i}{x - t_i} \mathrm{d}x\right) = \operatorname{res}_{P_i} \frac{lb_i}{n} \mathrm{d}(x - t_i) = \frac{lb_i}{n} \notin \mathbb{Z}, \quad i = -1, 0, \ldots, r.$$

Therefore

$$\operatorname{res}_{P_i} \left(\frac{\mathrm{d}R(x)}{R(x)} - \frac{l}{n} \sum_{-1}^{r} \frac{b_i}{x - t_i} \mathrm{d}x\right) \neq 0 \qquad (\in \mathbb{Q} \setminus \mathbb{Z}).$$

Hence, by (1.4.26), $\sum_{j=0}^{r} \alpha_j x^j \,\mathrm{d}x/R(x)$ has non-vanishing residues in P_i, $i = -1, 0, \ldots, r$. But this is only possible if $\left(\sum_{j=0}^{r} \alpha_j x^j \,\mathrm{d}x\right)\Big/ R(x)$ has poles in P_i. It follows that $R(x)$ has zeros there.

1.4.29. $R(x)$ is a polynomial in x of degree $\geqq r + 2$.

Indeed, by (1.4.25), the poles of $R(x)$ are located in the point set $\{P_{-1}, P_0, \ldots, P_r, P_\infty\}$. Now 1.4.29 follows from 1.4.27.

1.4.30. $$\operatorname{res}_\infty \left(\sum_{0}^{r} \alpha_i x^i \,\mathrm{d}x/R(x)\right) = 0.$$

Indeed, the differential form in the bracket is regular in ∞ because

$$\deg R(x) \geqq r + 2 \geqq \deg \sum_{0}^{r} \alpha_i x^i = r + 1.$$

On the other hand

$$\operatorname{res}_\infty \left(\frac{l}{n} \sum_{-1}^{r} \frac{b_i}{x - t_i} \mathrm{d}x\right) = \frac{l}{n} \sum_{-1}^{r} b_i \operatorname{res}_\infty \left(\frac{\mathrm{d}(x - t_i)}{x - t_i}\right)$$
$$= \frac{l}{n} \sum_{-1}^{r} b_i v_\infty(x - t_i) = -bl/n \notin \mathbb{Z}.$$

Together with (1.4.28) we have

1.4.31. $$\operatorname{res}_\infty \left(\frac{\mathrm{d}R(x)}{R(x)} - \frac{l}{n} \sum_{-1}^{r} \frac{b_i}{x - t_i} \mathrm{d}x\right) \neq 0 \qquad (\in \mathbb{Q} \setminus \mathbb{Z}).$$

The identities (1.4.26), (1.4.30) and (1.4.31) contradict each other. This proves Lemma 1.4.24.

Proof of Theorem 1.4.23. Diagram 1.3.a is compatible with localisations. Therefore

$$H^1_{DR}(\mathfrak{Y}/S) \otimes_{\mathbb{C}[S]} \mathbb{C}(t) = H^1_{DR}(\mathfrak{Y}/S)_{\mathbb{C}(t)} = H^1_{DR}(\mathfrak{Y}_{\mathbb{C}(t)}/\mathbb{C}(t)).$$

If $\overline{dx/y^l}, \overline{x\,dx/y^l}, \ldots, \overline{x^r\,dx/y^l}$ were $\mathbb{C}[S]$-linearly dependent in $H^1_{DR}(\mathfrak{Y}/S)$, then they would be $\mathbb{C}(t)$-linearly dependent in $H^1_{DR}(\mathfrak{Y}_{\mathbb{C}(t)}/\mathbb{C}(t))$. But this cannot be true by Lemma 1.4.24. So

$$\mathfrak{F}_l = \mathbb{C}[S] \cdot \overline{dx/y^l} \oplus \mathbb{C}[S] \cdot \overline{x\,dx/y^l} \oplus \cdots \oplus \mathbb{C}[S] \cdot \overline{x^r\,dx/y^l}$$

is a free $\mathbb{C}[S]$-submodule of $H^1_{DR}(\mathfrak{Y}/S)_l$ of rank $r + 1$. For $l' \gg 0$, $l' \equiv l \bmod n$, we know already that

$$\mathfrak{F}_l \subseteqq \overline{\mathbb{C}[S]\,[x]\,dx/y^l} \subseteqq H^1_{DR}(\mathfrak{Y}/S)_l = \mathfrak{F}_{l'} = \overline{\mathbb{C}[S]\,[x]\,dx/y^{l'}}$$

by Corollary 1.4.22. The localisations in the closed points s of S yield inclusions of $\mathbb{C}$-vector spaces $\mathfrak{F}_l(s) \subseteqq \mathfrak{F}_{l'}(s)$ because there are no non-trivial linear relations among our $r + 1$ generators of $\mathfrak{F}_l(s)$ in $H^1_{DR}(\mathfrak{Y}_{\mathbb{C}(s)}\mathbb{C}(s)) = \mathfrak{F}_{l'}(s)$ by Lemma 1.4.24. Both $\mathbb{C}$-vector spaces have dimension $r + 1$. So the inclusions are the identity maps. By the Nakayama Lemma (see [51], III, § 2)* we have the global identity $\mathfrak{F}_l = \mathfrak{F}_{l'}$. This proves (ii) of Theorem 1.4.23.

The local-global principle works also for

$$\varphi_l \colon \mathscr{E}_l \to H^1_{DR}(\mathfrak{Y}/S)_l.$$

The map φ_l is surjective because

$$\begin{aligned} H^1_{DR}(\mathfrak{Y}/S)_l &= \mathbb{C}[S]\,\overline{dx/y^l} + \cdots + \mathbb{C}[S]\,\overline{x^r\,dx/y^l} \\ &\subseteqq \mathbb{C}[S, D]\,\overline{\mathbb{C}[S]\,[x]\,dx/y^l} = \operatorname{Im} \varphi_l \end{aligned}$$

by 1.4.23(ii) and 1.4.17. So $\varphi_l(s)$ is a surjective linear map of $\mathbb{C}$-vector spaces of dimension $r + 1$. Therefore $\varphi_l(s)$ is bijective for all $s \in \operatorname{Spec}_{\max} \mathbb{C}[S]$. It follows that φ_l is bijective and that $\mathscr{E}_l$ is a free $\mathbb{C}[S]$-module of rank $r + 1$. This proves Theorem 1.4.23.

1.4.32. DEFINITION. For a given cycloelliptic curve family of type $(n, \mathfrak{b})$ we say that the natural number l' is *primitive* if and only if $n \nmid l'b$ and $n \nmid l'b_i$ for $i = -1, 0, \ldots, r$. We also say that $\overline{dx/y^{l'}}$ *is primitive*, if l' is primitive.

Obviously, for $l' \equiv l'' \bmod n$ it holds that l' is primitive if and only if l'' is. Now we can state the

1.4.33. AFFINE STRUCTURE THEOREM. *For any affine cycloelliptic curve family*

$$\ddot{\mathfrak{Y}}/S \colon y^n = \prod_{i=-1}^{r} (x - t_i)^{b_i}$$

there exists a direct sum decomposition of $\mathbb{C}[S, \mathrm{D}]$*-modules*

* See footnote on p. 145.

1.4.34. $H^1_{\mathrm{DR}}(\mathfrak{Y}/S) = H^1_{\mathrm{DR}}(\mathfrak{Y}/S)_{\text{prim.}} \oplus H^1_{\mathrm{DR}}(\mathfrak{Y}/S)_{\text{deg.}}$,

$$H^1_{\mathrm{DR}}(\mathfrak{Y}/S)_{\text{prim.}} \cong \bigoplus_{\substack{0<j<n\\ \text{prim.}}} \mathscr{E}_{l_j}, \qquad H^1_{\mathrm{DR}}(\mathfrak{Y}/S)_{\text{deg.}} = \bigoplus_{\substack{0\leqq k<n\\ \text{not prim.}}} \mathscr{E}'_{l'_k},$$

$l_j \equiv j \bmod n$ arbitrary, $l'_k \equiv k \bmod n$ sufficiently large; $\mathscr{E}_{l_j} \cong H^1_{\mathrm{DR}}(\mathfrak{Y}/S)_j$ is the principal primitive Euler-Picard module of type $(0, l_j, n, \mathfrak{b})$ with free generator $\overline{\mathrm{d}x/y^{l_j}}$ and $\mathscr{E}'_{l'_k} \cong H^1_{\mathrm{DR}}(\mathfrak{Y}/S)_k$ is a principal Euler-Picard module of type $(0, l'_k, n, \mathfrak{b})$ with generator $\overline{\mathrm{d}x/y^{l_k}}$.

Proof. By (1.3.18) we can define a unique decomposition (1.4.34) by setting

1.4.35. $$H^1_{\mathrm{DR}}(\mathfrak{Y}/S)_{\text{prim.}} = \bigoplus_{\substack{0<j<n\\ \text{prim.}}} H^1_{\mathrm{DR}}(\mathfrak{Y}/S)_j$$

and

$$H^1_{\mathrm{DR}}(\mathfrak{Y}/S)_{\text{deg.}} = \bigoplus_{\substack{0\leqq k<n\\ \text{not prim.}}} H^1_{\mathrm{DR}}(\mathfrak{Y}/S)_k .$$

We call the two summands of $H^1_{\mathrm{DR}}(\mathfrak{Y}/S)$ the *primitive part* and the *degenerate part*, respectively. The module $H^1_{\mathrm{DR}}(\mathfrak{Y}/S)_{\text{prim.}}$ is a primitive Euler-Picard module by Theorem 1.4.23(i). For the degenerate part we refer to Proposition 1.4.8.

1.5. Application to Some General Cycloelliptic Families

For a general cycloelliptic curve family

$$\ddot{\mathfrak{Y}}/S\colon Y^n = (X-1)\,X(X-t_1)\cdots(X-t_r)$$

we have the decomposition

$$H^1_{\mathrm{DR}}(\mathfrak{Y}/S) \cong \Bigl(\bigoplus_{\substack{0<j<n\\ n\nmid j(r+2)}} \mathscr{E}_j\Bigr) \oplus \Bigl(\bigoplus_{\substack{k=\frac{n\cdot m}{(n,r+2)}\\ 0\leqq m<n(n,r+2)}} \mathscr{E}'_{l'_k}\Bigr), \qquad l'_k \equiv k \bmod n, \qquad l'_k \gg 0,$$

where $(n, r+2) = \text{g.c.d.}\,(n, r+2)$. In particular there are general families with completely degenerate relative de Rham cohomology group:

$$H^1_{\mathrm{DR}}(\mathfrak{Y}/S) = H^1_{\mathrm{DR}}(\mathfrak{Y}/S)_{\text{deg.}} \quad \text{iff} \quad n \mid r+2, \quad \text{e.g.} \quad \text{if } r = n-2.$$

On the other hand for *primitive general families* $\bigl((n, r+2) = 1\bigr)$ the degenerate part of $H^1_{\mathrm{DR}}(\mathfrak{Y}/S)$ consists of $H^1_{\mathrm{DR}}(\mathfrak{Y}/S)_0$ only.

1.5.1.(i) $$H^1_{\mathrm{DR}}(\mathfrak{Y}/S) \cong \Bigl(\bigoplus_{l=1}^{n-1} \mathscr{E}_l\Bigr) \oplus \mathscr{E}'_{l'}, \qquad l' \equiv 0 \bmod n, \qquad l' \gg 0$$

*for primitive general cycloelliptic curve families;
furthermore we have a direct $\mathbb{C}[S]$-decomposition*

(ii) $$H^1_{\mathrm{DR}}(\mathfrak{Y}/S)_{\text{prim}} = \bigoplus_{l=1}^{n-1}\Bigl(\bigoplus_{i=0}^{r} \mathbb{C}[S]\,\overline{x^i\,\mathrm{d}x/y^l}\Bigr);$$

for the Euler-Picard operators $\varepsilon_{ij}^{(l)} = \varepsilon_{ij}^{0,l,n,(1,\ldots,1)}$, $0 < l < n$, *we have exact sequences* of $\mathbb{C}[S, \mathrm{D}]$*-modules*

$$\begin{array}{ll} & \qquad\qquad\qquad\qquad\quad \mathscr{E}_l \\ & \qquad\qquad\qquad\qquad\quad \| \wr \\ \text{(iii)} & \mathbb{C}[S, \mathrm{D}]^{r^2} \xrightarrow{\varepsilon^{(l)}} \mathbb{C}[S, \mathrm{D}] \to H^1_{\mathrm{DR}}(\mathfrak{Y}/S)_l \to 0 \\ & \quad 1_{ij} \mapsto \varepsilon_{ij}^{(l)}, \qquad 1 \mapsto \overline{\mathrm{d}x/y^l} \end{array}$$

Proof. (ii) follows from Theorem 1.4.23, (ii) and (iii) comes from part (iii) of 1.1.4.

Especially 1.5.1 is true for generalized Picard curve families

$$\ddot{\mathfrak{Y}}/S\colon y^n = (x-1)\,x(x-t_1)\cdots(x-t_{n-1})\,.$$

For $n = 2$ we obtain the family of elliptic curves

$$\ddot{\mathfrak{Y}}/S\colon y^2 = (x-1)\,x(x-t)\,.$$

Here we find

1.5.2. $$H^1_{\mathrm{DR}}(\mathfrak{Y}/S) = H^1_{\mathrm{DR}}(\mathfrak{Y}/S)_{\mathrm{prim.}} + H^1_{\mathrm{DR}}(\mathfrak{Y}/S)_{\mathrm{deg.}} = H^1_{\mathrm{DR}}(\mathfrak{Y}/S)_1 + H^1_{\mathrm{DR}}(\mathfrak{Y}/S)_0,$$

$$H^1_{\mathrm{DR}}(\mathfrak{Y}/S)_1 = \mathbb{C}[S, \mathrm{D}]\cdot\overline{\mathrm{d}x/y} \cong \mathbb{C}\left[t, \frac{1}{(t-1)\,t}, \mathrm{D}\right] \Big/ \mathbb{C}\left[t, \frac{1}{(t-1)\,t}, \mathrm{D}\right]\cdot\varepsilon,$$

$$\varepsilon = \mathrm{D}^2 + \frac{2t-1}{(t-1)\,t}\,\mathrm{D} + \frac{1}{4(t-1)\,t}, \qquad \mathrm{D} = \partial/\partial t\,.$$

Here ε is the Picard operator of type $(k, l, n, \mathfrak{b}) = (0, 1, 2, (1, 1, 1))$. For a 1-parameter family there is no Euler-operator. The relation $\varepsilon\cdot\overline{\mathrm{d}x/y} = 0$ is a very classical identity (see e.g. Fuchs [20], Manin [43]). We know also that the $\mathbb{C}[S, \mathrm{D}]$-left-multiples η of ε are the only elements of $\mathbb{C}[S, \mathrm{D}]$ satisfying $\eta\cdot\overline{\mathrm{d}x/y} = 0$.

For Picard curves

$$\ddot{\mathfrak{Y}}/S\colon y^3 = (x-1)\,x(x-t_1)\,(x-t_2)$$

one finds:

$$\begin{aligned} H^1_{\mathrm{DR}}(\mathfrak{Y}/S) &= H^1_{\mathrm{DR}}(\mathfrak{Y}/S)_{\mathrm{prim.}} \oplus H^1_{\mathrm{DR}}(\mathfrak{Y}/S)_{\mathrm{deg.}} \\ &= \left(H^1_{\mathrm{DR}}(\mathfrak{Y}/S)_1 \oplus H^1_{\mathrm{DR}}(\mathfrak{Y}/S)_2\right) \oplus H^1_{\mathrm{DR}}(\mathfrak{Y}/S)_0, \end{aligned}$$

$$\mathbb{C}[S] = \mathbb{C}[t_1, t_2, \left((t_1-1)\,(t_2-1)\,t_1t_2(t_2-t_1)\right)^{-1}],$$

$$\mathbb{C}[S, \mathrm{D}] = \mathbb{C}[S]\,[\mathrm{D}_1, \mathrm{D}_2], \qquad \mathrm{D}_1 = \partial/\partial t_1, \qquad \mathrm{D}_2 = \partial/\partial t_2,$$

$$H^1_{\mathrm{DR}}(\mathfrak{Y}/S)_1 = \mathbb{C}[S, \mathrm{D}]\cdot\overline{\mathrm{d}x/y} \cong \mathbb{C}[S, \mathrm{D}]/(\varepsilon_{11}^{(1)}, \varepsilon_{22}^{(1)}, \varepsilon_{12}^{(1)}],$$

$$H^1_{\mathrm{DR}}(\mathfrak{Y}/S)_2 = \mathbb{C}[S, \mathrm{D}]\,\overline{\mathrm{d}x/y^2} = \mathbb{C}[S, \mathrm{D}]/(\varepsilon_{11}^{(2)}, \varepsilon_{22}^{(2)}, \varepsilon_{12}^{(2)}],$$

where (, ,] denotes the left ideal in $\mathbb{C}[S, \mathrm{D}]$ generated by the elements in the "bracket". For the Euler-Picard operators $\varepsilon_{ij}^{(l)} = \varepsilon_{ij}^{0,l,3,(1,1,1,1)}$ one finds the explicit expressions

$$\varepsilon_{11}^{(1)} = \mathrm{D}_1^2 + \frac{1}{9(t_1 - 1)\, t_1(t_2 - t_1)} \times [3(-5t_1^2 + 4t_1t_2 + 3t_1 - 2t_2)\, \mathrm{D}_1 + 3(t_2 - 1)\, t_2\mathrm{D}_2 + (t_2 - t_1)],$$

$$\varepsilon_{22}^{(1)} = \mathrm{D}_2^2 + \frac{1}{9(t_2 - 1)\, t_2(t_1 - t_2)} \times [3(t_1 - 1)\, t_1\mathrm{D}_1 + 3(-5t_2^2 + 4t_1t_2 + 3t_2 - 2t_1)\, \mathrm{D}_2 + (t_1 - t_2)],$$

$$\varepsilon_{12}^{(1)} = \mathrm{D}_1\mathrm{D}_2 + \frac{1}{3(t_2 - t_1)} (\mathrm{D}_1 - \mathrm{D}_2),$$

$$\varepsilon_{11}^{(2)} = \mathrm{D}_1^2 + \frac{2}{9(t_1 - 1)\, t_1(t_2 - t_1)} \times [3(-5t_1^2 + 4t_1t_2 + 3t_1 - 2t_2)\, \mathrm{D}_1 + 3(t_2 - t_1)\, t_2\mathrm{D}_2 + 5(t_2 - t_1)],$$

$$\varepsilon_{22}^{(2)} = \mathrm{D}_2^2 + \frac{2}{9(t_2 - 1)\, t_2(t_1 - t_2)} \times [3(t_1 - 1)\, t_1\mathrm{D}_1 + 3(-5t_2^2 + 4t_1t_2 + 3t_2 - 2t_1)\, \mathrm{D}_2 + 5(t_1 - t_2)]$$

$$\varepsilon_{12}^{(2)} = \mathrm{D}_1\mathrm{D}_2 + \frac{2}{3(t_2 - t_1)} (\mathrm{D}_1 - \mathrm{D}_2).$$

The first time Euler-Picard operators for 2-parameter families appeared explicitly was in Picard's papers [54], [55]. In particular for the Picard curve family the system $\varepsilon_{11}^{(1)}$, $\varepsilon_{22}^{(1)}$, $\varepsilon_{12}^{(1)}$ can be found in [55]. But there are some sign errors.

Keeping the *Siegel modular threefold* in mind we consider the *general family of hyperelliptic curves of genus* 2

$$\ddot{\mathfrak{Y}}/S\colon y^2 = (x - 1)\, x(x - t_1)\, (x - t_2)\, (x - t_3).$$

Here we find

$$H_{\mathrm{DR}}^1(\mathfrak{Y}/S) = H_{\mathrm{DR}}^1(\mathfrak{Y}/S)_{\mathrm{prim.}} \oplus H_{\mathrm{DR}}^1(\mathfrak{Y}/S)_{\mathrm{deg.}} = H_{\mathrm{DR}}^1(\mathfrak{Y}/S)_1 \oplus H_{\mathrm{DR}}^1(\mathfrak{Y}/S)_0,$$

$$\mathbb{C}[S, \mathrm{D}] = \mathbb{C}[t_1, t_2, t_3, 1/(t_1 - 1)\, (t_2 - 1)\, (t_3 - 1)\, t_1t_2t_3(t_2 - t_1)\, (t_3 - t_1) \times (t_3 - t_2)]\, [\mathrm{D}_1, \mathrm{D}_2, \mathrm{D}_3],$$

$$H_{\mathrm{DR}}^1(\mathfrak{Y}/S)_1 = \mathbb{C}[S, \mathrm{D}] \cdot \overline{\mathrm{d}x/y} = \mathbb{C}\, [S, \mathrm{D}]/(\varepsilon_{11}, \varepsilon_{22}, \varepsilon_{33}, \varepsilon_{12}, \varepsilon_{13}, \varepsilon_{23}].$$

The Euler-Picard system is of type $(k, l, n, \mathfrak{b}) = \big(0, 1, 2, (1, 1, 1, 1, 1)\big)$. Explicitly it can be written as

$$\varepsilon_{11} = \mathrm{D}_1^2 + \frac{1}{2(t_1 - 1)\, t_1} \left[\left(6t_1 - 4 - \frac{(t_1 - 1)\, t_2}{t_2 - t_1} - \frac{(t_1 - 1)\, t_3}{t_3 - t_1}\right) \mathrm{D}_1 + \frac{(t_2 - 1)\, t_2}{t_2 - t_1}\, \mathrm{D}_2 + \frac{(t_3 - 1)\, t_3}{t_3 - t_1}\, \mathrm{D}_3 + \frac{3}{2}\right],$$

$$\varepsilon_{22} = \mathrm{D}_2^2 + \frac{1}{2(t_2-1)\,t_2}\left[\frac{(t_1-1)\,t_1}{t_1-t_2} + \left(6t_2 - 4 - \frac{(t_2-1)\,t_1}{t_1-t_2}\right.\right.$$
$$\left.\left. - \frac{(t_2-1)\,t_3}{t_3-t_2}\right)\mathrm{D}_2 + \frac{(t_3-1)\,t_3}{t_3-t_2}\,\mathrm{D}_3 + \frac{3}{2}\right],$$

$$\varepsilon_{33} = \mathrm{D}_3^2 + \frac{1}{2(t_3-1)\,t_3}\left[\frac{(t_1-1)\,t_1}{t_1-t_3}\,\mathrm{D}_1 + \frac{(t_2-1)\,t_2}{t_2-t_3}\,\mathrm{D}_2\right.$$
$$\left. + \left(6t_3 - 4 - \frac{(t_3-1)\,t_1}{t_1-t_3} - \frac{(t_3-1)\,t_2}{t_3-t_2}\right)\mathrm{D}_3 + \frac{3}{2}\right],$$

$$\varepsilon_{12} = \mathrm{D}_1\mathrm{D}_2 + \frac{1}{2(t_2-t_1)}\,(\mathrm{D}_1 - \mathrm{D}_2),$$

$$\varepsilon_{13} = \mathrm{D}_1\mathrm{D}_3 + \frac{1}{2(t_3-t_1)}\,(\mathrm{D}_1 - \mathrm{D}_3),$$

$$\varepsilon_{23} = \mathrm{D}_2\mathrm{D}_3 + \frac{1}{2(t_3-t_2)}\,(\mathrm{D}_2 - \mathrm{D}_3).$$

§ 2. EULER-PICARD CONNECTIONS

2.1. Projective Cycloelliptic Curve Families

As in the previous section we set

$$S = \operatorname{Spec} \mathbb{C}[S] \subset \mathbb{A}^r = \operatorname{Spec} \mathbb{C}[t], \qquad \mathbb{C}[t] = \mathbb{C}[t_1, t_2, \ldots, t_r],$$
$$\mathbb{C}[S] = \mathbb{C}[t_1, \ldots, t_r, \prod_{-1 \leqq i < j \leqq r} (t_j - t_i)^{-1}], \qquad t_{-1} = 1, \qquad t_0 = 0.$$

2.1.1. DEFINITION. A *projective cycloelliptic curve family* $\overline{\mathfrak{Y}}/S$ is a family of equations of the type

2.1.2. $$\overline{\mathfrak{Y}}/S\colon\; W^{b-n}Y^n = \prod_{i=-1}^{r} (X - t_i W)^{b_i},$$

$$0 < b_i \text{ natural numbers}, \qquad n \geqq 2, \qquad b = \sum_{i=-1} b_i.$$

(If $b - n < 0$, then the W-power is to be interpreted as W^{n-b} written on the right-hand side of equation (2.1.2).) More precisely, $\overline{\mathfrak{Y}}$ is the divisor on $S \times \mathbb{P}^2 \subset \mathbb{A}^r \times \mathbb{P}^2$, $\mathbb{P}^2 = \operatorname{Proj} \mathbb{C}[W, X, Y]$, described by equation (2.1.2), and $\overline{\mathfrak{Y}}/S$ is the morphism $\bar{\pi}\colon \overline{\mathfrak{Y}} \to S$ defined by the following commutative diagram

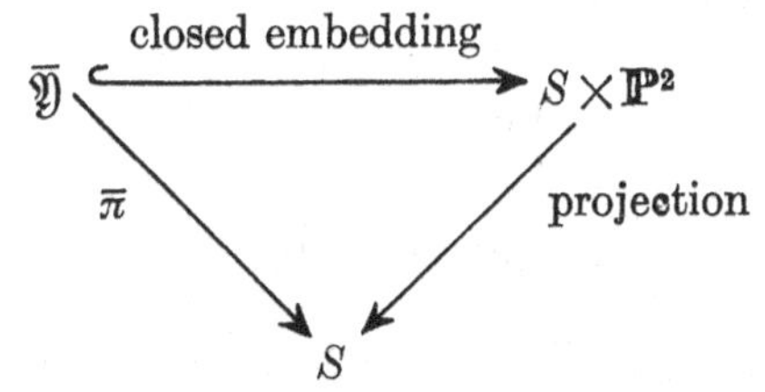

With the notations of § 1 we have a commutative Diagram 2.1.a with cartesian

2.1.a.

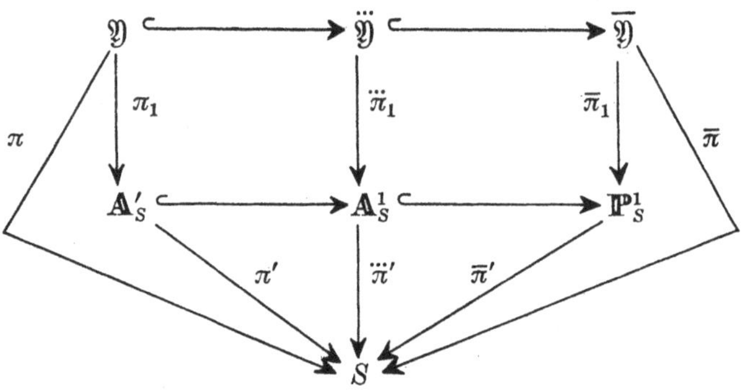

squares, horizontal open embeddings, and vertical cyclic (finite) coverings $\pi_1, \dddot{\pi}_1, \bar{\pi}_1$ with Galois group $G \simeq \mathbb{Z}/n\mathbb{Z}$; $\pi_1, \dddot{\pi}_1, \pi, \dddot{\pi} = \dddot{\pi}' \circ \dddot{\pi}_1$ affine; $\bar{\pi}, \bar{\pi}'$ projective, hence proper; π_1 étale; $\pi, \pi', \pi_1, \dddot{\pi}', \bar{\pi}'$ smooth.

We are interested in the singular locus $\operatorname{Sg} \overline{\mathfrak{Y}}$ of $\overline{\mathfrak{Y}}$. For this purpose we divide the set of all projective cycloelliptic curve families into three classes:

$$\text{type A)}: \ b > n, \qquad \text{type B)}: \ b = n, \qquad \text{type C)}: \ b < n.$$

Clearly, $\operatorname{Sg} \overline{\mathfrak{Y}}$ lies on $\overline{\mathfrak{Y}} \setminus \mathfrak{Y}$. We set

$$S_\infty = \overline{\mathfrak{Y}} \setminus \dddot{\mathfrak{Y}} \quad \text{and} \quad S_f = \dddot{\mathfrak{Y}} \setminus \mathfrak{Y}.$$

S_∞ is characterized on $\overline{\mathfrak{Y}}$ by the equation $W = 0$. For the three different types of families we find that S_∞ is the divisor

$$S_\infty = \begin{cases} S \times (0:0:1), & \text{type A)}, \\ \sum_{i=0}^{n-1} S_\infty^{(i)} \text{ (disjoint sum)}, & \text{type B)}, \\ S \times (0:1:0), & \text{type C)}, \end{cases}$$

where $S_\infty^{(i)}$ is the image of the following section of $\bar{\pi}$

$$\sigma_\infty^{(i)}: S \to \overline{\mathfrak{Y}}, \qquad s \mapsto s \times (0:1:\zeta^i), \qquad \zeta = e^{2\pi i/n}.$$

S_f is described by $W \neq 0$, $Y = 0$ because $\mathfrak{Y} = \operatorname{Spec} \mathbb{C}[S][x, y, y^{-1}]$ (see 1.2). The divisor S_f is the disjoint sum of sections

$$S_f = S_{-1} + S_0 + \sum_{i=1}^{r} S_i, \qquad S_i = \operatorname{Im} \sigma_i,$$

$$\sigma_{-1}: s \mapsto s \times (1:1:0), \qquad \sigma_0: s \mapsto s \times (1:0:0),$$

$$\sigma_i: s = (s_1, \ldots, s_r) \mapsto s \times (1:s_i:0), \qquad i = 1, \ldots, r.$$

Altogether we have

2.1.3. $\operatorname{Sg} \overline{\mathfrak{Y}} \subseteq \overline{\mathfrak{Y}} \setminus \mathfrak{Y} = S_\infty + \sum_{i=-1}^{r} S_i$ (disjoint sum of divisors),

$S_\infty = \sum_{i=0}^{n-1} S_\infty^{(i)}$ in case B) and S_∞ irreducible in the cases A) and C).

2.1.4. LEMMA. *Assume that $P \in S_i \subset S_f$, $i \in \{-1, 0, \ldots, r\}$, and let $\overline{\mathfrak{Y}}_P$ be the fibre of $\bar{\pi}$ through P. Then the following five conditions are equivalent:*

(i) *P is a singular point of $\overline{\mathfrak{Y}}$;*

(ii) *P is a singular point of $\overline{\mathfrak{Y}}_P$;*

(iii) *Q is a singular point of $\overline{\mathfrak{Y}}$ for all $Q \in S_i$;*

(iv) *Q is a singular point of $\overline{\mathfrak{Y}}_Q$ for all $Q \in S_i$;*

(v) $b_i > 1$.

Two singularities $(\overline{\mathfrak{Y}}_P, P)$ and $(\overline{\mathfrak{Y}}_Q, Q)$ are isomorphic for $P, Q \in S_i$.

Proof. We set

$$F = Y^n - \prod_{i=-1}^{r} (X - t_i)^{b_i}$$

and look for common zeros of F, $\frac{\partial F}{\partial X}$, $\frac{\partial F}{\partial Y}$, $\frac{\partial F}{\partial t_i}$, $i = 1, \ldots, r$. These are the singularities of $\overline{\mathfrak{Y}}$. We have

$$\frac{\partial F}{\partial X} = -\prod_{-1}^{r} (X - t_i)^{b_i} \left(\sum_{-1}^{r} \frac{b_i}{X - t_i}\right), \qquad \frac{\partial F}{\partial Y} = nY^{n-1},$$

$$\frac{\partial F}{\partial t_i} = \left(\prod_{-1}^{r} (X - t_i)^{b_i}\right) \cdot \frac{b_i}{X - t_i}, \qquad i = 1, \ldots, r.$$

The point $P \in S_i$ is a singular point of $\overline{\mathfrak{Y}}$ if and only if $b_i > 1$. This condition is equivalent to $F(P) = \frac{\partial F}{\partial X}(P) = \frac{\partial F}{\partial Y}(P) = 0$, that means that P is a singular point of the fibre $\overline{\mathfrak{Y}}_P$. The type of the (embedded) singularity is described by $Y^n = U^{b_i} \cdot \varepsilon$, ε a unit in the local ring of the point $(Y, U) = (0, 0)$.

2.1.5. COROLLARY. *$\overline{\mathfrak{Y}}$ is smooth if and only if $\overline{\mathfrak{Y}}/S$ is a general cycloelliptic curve family* (see Definition 1.2.3).

It is not difficult to investigate in the same manner the situation along (the components of) S_∞. In this way one can prove

2.1.6. LEMMA. *For a cycloelliptic curve family $\overline{\mathfrak{Y}}/S$ the following conditions are equivalent:*

(i) *$\overline{\mathfrak{Y}}$ is smooth;*

(ii) *the fibres $\overline{\mathfrak{Y}}_s = \bar{\pi}^{-1}(s)$ are smooth for each $s \in S$;*

(iii) $\overline{\mathfrak{Y}}/S$ *is a general cycloelliptic family of one of the following types:*

type A): $b = n + 1$ *(generalized Picard curves)*,

type B): *all general families,*

type C): $b = n - 1$.

If these conditions are satisfied, then $\bar{\pi}: \overline{\mathfrak{Y}} \to S$ *is a smooth morphism.*

For the proof we need remark only that one can obtain equations for the embedded singularities along S_∞ in the following manner: For type A) we set $Y = 1$ in (2.1.2), and for the types B), C) we set $X = 1$ in (2.1.2). The study of these equations and of their derivatives is left to the reader as a simple exercise. Now let $\overline{\mathfrak{Y}}$ and S be two smooth algebraic varieties over $\mathbb{C}$ and $\bar{\pi}: \overline{\mathfrak{Y}} \to S$ an arbitrary morphism. It is well-known that $\bar{\pi}$ is smooth if and only if for each closed point $P \in \overline{\mathfrak{Y}}$ the induced map $T_{\bar{\pi}}(P): T_P \to T_s$, $s = \bar{\pi}(P)$, of Zariski tangent spaces is surjective (see e.g. [27], III, § 10, Prop. 10.4, (iii)). The morphism $\pi: \mathfrak{Y} \to S$ is smooth for each cycloelliptic curve family (see 2.1.a). Since $\overline{\mathfrak{Y}} \setminus \mathfrak{Y}$ is a sum of sections over S, $T_{\bar{\pi}}(P)$ is surjective for each $P \in \overline{\mathfrak{Y}} \setminus \mathfrak{Y}$, if $\mathfrak{Y}/S$ is one of the families satisfying conditions (i), (ii), (iii).

Now we return to arbitrary cycloelliptic families. Let $P \in \overline{\mathfrak{Y}}_s \subset \mathbb{P}^2$ be an (embedded) singularity of the fibre $\overline{\mathfrak{Y}}_s$. Then one can remove the curve singularity by means of σ-processes over $P \in \mathbb{P}^2$. This procedure works simultaneously for all $Q \in \overline{\mathfrak{Y}}$ lying in the same component S_i of $\overline{\mathfrak{Y}} \setminus \mathfrak{Y}$ as P, $i = -1, 0, \ldots, r, \infty$. (In case B) $\overline{\mathfrak{Y}}, \overline{\mathfrak{Y}}_s$ is already smooth along $S_\infty = \sum_{0}^{n-1} S_\infty^{(i)}$.) After performing these resolutions along the singular sections S_i we obtain a family $\tilde{\mathfrak{Y}}/S$ with good properties as in (i), (ii) of Lemma 2.1.6. We have a commutative diagram

$$\begin{array}{ccc} \tilde{\mathfrak{Y}} & \xrightarrow{\text{sing. resol.}} & \overline{\mathfrak{Y}} \\ & {\scriptstyle\tilde{\pi}}\searrow \quad \swarrow{\scriptstyle\bar{\pi}} & \\ & S & \end{array}$$

2.1.7. PROPOSITION-DEFINITION.

a) $\tilde{\mathfrak{Y}}/S$ *is a smooth family,* i.e. $\tilde{\pi}: \tilde{\mathfrak{Y}} \to S$ *is smooth;*

b) *in particular* $\tilde{\mathfrak{Y}}_s = \tilde{\pi}^{-1}(s)$ *is smooth for each* $s \in S$;

c) $\tilde{\mathfrak{Y}} \setminus \mathfrak{Y}$ *is a disjoint sum of sections over* S.

$\tilde{\mathfrak{Y}}/S$ is called the *projective cycloelliptic smooth curve family of type* $(n, \mathfrak{b})$, if $\overline{\mathfrak{Y}}/S$ is of equation type (2.1.2).

2.2. Simple Quadratic Connections

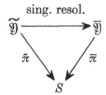

Let U be an open analytic subset of $\mathbb{C}^r$. We restrict our attention to Stein subsets U, especially to small polycylindres and to Zariski-open affine subsets of $S(\mathbb{C})$. The symbol $\mathcal{D}(U)$ denotes the *ring of (analytic) differential operators*

over U,

$$\mathcal{D}(U) = \left\{ \sum_{i_1,\ldots,i_r} f_{i_1\ldots i_r} \cdot \mathrm{D}_1^{i_1} \cdots \mathrm{D}_r^{i_r};\ f_{i_1\ldots i_r} \in \mathcal{O}(U) = \Gamma(U, \mathcal{O}_U^{\mathrm{an}}) \right\},$$

where $\mathcal{O}_U^{\mathrm{an}}$ is the sheaf of germs of holomorphic functions on U and $\mathrm{D}_k = \partial/\partial t_k$ is the kth partial derivative. Taking limits over open neighbourhoods of points we obtain a non-commutative sheaf of rings $\mathcal{D}_U$ over U. The structure sheaf $\mathcal{O}_U$ is a sheaf of $\mathcal{D}_U$-modules. Both $\mathcal{O}_U$ and $\mathcal{D}_U$ are coherent sheaves of rings (see Pham [53], 6.3). A *system of (partial) differential equations* over U is a homomorphism of sheaves of $\mathcal{D}_U$-modules

$$\delta\colon \mathcal{D}_U^p \to \mathcal{D}_U^q.$$

A homomorphism of sheaves of $\mathcal{D}_U$-modules $\varphi\colon \mathcal{D}_U^q \to \mathcal{F}$ is called a *solution* of δ (in $\mathcal{F}$), if

2.2.1. $$\mathcal{D}_U^p \xrightarrow{\delta} \mathcal{D}_U^q \xrightarrow{\varphi} \mathcal{F}$$

is a complex; i.e. $\varphi \circ \delta = 0$. The homomorphism φ is called a *general solution* if and only if (2.2.1) is an exact sequence.

We translate these notions in classical language. Let $e_1, \ldots, e_p$ be the canonical basis of $\mathcal{D}(U)^p$. Then $\delta_i = \delta(e_i) \in \mathcal{D}(U)^q$, $\delta_i = \sum_{j=1}^{q} \delta_{ij} e_j$. So δ can be identified with a system of p elements of $\mathcal{D}(U)^q$ or with a matrix (δ_{ij}) of differential operators over U. If we denote the φ-images of e_i by φ_i, $i = 1, \ldots, q$, then φ is a solution of δ if and only if

$$\begin{pmatrix} \delta_{11} & \cdots & \delta_{1q} \\ \vdots & & \vdots \\ \delta_{p1} & \cdots & \delta_{pq} \end{pmatrix} \begin{pmatrix} \varphi_1 \\ \vdots \\ \varphi_q \end{pmatrix} = \begin{pmatrix} 0 \\ \vdots \\ 0 \end{pmatrix}.$$

φ is a general solution if and only if it is a solution and

2.2.2. $$(\delta_1, \ldots, \delta_q) \begin{pmatrix} \varphi_1 \\ \vdots \\ \varphi_q \end{pmatrix} = 0 \Rightarrow (\delta_1, \ldots, \delta_q) \in (\delta]\,(V)$$

for each system of differential operators $\delta_i \in \mathcal{D}(V)$, V an open Stein submanifold of U; here $(\delta] = (\delta_1, \ldots, \delta_p]$ denotes the sheaf of left $\mathcal{D}_V$-submodules of $\mathcal{D}_V^q$ generated by $\delta_1, \ldots, \delta_p$.

2.2.3. EXAMPLE. The analytic de Rham cohomology groups

$$\mathcal{H}_{\mathrm{DR}}^{1,\mathrm{an}}(\mathfrak{Y}/S) = \mathcal{H}_{\mathrm{DR}}^{1}(\mathfrak{Y}/S) \otimes_{\mathcal{O}_S} \mathcal{O}_S^{\mathrm{an}}$$

are sheaves of $\mathcal{D}_S$-modules. The homomorphism $\mathcal{D}_S \to \mathcal{H}_{\mathrm{DR}}^{1,\mathrm{an}}(\mathfrak{Y}/S)$, $1 \to \overline{\mathrm{d}x/y^l}$, is a general solution of an Euler-Picard system of differential operators, if $\overline{\mathrm{d}x/y^l}$ is primitive (see 1.5.1).

We are interested in *systems of holomorphic solutions*; i.e. in solutions in

$\mathcal{F} = (\mathcal{O}_U^{\text{an}})^{m+1}$. We will reduce the study of simple quadratic systems of differential equations (see 1.1) to the study of simple first order systems.

We consider square matrices of holomorphic functions on U or of first order differential operators, respectively:

2.2.4. $\mathsf{G}_k = (g_{\beta k}^{\alpha})_{\alpha,\beta=0,\ldots,m}, \qquad \Lambda_k = \mathrm{D}_k I_{m+1} + \mathsf{G}_k, \qquad k = 1, \ldots, r.$

Here I_{m+1} denotes the unit matrix. Assign to the elements $1_{\alpha k}$ of the canonical basis of $\mathcal{D}(U)^{r\cdot(m+1)}$ the αth row of the matrix Λ_k:

2.2.5. $\delta\colon 1_{\alpha k} \mapsto (g_{0k}^{\alpha}, \ldots, \mathrm{D}_k + g_{\alpha k}^{\alpha}, \ldots, g_{mk}^{\alpha}).$

2.2.6. DEFINITION. The $\mathcal{D}_U$-homomorphism

$$\delta\colon \mathcal{D}_U^{r(m+1)} \to \mathcal{D}_U^{m+1}$$

defined by (2.2.5) is called a *simple first order system of differential operators* (or *equations*) over U.

The definition becomes clear if one considers solutions of δ in $\mathcal{O}_U$. These are $(m+1)$-tuples $\mathfrak{F} = {}^t(F^0, F^1, \ldots, F^m)$ of holomorphic functions on U such that

2.2.7. $\mathrm{D}_k F^{\alpha} = \dfrac{\partial}{\partial t_k} F^{\alpha} = -\sum\limits_{\beta=0}^{m} g_{\beta k}^{\alpha} \cdot F^{\beta}, \qquad k = 1, \ldots, r, \qquad \alpha = 0, \ldots, m.$

A solution in $\mathcal{O}_U^{m+1}$ consists of $m+1$ solutions in $\mathcal{O}_U$. These solutions can be seen as columns in a square matrix of holomorphic functions

2.2.8. $\mathsf{F} = \begin{pmatrix} F_0^0 & \cdots & F_l^0 & \cdots & F_m^0 \\ \vdots & & \vdots & & \vdots \\ F_0^m & \cdots & F_l^m & \cdots & F_m^m \end{pmatrix} = \begin{pmatrix} \vdots & & \vdots & & \vdots \\ \mathfrak{F}_0 & \cdots & \mathfrak{F}_l & \cdots & \mathfrak{F}_m \\ \vdots & & \vdots & & \vdots \end{pmatrix}$

satisfying $\Lambda_k \cdot \mathsf{F} = 0$ for $k = 1, \ldots, r$.

2.2.9. DEFINITION. The simple first order system δ is called *integrable,* if

2.2.10. $-\mathrm{D}_l(\mathsf{G}_k) + \mathsf{G}_k \cdot \mathsf{G}_l = -\mathrm{D}_k(\mathsf{G}_l) + \mathsf{G}_l \cdot \mathsf{G}_k \qquad \text{for} \qquad k, l = 1, \ldots, r,$

where $\mathrm{D}_l(\mathsf{G}_k) = \big(\mathrm{D}_l(g_{\beta k}^{\alpha})\big)_{\alpha,\beta=0,\ldots,m}$. The system δ is called *primitive* (over U), if the $\mathcal{D}(U)$-left submodule $(\delta]\,(U)$ of $\mathcal{D}(U)^{m+1}$ generated by all the rows $\delta(1_{\alpha k})$ of the matrices Λ_k does not contain a non-trivial element of $(a_0, a_1, \ldots, a_m)$, $a_i \in \mathcal{O}(U)$ of pure order 0.

2.2.11. PROPOSITION. *Let $U \subset \mathbb{C}^r$ be a polycylindre and δ a simple first order system of differential equations defined by the system of matrices G_k in* (2.2.4). *Then the following conditions* (i), (ii), (iii) *are equivalent and* (iv) *is a consequence of them.*

(i) *δ is integrable;*

(ii) *δ is primitive;*

(iii) *For each $P \in U$ and each $(m+1)$-tuple of complex numbers $(c^0, c^1, \ldots, c^m)$*

there exists one and only one solution ${}^t(F^0, \ldots, F^m)$ *of* δ *in* $\mathcal{O}_U$ *satisfying the initial condition* $F^\beta(P) = c^\beta$, $\beta = 0, \ldots, r$.

(iv) *δ has a general solution* F *in* $\mathcal{O}_U^{m+1}$ *with linearly independent* $\mathfrak{F}_0, \mathfrak{F}_1, \ldots, \mathfrak{F}_m$.

Proof. The central point is the implication (i) $\Rightarrow$ (iii). For a simple proof we refer the reader to Pham's monograph [53], 7.3. For the other implications we need some of the following characterizations of a general solution F of δ in $\mathcal{O}_U^{m+1}$.

2.2.12. LEMMA. *With the notations of* (2.2.8) *and* 2.2.11 *let* F *be a solution in* $\mathcal{O}_U^{m+1}$ *of the integrale system* δ. *Then the following conditions are equivalent:*

(a) F *is a general solution of* δ *with linearly independent* $\mathfrak{F}_0, \ldots, \mathfrak{F}_m$;

(b) $\det \mathsf{F}(P) \neq 0$ *for some point* $P \in U$;

(c) $\det \mathsf{F}(Q) \neq 0$ *for each* $Q \in U$;

(d) $(a_0, \ldots, a_m) \cdot \mathsf{F} = 0 \Rightarrow a_0 = \cdots = a_m = 0$
for each $(m+1)$-*tuple of holomorphic functions* a_i *on* U;

(e) *each solution of* δ *in* $\mathcal{O}_U$ *is a (unique)* $\mathbb{C}$-*linear combination of the solutions* $\mathfrak{F}_0, \ldots, \mathfrak{F}_m$;

(f) *the solutions* $\mathfrak{F}_0, \ldots, \mathfrak{F}_m$ *are* $\mathbb{C}$-*linearly independent.*

Proof of the lemma. (b) $\Rightarrow$ (c): Assume that $\det \mathsf{F}(Q) = 0$. Then there is a non-trivial relation

$$c_0\mathfrak{F}_0(Q) + \cdots + c_m\mathfrak{F}_m(Q) = 0, \qquad c_i \in \mathbb{C}.$$

$\mathfrak{F}_{m+1} = c_0\mathfrak{F}_0 + \cdots + c_m\mathfrak{F}_m$ is a solution of δ in $\mathcal{O}_U$ with the initial condition $\mathfrak{F}_{m+1}(Q) = 0$. From the uniqueness assertion of (iii) in Proposition 4.2 it follows that $\mathfrak{F}_{m+1} = 0$. So in particular, the vectors $\mathfrak{F}_0(P), \ldots, \mathfrak{F}_m(P)$ are linearly dependent, hence $\det \mathsf{F}(P) = 0$. This contradicts our hypothesis (b).

(c) $\Rightarrow$ (d): If $(a_0, \ldots, a_m) \cdot \mathsf{F} = 0$, then $\big(a_0(Q), \ldots, a_m(Q)\big) \cdot \mathsf{F}(Q) = 0$ for each $Q \in U$. From (c) follows that $a_0(Q) = \cdots = a_m(Q) = 0$ for all $Q \in U$.

(d) $\Rightarrow$ (b): Assume that $\det \mathsf{F}(P) = 0$. Then there is a non-trivial linear relation $(c_0, \ldots, c_m) \cdot \mathsf{F}(P) = 0$, $c_i \in \mathbb{C}$, $\mathfrak{F}_{m+1} = \sum_{i=0}^{m} c_i\mathfrak{F}_i$ is a solution of δ, and $\mathfrak{F}_{m+1}(P) = 0$. Therefore $\mathfrak{F}_{m+1} = 0$ by (iii), hence $(c_0, \ldots, c_m) \cdot \mathsf{F} = 0$. This contradicts hypothesis (d).

(b) $\Rightarrow$ (e): Let $\mathfrak{F}_{m+1}$ be a solution of δ in $\mathcal{O}_U$. Then there must be a relation $\mathfrak{F}_{m+1}(P) = \sum_{i=0}^{m} c_i\mathfrak{F}_i(P)$, $c_i \in \mathbb{C}$. The solutions $\mathfrak{F}_{m+1}$ and $\sum c_i\mathfrak{F}_i$ have the same initial value in P. From (iii) it follows that $\mathfrak{F}_{m+1} = \sum c_i\mathfrak{F}_i$. Obviously, the c_i's are uniquely determined.

(e) $\Rightarrow$ (b): Assume that $\det \mathsf{F}(P) = 0$. By (iii) there exist $m + 1$ solutions $\mathfrak{F}'_0, \ldots, \mathfrak{F}'_m$ of δ in $\mathcal{O}_U$ such that $\det \mathsf{F}'(P) \neq 0$. So not all of the vectors $\mathfrak{F}'_i(P)$ are linear combinations of the vectors $\mathfrak{F}_i(P)$. Therefore not all of the solutions $\mathfrak{F}'_i$ are linear combinations of the solutions $\mathfrak{F}_i$. This contradicts hypothesis (e).

(d) $\Rightarrow$ (a): We check criterion (2.2.2) for δ, F. Assume that

$$(\delta_0, \dots, \delta_m) \cdot \mathsf{F} = 0, \qquad \delta_i \in \mathcal{D}(U).$$

By considering the rows (2.2.5) of the matrices $\Lambda_k = \mathrm{D}_k I_{m+1} + \mathsf{G}_k$ we can find an $(m+1)$-tuple $(a_0, \dots, a_m)$ of holomorphic functions on U such that

$$(a_0, \dots, a_m) \equiv (\delta_0, \dots, \delta_m) \bmod (\delta] \quad \text{and} \quad (a_0, \dots, a_m) \cdot \mathsf{F} = 0.$$

Then from (d) it follows immediately that $(\delta_0, \dots, \delta_m) \equiv (0, \dots, 0) \bmod (\delta]$, i.e. $(\delta_0, \dots, \delta_m) \in (\delta]$. If $\mathfrak{F}_0, \dots, \mathfrak{F}_m$ were be linearly dependent, then $\det \mathsf{F}(P) = 0$. This contradicts to (b), which is equivalent to (d).

(a) $\Rightarrow$ (b): Assume that $\det \mathsf{F}(P) = 0$. Using a non-trivial linear combination of $\mathfrak{F}_0(P), \dots, \mathfrak{F}_m(P)$ which is equal to 0 we find by means of (iii) a non-trivial linear combination $\mathfrak{F}_0, \dots, \mathfrak{F}_m$ which is the zero function. This contradicts the second part of (a).

(b) $\Rightarrow$ (f) is trivial.

(f) $\Rightarrow$ (b) is also clear (use (iii) again).

The lemma is proved.

We now continue the proof of Proposition 2.2.11.

(iii) $\Rightarrow$ (iv): Let M be a regular square matrix with $m+1$ rows and coefficients in $\mathbb{C}$, $P \in U$. By (iii) there is a solution $\mathfrak{F}_i$ of δ in $\mathcal{O}_U$ such that $\mathfrak{F}_i(P)$ is the ith column of M. The $\mathfrak{F}_i$'s are the columns of a solution F of δ in $\mathcal{O}_U^{m+1}$ satisfying condition (b) in Lemma 5.1, and (b) is equivalent with (a).

(iii), (iv) $\Rightarrow$ (ii): Assume that

$$(a_0, \dots, a_m) \in (\delta]\,(U), \qquad a_i \text{ holomorphic functions on } U,$$

and let F be a general solution of δ as in (iv) or Lemma 2.2.12(a). We know that

$$(\delta_0, \dots, \delta_m) \cdot \mathsf{F} = 0 \quad \text{for all} \quad (\delta_0, \dots, \delta_m) \in (\delta]\,(U).$$

(ii) $\Rightarrow$ (i): The αth row of G_k is denoted by $\mathfrak{G}^{\alpha}_{.k}$. The element

2.2.13. $\mathrm{D}_k \cdot \delta(1_{\alpha l}) - \mathrm{D}_l \cdot \delta(1_{\alpha l}) = \mathrm{D}_k(\mathfrak{G}^{\alpha}_{.l}) - \mathrm{D}_l(\mathfrak{G}^{\alpha}_{.k}) + \mathfrak{G}^{\alpha}_{.l} \cdot \mathrm{D}_k - \mathfrak{G}^{\alpha}_{.k} \cdot \mathrm{D}_l$

lies in $(\delta]$. Using $g^{\alpha}_{\beta l}$- or $g^{\alpha}_{\beta k}$-multiples of the rows of $\mathrm{D}_k I_{m+1} + \mathsf{G}_k$ and $\mathrm{D}_l I_{m+1} + \mathsf{G}_l$, respectively, we can eliminate the operators D_k, D_l on the right-hand side of (2.2.13) and stay within $(\delta]$. It turns out that the rows of $\mathrm{D}_k(\mathsf{G}_l) - \mathrm{D}_l(\mathsf{G}_k) + \mathsf{G}_k \cdot \mathsf{G}_l - \mathsf{G}_l \cdot \mathsf{G}_k$ lie in $(\delta]$. Since $(\delta]$ is primitive the rows have to be 0, q.e.d.

REMARK. Originally the integrability condition (2.2.10) comes from

$$\mathrm{D}_k \mathrm{D}_l F^{\alpha} = \mathrm{D}_l \mathrm{D}_k F^{\alpha}, \qquad k, l = 1, \dots, r, \qquad \alpha = 0, \dots, m,$$

where the F^{α}'s satisfy (2.2.7). After substitutions of type (2.2.7) and the formal elimination of the F^{α}'s one obtains (2.2.10).

In 1.1 we defined (primitive) simple quadratic systems of differential operators.

These definitions can be translated in an obvious manner from the algebraic to the analytic situation. So let

2.2.14. $\varepsilon\colon \mathcal{D}_U^{r^2} \to \mathcal{D}_U, \qquad \varepsilon_{k\alpha} = \varepsilon(1_{k\alpha}) = \mathrm{D}_k\mathrm{D}_\alpha + \sum_{\beta=0}^{r} g_{\beta k}^{\alpha}\mathrm{D}_\beta,$

$\mathrm{D}_0 = 1$, $\varepsilon_{k\alpha} = \varepsilon_{\alpha k}$, be a simple quadratic system of differential equations (operators) over U.

2.2.15. DEFINITION. The *linealisation* ε' of ε is the simple linear system of differential equations of first order

$$\varepsilon'\colon \mathcal{D}_U^{r(r+1)} \to \mathcal{D}_U^{r+1}$$

which is represented by the system of matrices of first order differential operators

$$\Lambda_k = \mathrm{D}_k I_{r+1} + \mathsf{G}_k, \qquad k = 1, \ldots, r,$$

$$\mathsf{G}_k = \left(\frac{0\ldots 0\ -1\ 0\ldots 0}{g_{\beta k}^{\alpha}}\right), \qquad g_{\beta k}^{0} = \begin{cases} -1 & \text{if } \beta = k, \\ 0 & \text{if } \beta \neq k. \end{cases}$$

2.2.16. *${}^t(F, F^1, \ldots, F^r)$ is a solution of ε' if and only if F is a solution of ε and $F^\alpha = \mathrm{D}_\alpha(F)$ for $\alpha = 1, \ldots, r$.*

This is immediately clear from the definition of ε'.

2.2.17. *ε is primitive if and only if ε' is primitive.*

Proof. Assume that ε is primitive and

$$(a_0, \ldots, a_r) \in (\varepsilon'], \qquad a_i \in \mathcal{O}(U),$$

this means that

$$(a_0, \ldots, a_r) = \sum_{\alpha,k} \delta_\alpha^k \cdot \Lambda_{.k}^{\alpha}, \qquad \delta_\alpha^k \in \mathcal{D}(U).$$

We multiply this identity from the right-hand side with ${}^t(\mathrm{D}_0, \mathrm{D}_1, \ldots, \mathrm{D}_r)$. Since

$$\Lambda_{.k}^{\alpha} \cdot \begin{pmatrix} \mathrm{D}_0 \\ \vdots \\ \mathrm{D}_r \end{pmatrix} = \begin{cases} 0, & \alpha = 0, \\ \varepsilon_{\alpha k}, & \alpha > 0, \end{cases}$$

we get $a_0 + a_1\mathrm{D}_1 + \cdots + a_r\mathrm{D}_r \in (\varepsilon]$, hence $a_0 = a_1 = \cdots = a_r = 0$ because $(\varepsilon]$ is primitive. The inverse implication can be easily brought out from the proof of the following

2.2.18. PROPOSITION. *If $\varepsilon\colon \mathcal{D}_U^{r^2} \to \mathcal{D}_U$ is a primitive simple quadratic system of differential equations over a polycylindre U, then it has a general linearly independent solution $(F_0, F_1, \ldots, F_r)$ in $\mathcal{O}_U^{r+1}$. Such a solution is characterized as a solution with one of the following equivalent properties:*

(o) *$(F_0, \ldots, F_r)$ is a general linearly independent solution of $(\varepsilon]$;*

(i) *$F_0, \ldots, F_r$ are linearly independent;*

(ii) *each solution of ε in $\mathcal{O}_U$ is a $\mathbb{C}$-linear combination of $F_0, F_1, \ldots, F_r$;*

(iii) $$\mathsf{F} = \begin{pmatrix} F_0 & \cdots & F_r \\ \mathrm{D}_1F_0 & \cdots & \mathrm{D}_1F_r \\ \vdots & & \vdots \\ \mathrm{D}_rF_0 & \cdots & \mathrm{D}_rF_r \end{pmatrix}$$

is a general linearly independent solution of ε' in $\mathcal{O}_U^{r+1}$.

Proof. ε' is primitive by the proof of 2.2.17. Therefore ε' has a general solution F in $\mathcal{O}_U^{r+1}$ by Proposition 2.2.11. If F is written as in (2.2.8), then the columns are solutions of ε' in $\mathcal{O}_U$. By 2.2.16 F is of type described in (iii), each F_i is a solution of ε in $\mathcal{O}_U$, and $(F_0, \ldots, F_r)$ is a solution of ε in $\mathcal{O}_U^{r+1}$. Assume that $\delta \cdot (F_0, \ldots, F_r) = 0$ for an element δ of $\mathcal{D}(U)$. Since also $\varepsilon_{\alpha\beta} \cdot (F_0, \ldots, F_r) = 0$ for $\alpha, \beta = 1, \ldots, r$, there exists a linear operator

2.2.19. $\delta' = a_0\mathrm{D}_0 + \cdots + a_r\mathrm{D}_r = (a_0, \ldots, a_r) \cdot {}^t(\mathrm{D}_0, \ldots, \mathrm{D}_r) \in \mathcal{D}(U)$

such that

$$\delta \equiv \delta' \bmod (\varepsilon] \quad \text{and} \quad \delta' \cdot (F_0, \ldots, F_r) = 0.$$

The latter identity can be written as

2.2.20. $(a_0, \ldots, a_r) \cdot \mathsf{F} = 0, \qquad a_i \in \mathcal{O}(U).$

From Lemma 2.2.12(d) and assumption (iii) it follows that $\delta' = 0$, hence $\delta \in (\varepsilon]$. So we have proved the existence of a general solution of ε and the implication (iii) $\Rightarrow$ (o).

(o) $\Rightarrow$ (iii): We check property (d) of Lemma 2.2.12 with F as in (iii). Equation (2.2.20) can be written as $\delta' \cdot (F_0, \ldots, F_r) = 0$ with the linear operator as δ' in (2.2.19). From (o) and the primitivity of ε it follows successively that $\delta' \in (\varepsilon]$, $\delta' = 0$, and $a_0 = \cdots = a_r = 0$.

(iii) $\Leftrightarrow$ (i): $F_0, \ldots, F_r$ are linearly independent if and only if the columns of F are linearly independent. Now the equivalence follows from Lemma 2.2.12, (a) $\Leftrightarrow$ (f).

(iii) $\Leftrightarrow$ (ii): Each solution of ε in $\mathcal{O}_U$ is a $\mathbb{C}$-linear combination of $F_0, \ldots, F_r$ if and only if each solution of ε' is a $\mathbb{C}$-linear combination of the columns of F in (iii). But this is characterization (e) of Lemma 2.2.12 for general solutions of ε'.

REMARK. It is very natural to call the members $F_0, \ldots, F_r$ of our general solutions of ε a *fundamental system of solutions*. Dually the differential operators $\varepsilon_{k\alpha}$ of a primitive simple quadratic system can be understood as a *fundamental system of differential operators* for the functions $F_0, \ldots, F_r$, because $(\varepsilon_{11}, \varepsilon_{12}, \ldots, \varepsilon_{rr}]$ contains exactly all differential operators annihilating these functions and the $\varepsilon_{k\alpha}$'s are the only simple quadratic differential operators in $(\varepsilon]$.

We close this section with some definitions, which are useful for the next sections.

2.2.21. DEFINITIONS. We say that the sheaf of $\mathcal{D}_U$-modules $\mathcal{M}$ carries the structure of a *principal Euler-Picard sheaf of modules* over U, if there exists a complex of $\mathcal{D}_U$-module homomorphisms

2.2.22. $\mathcal{D}^{r^2} \xrightarrow{\varepsilon} \mathcal{D} \xrightarrow{\varphi} \mathcal{M} \to 0$

such that ε is an Euler-Picard system of differential operators (see 1.1) and φ is surjective. The Euler-Picard structure is not uniquely determined by $\mathcal{M}$; e.g. it depends on the choice of a suitable element $m = \varphi(1) \in \mathcal{M}(U)$. An *Euler-Picard sheaf of modules* over U is a sheaf of $\mathcal{D}_U$-modules, which is a finite sum of principal Euler-Picard sheaves of submodules. Following Pham [53], 7, we call a coherent sheaf of $\mathcal{D}_U$-modules $\mathcal{M}$, which is at the same time coherent as sheaf of $\mathcal{O}_U$-modules ($\mathcal{O}_U \subset \mathcal{D}_U$), a *connection*, if it is a locally free sheaf of $\mathcal{O}_U$-modules. If $\mathcal{M}$ is simultaneously a (principal) Euler-Picard sheaf of modules and a connection over U, then it is called a (*principal*) *Euler-Picard connection over* U. We call $\mathcal{M}$ a *free* principal Euler-Picard connection, if there exists an exact sequence (2.2.22) defining the Euler-Picard structure of $\mathcal{M}$. If moreover $\mathcal{M}$ is the direct sum of the sheaves of $\mathcal{O}_U$-submodules generated by $m = \varphi(1)$, $D_1 m, \ldots, D_r m$, then it is called a *primitive principal Euler-Picard connection*. Obviously, in this case the Euler-Picard system ε in (2.2.22) is a primitive simple quadratic system of differential operators over U. A *primitive Euler-Picard connection* is a $\mathcal{D}_U$-direct sum of primitive principal Euler-Picard connections.

2.2.23. EXAMPLE. Let $\mathfrak{Y}/S$ be a cycloelliptic curve family. Then $H^{1,\mathrm{an}}_{\mathrm{DR}}(\mathfrak{Y}/S)_{\mathrm{prim.}}$ *is a primitive Euler-Picard sheaf of modules over* $S(\mathbb{C})$, hence over all open $U \subset S(\mathbb{C})$; see Example 2.2.3, Definition 1.4.32, Theorem 1.4.23(i), Theorem 1.4.33 and 1.5.1(iii).

2.3. Hypercohomology

Let $\bar{\mathfrak{Y}}/S$ be a smooth algebraic family, $\bar{\mathfrak{Y}} = \bar{\mathfrak{Y}}/\mathbb{C}$, $S = S/\mathbb{C}$ smooth, $\mathfrak{S}$ a divisor on $\bar{\mathfrak{Y}}$ with smooth irreducible normal crossing components relative to S, $\mathfrak{Y} = \bar{\mathfrak{Y}} \setminus \mathfrak{S}$. For simplicity we assume that S, $\mathfrak{Y}$ are affine. We allow that $\bar{\mathfrak{Y}}/S$ is not proper and that $\mathfrak{S} = \emptyset$. For $s \in S$ we fix notations by means of the commutative Diagram 2.3.a.

2.3.a.

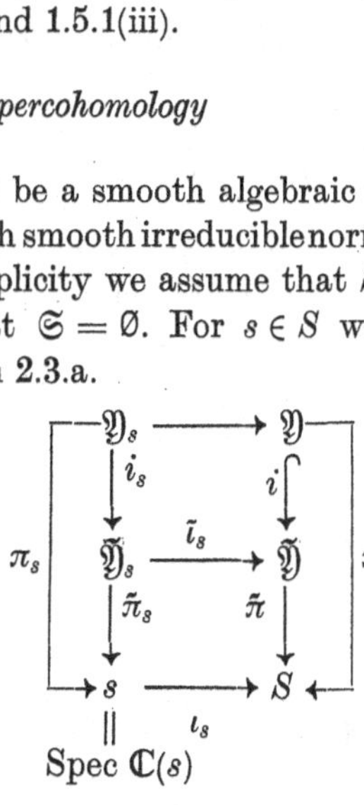

It has cartesian squares and i is affine. Therefore

$$i_*: \mathbf{mod}_{\text{q.c.}}\, \mathcal{O}_{\mathfrak{Y}} \to \mathbf{mod}_{\text{q.c.}}\, \mathcal{O}_{\tilde{\mathfrak{Y}}}$$

is an exact functor on categories of quasicoherent sheaves of modules (see e.g. [27], II, § 5, 5.17(e)). There is a commutative diagram of quasicoherent $\mathcal{O}_{\tilde{\mathfrak{Y}}}$-modules with exact rows

$$\begin{array}{ccccccc}
0 \to \tilde{\pi}^*\Omega^1_S & \longrightarrow & \Omega^1_{\tilde{\mathfrak{Y}}} & \longrightarrow & \Omega^1_{\tilde{\mathfrak{Y}}/S} & \longrightarrow & 0 \\
\uparrow & & \uparrow & & \uparrow & & \\
0 \to \tilde{\pi}^*\Omega^1_S \cap \Omega^1_{\tilde{\mathfrak{Y}}}(\log \mathfrak{S}) & \to & \Omega^1_{\tilde{\mathfrak{Y}}}(\log \mathfrak{S}) & \to & \Omega^1_{\tilde{\mathfrak{Y}}/S}(\log \mathfrak{S}) & \to & 0 \\
\uparrow & & \uparrow & & \uparrow & & \\
0 \to i_*(\pi^*\Omega^1_S) & \longrightarrow & i_*\Omega^1_{\mathfrak{Y}} & \longrightarrow & i_*\Omega^1_{\mathfrak{Y}/S} & \longrightarrow & 0 \\
\| & & \| & & & & \\
i_*(i^*\tilde{\pi}^*\Omega^1_S) & & i_*(i^*\Omega^1_{\tilde{\mathfrak{Y}}}) & & & &
\end{array}$$

which defines $\Omega^1_{\tilde{\mathfrak{Y}}/S}(\log \mathfrak{S})$. In the above diagram $\Omega^1_{\tilde{\mathfrak{Y}}}(\log \mathfrak{S})$ is the sheaf of germs of differential forms of degree 1 with at most logarithmic poles along $\mathfrak{S}$ (see Griffiths/Harris [23], III, 5. (analytic language); Deligne [12], 3.1 (algebraic language)). The sheaf $\Omega^1_{\tilde{\mathfrak{Y}}/S}(\log \mathfrak{S})$ is a locally free coherent $\mathcal{O}_{\tilde{\mathfrak{Y}}}$-module. We set

$$\Omega^p_{\tilde{\mathfrak{Y}}/S}(\log \mathfrak{S}) = \Lambda^p_{\mathcal{O}_{\tilde{\mathfrak{Y}}}}\big(\Omega^1_{\tilde{\mathfrak{Y}}/S}(\log \mathfrak{S})\big).$$

Let $\tilde{\pi}^{-1}(\mathcal{O}_S)$ be the sheaf-theoretic preimage of $\mathcal{O}_S$ on $\tilde{\mathfrak{Y}}$. We have available the exterior differential maps

$$\mathrm{d} = \mathrm{d}_p: \Omega^p_{\tilde{\mathfrak{Y}}/S}(\log \mathfrak{S}) \to \Omega^{p+1}_{\tilde{\mathfrak{Y}}/S}(\log \mathfrak{S})$$

in the category of $\tilde{\pi}^{-1}(\mathcal{O}_S)$-modules coming from $d_0: \tilde{\pi}^*(\mathcal{O}_S) \to \Omega^1_{\tilde{\mathfrak{Y}}/S}(\log \mathfrak{S})$ and exterior products. The complex of $\tilde{\pi}^{-1}(\mathcal{O}_S)$-modules

$$\tilde{\pi}^*\mathcal{O}_S \xrightarrow{\mathrm{d}} \Omega^1_{\tilde{\mathfrak{Y}}/S}(\log \mathfrak{S}) \xrightarrow{\mathrm{d}} \cdots \xrightarrow{\mathrm{d}} \Omega^p_{\tilde{\mathfrak{Y}}/S}(\log \mathfrak{S}) \xrightarrow{\mathrm{d}} \cdots$$

is called the *logarithmic relative de Rham complex* of $\tilde{\mathfrak{Y}}/S$ with respect to $\mathfrak{S}$. If $\mathfrak{S} = \emptyset$, then it is called the *relative de Rham complex*. These complexes are denoted by $\Omega^{\cdot}_{\tilde{\mathfrak{Y}}/S}(\log \mathfrak{S})$ and $\Omega^{\cdot}_{\tilde{\mathfrak{Y}}/S}$, respectively. We have inclusions

2.3.1. $$\Omega^{\cdot}_{\tilde{\mathfrak{Y}}/S} \hookrightarrow \Omega^{\cdot}_{\tilde{\mathfrak{Y}}/S}(\log \mathfrak{S}) \hookrightarrow i_*\Omega^{\cdot}_{\mathfrak{Y}/S}$$

of complexes of $\tilde{\pi}^{-1}(\mathcal{O}_S)$-modules.

Each complex $(C^{\cdot}, \mathrm{d})$ of $\mathcal{O}_Y$-quasicoherent $\tilde{\pi}^{-1}(\mathcal{O}_S)$-modules defines *Čech-bicomplexes* $(C^{\cdot\cdot}, \mathrm{d}, \delta)$ of quasicoherent $\mathcal{O}_S$-modules:

$$C^{\cdot\cdot} = C^{\cdot\cdot}(\{U_\alpha\}, C^{\cdot}),$$

$$C^{p,q} = C^p(\{U_\alpha\}, C^q) = \bigoplus_{A_p} (\tilde{\pi} \mid A_p)_* \, (C^q \mid A_p).$$

Here

$$\mathrm{d}: C^{\cdot,q} \to C^{\cdot,q+1}; \qquad \delta: C^{p,\cdot} \to C^{p+1,\cdot}$$

are the Čech-differentials with respect to an affine covering $\{U_\alpha\}$ of $\mathfrak{Y}$, $i_\alpha: U_\alpha \hookrightarrow \mathfrak{Y}$, $\bar{\pi}_\alpha = \bar{\pi} \circ i_\alpha: U_\alpha \to S$, and A_p runs over all p-simplices $U_{\alpha_0} \cap \cdots \cap U_{\alpha_p}$ of the nerve of the covering $\{U_\alpha\}$.

Each bicomplex $C^{\cdot\cdot}$ defines a total complex $(K^\cdot, \partial)$:

$$K^m = \bigoplus_{p+q=m} C^{p,q}, \qquad \partial = \mathrm{d} + \delta.$$

DEFINITION. The homology sheaves of the associated total complex $K^\cdot$ of $C^{\cdot\cdot}$ are called the *hypercohomology sheaves* of $C^\cdot$. They are denoted by $\underline{\mathscr{H}}^m(C^\cdot)$.

They do not depend on the choice of the open affine covering $\{U_\alpha\}$ of $\mathfrak{Y}$ (see e.g. Katz [36], p. 38). The sheaf $\underline{\mathscr{H}}^m(C^\cdot)$ is a quasicoherent $\mathcal{O}_S$-module. It is easy to see that

2.3.2. if $C^\cdot = i_* C^\cdot_{\mathfrak{Y}}$, then $\underline{\mathscr{H}}^m(C^\cdot_{\mathfrak{Y}}) = \underline{\mathscr{H}}^m(i_* C^\cdot_{\mathfrak{Y}})$.

The $\mathbb{C}[S]$-module of global sections of $\underline{\mathscr{H}}^m(C^\cdot)$ is denoted by $\mathbf{H}^m(C^\cdot)$. It is called the mth *hypercohomology group* of $C^\cdot$. Furthermore we set

$$\underline{\mathscr{H}}^m_{\mathrm{DR}}(\bar{\mathfrak{Y}}/S) = \underline{\mathscr{R}}^m\bar{\pi}_*(\Omega^\cdot_{\bar{\mathfrak{Y}}/S}) = \underline{\mathscr{H}}^m(\Omega^\cdot_{\bar{\mathfrak{Y}}/S}),$$

$$\mathbf{H}^m_{\mathrm{DR}}(\bar{\mathfrak{Y}}/S) = \mathbf{R}^m\bar{\pi}_*(\Omega^\cdot_{\bar{\mathfrak{Y}}/S}) = \mathbf{H}^m(\Omega^\cdot_{\bar{\mathfrak{Y}}/S}),$$

$$\underline{\mathscr{H}}^m_{\mathrm{DR}}\big(\bar{\mathfrak{Y}}/S(\log \mathfrak{S})\big) = \underline{\mathscr{R}}^m\bar{\pi}_*\big(\Omega^\cdot_{\bar{\mathfrak{Y}}/S}(\log \mathfrak{S})\big) = \underline{\mathscr{H}}^m\big(\Omega^\cdot_{\bar{\mathfrak{Y}}/S}(\log \mathfrak{S})\big),$$

$$\mathbf{H}^m_{\mathrm{DR}}\big(\bar{\mathfrak{Y}}/S(\log \mathfrak{S})\big) = \mathbf{R}^m\bar{\pi}_*\big(\Omega^\cdot_{\bar{\mathfrak{Y}}/S}(\log \mathfrak{S})\big) = \mathbf{H}^m\big(\Omega^\cdot_{\bar{\mathfrak{Y}}/S}(\log \mathfrak{S})\big).$$

These groups and sheaves are called the *relative* (*logarithmic*) *de Rham cohomology groups and sheaves*, respectively, of the family $\bar{\mathfrak{Y}}/S$.

The formation $\underline{\mathscr{H}}^m$ is a functor. So we obtain from (2.3.1) natural morphisms

2.3.3.
$$\begin{array}{ccccc} \underline{\mathscr{H}}^m_{\mathrm{DR}}(\bar{\mathfrak{Y}}/S) & \to & \underline{\mathscr{H}}^m_{\mathrm{DR}}\big(\mathfrak{Y}/S(\log \mathfrak{S})\big) & \xrightarrow{\sim} & \underline{\mathscr{H}}^m(i_*\Omega^\cdot_{\mathfrak{Y}/S}) \\ & & & & \| \wr \\ & & \mathscr{H}^m_{\mathrm{DR}}(\mathfrak{Y}/S) = \underline{\mathscr{H}}^m_{\mathrm{DR}}(\mathfrak{Y}/S) & = & \underline{\mathscr{H}}^m(\Omega^\cdot_{\mathfrak{Y}/S}). \end{array}$$

For the vertical isomorphism on the right-hand side we refer to 2.3.2. It follows immediately that $\underline{\mathscr{H}}^m(i_*\Omega^\cdot_{\mathfrak{Y}/S}) = \underline{\mathscr{H}}^m_{\mathrm{DR}}(\mathfrak{Y}/S)$. Since $\mathfrak{Y}/S$ is an affine family we can use the covering $\{U_\alpha\} = \{\mathfrak{Y}\}$ for the construction of $\underline{\mathscr{H}}^m_{\mathrm{DR}}(\mathfrak{Y}/S) = \underline{\mathscr{H}}^m(\Omega^\cdot_{\mathfrak{Y}/S})$. Obviously, the total complex of the Čech-bicomplex of $\Omega^\cdot_{\mathfrak{Y}/S}$ coincides with $\pi_*\Omega^\cdot_{\mathfrak{Y}/S}$. Therefore the hypercohomology sheaves $\underline{\mathscr{H}}^m_{\mathrm{DR}}(\mathfrak{Y}/S)$ coincide with the ordinary cohomology sheaves $\mathscr{H}^m_{\mathrm{DR}}(\mathfrak{Y}/S) = \mathscr{H}^m(\pi_*\Omega^\cdot_{\mathfrak{Y}/S})$ of the complex $\pi_*\Omega^\cdot_{\mathfrak{Y}/S}$. We take also note of the fact that $\mathbf{H}^m_{\mathrm{DR}}(\mathfrak{Y}/S) = H^m_{\mathrm{DR}}(\mathfrak{Y}/S)$.

Now we prove that the second horizontal morphism in (2.3.3) is an isomorphism. We use the following fact (Deligne [13], Prop. 3.1.8): The complexes $\Omega^\cdot_{\bar{\mathfrak{Y}}/S}(\log \mathfrak{S})$ and $i_*\Omega^\cdot_{\mathfrak{Y}/S}$ are quasiisomorphic; i.e. the cohomology sheaves of these complexes are naturally isomorphic. Then it follows that the corresponding relative hypercohomology sheaves coincide (see [13], (3.1.8.2)).

For the study of the first horizontal homomorphism in (2.3.3) we need some further results of algebraic geometry.

2.3.4. PROPOSITION (Deligne [11]). *$\underline{\mathscr{H}}^m_{\mathrm{DR}}(\mathfrak{Y}/S)$ and $\underline{\mathscr{H}}^m_{\mathrm{DR}}(\bar{\mathfrak{Y}}/S(\log \mathfrak{S}))$ are locally free coherent sheaves of $\mathcal{O}_S$-modules.*

2.3.5. PROPOSITION (Deligne [11], see also Katz [36]). *The formations $\underline{\mathscr{H}}^m_{\mathrm{DR}}(\bar{\mathfrak{Y}}/S)$, $\underline{\mathscr{H}}^m_{\mathrm{DR}}(\bar{\mathfrak{Y}}/S(\log \mathfrak{S}))$, $\underline{\mathscr{H}}^m_{\mathrm{DR}}(\mathfrak{Y}/S)$ commute with arbitrary base changes. In particular, for the closed points $s \in S$ one has*

$$\underline{\mathscr{H}}^m_{\mathrm{DR}}(\bar{\mathfrak{Y}}_s) = \iota_s^* \underline{\mathscr{H}}^m_{\mathrm{DR}}(\bar{\mathfrak{Y}}/S), \qquad \underline{\mathscr{H}}^m_{\mathrm{DR}}\big(\bar{\mathfrak{Y}}_s(\log \iota_s^*(\mathfrak{S}))\big) = \iota_s^* \underline{\mathscr{H}}^m_{\mathrm{DR}}(\bar{\mathfrak{Y}}/S(\log \mathfrak{S}))$$

with the notations of Diagram 2.3.a.

This lead us to a local study of the homomorphisms in (2.3.3). For localizations we use the following notations: Let $\mathscr{F}$ be a sheaf of $\mathcal{O}_S$-modules, then $\mathscr{F}_s$ denotes the stalk of $\mathscr{F}$ in $s \in S$. The stalk $\mathscr{F}_s$ is a module over the local ring $\mathcal{O}_s = \mathcal{O}_{S,s}$ with maximal ideal $\mathfrak{m}_s$. Furthermore we set

$$\mathscr{F}(s) = \mathbb{C}(s) \otimes_{\mathcal{O}_s} \mathscr{F}_s \cong \mathscr{F}_s/\mathfrak{m}_s \cdot \mathscr{F}_s = \iota_s^*(\mathscr{F}), \qquad \mathbb{C}(s) = \mathcal{O}_s/\mathfrak{m}_s \cong \mathbb{C},$$

$$M_s = (\tilde{M})_s, \qquad M(s) = \tilde{M}(s) \quad \text{for a } \mathbb{C}[S]\text{-module } M.$$

2.3.6. PROPOSITION. *Let $\mathscr{F}$ and $\mathscr{G}$ be locally free coherent sheaves of $\mathcal{O}_S$-modules and $u: \mathscr{F} \to \mathscr{G}$ an $\mathcal{O}_S$-homomorphism. Then:*

(i) *u is injective (surjective, an isomorphism) if and only if $u_s: \mathscr{F}_s \to \mathscr{G}_s$ is injective (surjective, an isomorphism) for all closed points $s \in S$.*

(ii) *u is surjective (an isomorphism) if and only if $u(s) = 1_{\mathbb{C}(s)} \otimes u_s$ is a surjective homomorphism (an isomorphism) from $\mathscr{F}(s)$ into $\mathscr{G}(s)$ for all closed points $s \in S$.*

(iii) *$u(s)$ is injective if and only if u_s is injective and $\mathscr{F}_s$ is a direct summand of $\mathscr{G}_s$.*

For the proof we refer the reader to Bourbaki [7], II, § 3, Cor. 1 of Prop. 4, Prop. 6 and Theorem 1.

Now we look at the commutative diagrams

2.3.b.

$$\begin{array}{ccccccc}
\mathbf{H}^m_{\mathrm{DR}}(\bar{\mathfrak{Y}}/S) & \longrightarrow & \mathbf{H}^m_{\mathrm{DR}}(\bar{\mathfrak{Y}}/S(\log \mathfrak{S})) & \xrightarrow{\sim} & \mathbf{H}^m(i_* \Omega^{\cdot}_{\mathfrak{Y}/S}) = H^m_{\mathrm{DR}}(\mathfrak{Y}/S) & & \bar{\omega} \\
\downarrow & & \downarrow & & \downarrow & & \downarrow \\
\mathbf{H}^m_{\mathrm{DR}}(\bar{\mathfrak{Y}}_s) & \longrightarrow & \mathbf{H}^m_{\mathrm{DR}}(\bar{\mathfrak{Y}}_s(\log \mathfrak{S}_s)) & \xrightarrow{\sim} & \mathbf{H}^m_{\mathrm{DR}}(i_{s*} \Omega^{\cdot}_{\mathfrak{Y}_s}) = H^m_{\mathrm{DR}}(\mathfrak{Y}_s) & & \bar{\omega}(s) \\
\| & & \| & & \| & & \\
\underline{\mathscr{H}}^m_{\mathrm{DR}}(\bar{\mathfrak{Y}}/S)(s) & & \underline{\mathscr{H}}^m_{\mathrm{DR}}(\bar{\mathfrak{Y}}/S(\log \mathfrak{S}))\,(s) & & \mathscr{H}^m_{\mathrm{DR}}(\mathfrak{Y}/S)\,(s) & &
\end{array}$$

$$\mathfrak{S}_s = \iota_s^*(\mathfrak{S})$$

and make the following

2.3.7. REMARK. There is a construction of analytic (relative, logarithmic) de Rham cohomology sheaves analogous to our algebraic construction. For the

Čech-bicomplexes one uses analytic open coverings. The limit over all refinements of such a covering leads to the analytic de Rham cohomology groups (see Griffiths/Harris [23], III, § 5). The algebraic construction is compatible with the analytic construction. More precisely, for $\tilde{\pi}\colon \mathfrak{Y} \to S$ projective and smooth one has that

$$\begin{aligned}&\underline{\mathscr{H}}^{m,\mathrm{an}}_{\mathrm{DR}}\big((\mathfrak{Y}/S(\log \mathfrak{S}))^{\mathrm{an}}\big)\\ &= \underline{\mathscr{R}}^m\tilde{\pi}^{\mathrm{an}}_*\big((\Omega^{\bullet}_{\mathfrak{Y}/S}(\log \mathfrak{S}))^{\mathrm{an}}\big)\\ &= \underline{\mathscr{R}}^m\tilde{\pi}^{\mathrm{an}}_*\big(\Omega^{\bullet}_{\mathfrak{Y}^{\mathrm{an}}/S^{\mathrm{an}}}(\log \mathfrak{S}^{\mathrm{an}})\big)\\ &= \big(\underline{\mathscr{R}}^m\tilde{\pi}_*(\Omega^{\bullet}_{\mathfrak{Y}/S}(\log \mathfrak{S}))\big)^{\mathrm{an}}\\ &= \big(\underline{\mathscr{H}}^m_{\mathrm{DR}}(\mathfrak{Y}/S(\log \mathfrak{S}))\big)^{\mathrm{an}} = \underline{\mathscr{H}}^m_{\mathrm{DR}}(\mathfrak{Y}/S(\log \mathfrak{S})) \otimes_{\mathcal{O}_S} \mathcal{O}^{\mathrm{an}}_S\,,\end{aligned}$$

where the index "an" denotes the corresponding analytic object. For these identities we refer to Katz [36], 4.3, and Deligne [12], Theorem 6.13. In particular, if $S \ni s \cong \operatorname{Spec} \mathbb{C}$ is a closed point, then the algebraic and the analytic constructions lead to the same vector spaces $\mathbf{H}^m_{\mathrm{DR}}\big(\mathfrak{Y}_s(\log \mathfrak{S}_s)\big)$ or $\mathbf{H}^m_{\mathrm{DR}}(\mathfrak{Y}_s)$ of finite dimension.

2.3.8. LEMMA. *If $\mathfrak{Y}/S$ is projective and smooth, then the horizontal homomorphisms of the Diagrams* 2.3.b *and* 2.3.3 *are injective for $m = 1$. In particular, we have a natural embedding*

$$\mathbf{H}^1_{\mathrm{DR}}(\mathfrak{Y}/S) \hookrightarrow H^1_{\mathrm{DR}}(\mathfrak{Y}/S)\,.$$

Proof. By Proposition 2.3.6 it suffices to prove that the local homomorphisms

$$\mathbf{H}^1_{\mathrm{DR}}(\mathfrak{Y}_s) \to \mathbf{H}^1_{\mathrm{DR}}\big(\mathfrak{Y}_s(\log \mathfrak{S}_s)\big) = \mathbf{H}^1_{\mathrm{DR}}(i_{s*}\Omega^{\bullet}_{\mathfrak{Y}_s})$$

are injective. Let $\Omega^{\bullet}_{\mathfrak{Y}_s}(*)$ be the complex consisting of

$$\Omega^p_{\mathfrak{Y}_s}(*) = \bigcup_{D\in \mathrm{Div}_r\mathfrak{Y}_s} \Omega^p_{\mathfrak{Y}_s}(*D)\,, \qquad \Omega^p_{\mathfrak{Y}_s}(*D) = i_{D*}(\Omega^p_{\mathfrak{Y}_s\setminus D})\,,$$

where $\mathrm{Div}_r\,\mathfrak{Y}_s$ denotes the set of reduced divisors on $\mathfrak{Y}_s$ and i_D is the open em- bedding $i_D\colon \mathfrak{Y}_s \setminus D \hookrightarrow \mathfrak{Y}_s$. Changing over to analytic objects one has accord to Remark 2.3.7 an exact sequence

$$0 \to \mathbf{H}^1_{\mathrm{DR}}(\mathfrak{Y}_s) \xrightarrow{\lambda} \mathbf{H}^1_{\mathrm{DR}}\big(\Omega^{\bullet}_{\mathfrak{Y}_s}(*)\big) \xrightarrow{R} H^0\Big(\bigoplus_{D\in \mathrm{Div}\mathfrak{Y}_s} \mathbb{C}_D\Big)$$

where $\mathbb{C}_D$ denotes the sheaf with stalks isomorphic to $\mathbb{C}$ on the supp of D and which is zero outside of D. Here R is the "*residue map*" (see Griff [23], III, § 5, p. 458). Clearly λ factorizes through $\mathbf{H}^1_{\mathrm{DR}}(i_{s*}\Omega^{\bullet}_{\mathfrak{Y}_s})$. Thi lemma.

Lemma 2.3.8 together with Remark 2.3.7 allows an affine algeb of the analytic de Rham cohomology sheaf $\underline{\mathscr{H}}^{1,\mathrm{an}}_{\mathrm{DR}}(\bar{\mathfrak{Y}}/S)$. For a smooth projective curve family $\bar{\mathfrak{Y}}/S$ one has locally the very classical interpretation:

$$\text{2.3.9.} \quad H^1(\bar{\mathfrak{Y}}_s, \mathbb{C}) = \mathbf{H}^1_{\mathrm{DR}}(\bar{\mathfrak{Y}}_s) = \frac{\{\text{differential forms of 2. kind on } \bar{\mathfrak{Y}}_s\}}{\{\text{exact forms}\}}$$

$$\text{2.3.10.} \quad \mathbf{H}^1\big(\Omega^{\cdot}_{\bar{\mathfrak{Y}}_s}(*)\big) = \frac{\{\text{merom. diff. forms on } \bar{\mathfrak{Y}}_s\}}{\{\text{exact forms}\}}$$

(see Griffiths/Harris [23], III.5). *Differential forms of the second kind* are those forms which have no residues (i.e. everywhere vanishing residues). Now we define the sheaf of $\mathcal{O}_S$-submodules $\mathscr{H}^1_{\mathrm{DR}}(\mathfrak{Y}/S)_{2.\mathrm{kind}}$ of $\mathscr{H}^1_{\mathrm{DR}}(\mathfrak{Y}/S)$ by

$$U \mapsto \{\overline{\omega} \in \Gamma\big(U, \mathscr{H}^1_{\mathrm{DR}}(\mathfrak{Y}/S)\big);\ \overline{\omega}(s) \text{ of 2. kind on } \bar{\mathfrak{Y}}_s \text{ for all } s \in U\}$$

where U is an open subset of S. We set

$$H^1_{\mathrm{DR}}(\mathfrak{Y}/S)_{2.\mathrm{kind}} = \Gamma\big(S, \mathscr{H}^1_{\mathrm{DR}}(\mathfrak{Y}/S)_{2.\mathrm{kind}}\big).$$

Then we obtain from Diagram 2.3.b, Remark 2.3.7, Lemma 2.3.8 and (2.3.9) the following

2.3.11. PROPOSITION. *Let $\bar{\mathfrak{Y}}/S$ be a projective curve family, $\mathfrak{Y} = \bar{\mathfrak{Y}} \setminus \mathfrak{S}$, with the good properties stated at the beginning of this section. Then*:

(i) $\mathscr{H}^{1,\mathrm{an}}_{\mathrm{DR}}(\bar{\mathfrak{Y}}/S) = \widetilde{H^1_{\mathrm{DR}}(\mathfrak{Y}/S)_{2.\mathrm{kind}}} \otimes_{\mathcal{O}_S} \mathcal{O}^{\mathrm{an}}_S = \mathscr{H}^1_{\mathrm{DR}}(\mathfrak{Y}/S)_{2.\mathrm{kind}} \otimes_{\mathcal{O}_S} \mathcal{O}^{\mathrm{an}}_S$,

(ii) $\mathrm{rk}_{\mathcal{O}_S} \mathscr{H}^1_{\mathrm{DR}}(\bar{\mathfrak{Y}}/S) = \dim_{\mathbb{C}} H^1(\bar{\mathfrak{Y}}_s, \mathbb{C}) = 2g$,

where g denotes the genus of any fibre curve $\bar{\mathfrak{Y}}_s$.

For the understanding of (ii) we remark that g does not depend on the special choice of $s \in S$ (see e.g. [27], III, 12.9.3), that $\mathbf{H}^1_{\mathrm{DR}}(\bar{\mathfrak{Y}}/S)$ is locally free over $\mathcal{O}_S$ with localizations

$$\mathbf{H}^1_{\mathrm{DR}}(\bar{\mathfrak{Y}}/S)(s) = \mathbf{H}^1_{\mathrm{DR}}(\bar{\mathfrak{Y}}_s) = H^1(\bar{\mathfrak{Y}}_s, \mathbb{C})$$

and that $\dim_{\mathbb{C}} H^1(C, \mathbb{C}) = h^{1,0}(C) + h^{0,1}(C) = 2h^{0,1} = 2 \dim H^0(\Omega^1_C) = 2g$ for compact Riemann surfaces C.

2.3.12. EXAMPLE. We apply Proposition 2.3.11 to generalized projective Picard curve families

$$\bar{\mathfrak{Y}}/S = \bar{\mathfrak{Y}}/S : {}^{\cdot}WY^n = \prod_{i=-1}^{n-1} (X - t_i W) \qquad \text{(see 2.1.6 (iii) A))}.$$

We assert that

$$\underline{\mathscr{H}}^1_{\mathrm{DR}}(\bar{\mathfrak{Y}}/S) = \mathscr{H}^1_{\mathrm{DR}}(\mathfrak{Y}/S)_{2.\mathrm{kind}} = \bigoplus_{\substack{0 \le k \le n \\ 0 < l < n}} \mathcal{O}_S \cdot \overline{x^k\, \mathrm{d}x/y^l}.$$

Proof. Choose $s = (s_1, \ldots, s_r) \in S(\mathbb{C})$. Now $\overline{(x^k\, \mathrm{d}x/y^l)}\,(s)$ is $x^k\, \mathrm{d}x/y^l$ considered s an element of $\Omega^1_{\mathfrak{Y}_s}$. The affine equation for $\ddot{\mathfrak{Y}}_s = \bar{\mathfrak{Y}}_s \setminus \{\infty\}$ is

$$F = F(X, Y) = Y^n - (X - 1)\, X(X - s_1) \cdots (X - s_{n-1}) = 0.$$

2.3.13. *$x^k\,dx/y^l$ is regular on $\ddot{\mathfrak{Y}}_s$ for $k \geqq 0$, $l < n$.*

Proof. Let f be the polynomial $(X-1)\,X(X-s_1)\cdots(X-s_r)$. Taking differentials on both sides of the identity $y^n = f(x)$ on $\overline{\mathfrak{Y}}_s$ we obtain with $f_x = (\partial F/\partial X)\,(x)$

$$ny^{n-1}\,dy = f_x\,dx \qquad \text{or} \qquad dx/y^{n-1} = n\,dy/f_x.$$

The poles of dx/y^{n-1} lie in the intersection of the point sets on $\ddot{\mathfrak{Y}}_s$ described by the equations $y = f(x) = 0$ and $f_x = 0$, respectively. But $f(x)$ and f_x have no common zeros on $\ddot{\mathfrak{Y}}_s$ because the polynomial f has only zeros of order 1. This proves 2.3.13.

From 2.3.13 it follows that the differential forms $x^k\,dx/y^l$, $l \leqq n-1$, $k \geqq 0$, have no residues on $\ddot{\mathfrak{Y}}_s$ because they are regular there. By the residue theorem (see e.g. [23], II.1) the sum of all residues of a meromorphic differential form on $\overline{\mathfrak{Y}}_s$ has to be zero. Since $\overline{\mathfrak{Y}}_s \setminus \ddot{\mathfrak{Y}}_s$ consists of one point only, $x^k\,dx/y^l$ has no residues on $\overline{\mathfrak{Y}}_s$. Using 1.5.1(ii) we find

$$\textbf{2.3.14.} \quad \mathcal{M} = \bigoplus_{\substack{0 \leqq k < n \\ 0 < l < n}} \mathcal{O}_S \cdot \overline{\frac{x^k\,dx}{y^l}} \subseteqq \mathcal{H}^1_{\mathrm{DR}}(\mathfrak{Y}/S)_{2.\mathrm{kind}}.$$

On the other hand the genus of $\overline{\mathfrak{Y}}_s \subset \mathbb{P}^2(\mathbb{C})$ is $g = (d-1)\,(d-2)/2 = n(n-1)/2$, where $d = n+1$ is the degree of the plane curve $\overline{\mathfrak{Y}}_s$ (see e.g. [23], II.1, p. 220). Therefore, by Proposition 2.3.11 and (2.3.9)

$$\begin{aligned} n(n-1) = 2g &= \dim H^1(\overline{\mathfrak{Y}}_s, \mathbb{C}) = \dim \mathbf{H}^1_{\mathrm{DR}}(\overline{\mathfrak{Y}}_s) \\ &= \dim \left((H^1_{\mathrm{DR}}(\mathfrak{Y}/S)_{2.\mathrm{kind}})\,(s)\right) = \dim \bigoplus_{\substack{0 \leqq k < n \\ 0 < l < n}} \mathbb{C} \cdot x^k\,dx/y^l = \dim \mathcal{M}(s). \end{aligned}$$

Consequently

$$\mathcal{M}(s) = \left(H^1_{\mathrm{DR}}(\mathfrak{Y}/S)_{2.\mathrm{kind}}\right)(s) \qquad \text{for all } s \in S.$$

By Proposition 2.3.6 the embedding (2.3.14) is an isomorphism.

2.4. The Gauss-Manin Connection

Katz and Oda proved in [37] that for smooth algebraic families $\mathfrak{X}/S$, $\mathfrak{X}$, S smooth over a field k, there exists an $\mathcal{O}_S$-homomorphism

$$Д: \underline{\mathcal{H}}^1_{\mathrm{DR}}(\mathfrak{X}/S) \to \Omega^1_{S/k} \otimes_{\mathcal{O}_S} \underline{\mathcal{H}}^1_{\mathrm{DR}}(\mathfrak{X}/S)$$

with the property

$$Д(f\omega) = f \cdot Д(\omega) + df \otimes \omega,$$

f, ω sections in $\mathcal{O}_S$ and $\underline{\mathcal{H}}^1_{\mathrm{DR}}(\mathfrak{X}/S)$, respectively.

Д gives rise to an $\mathcal{O}_S$-homomorphism

$$\nabla: \mathcal{D}er_k(\mathcal{O}_S) \to \mathcal{E}nd_k(\underline{\mathcal{H}}^1_{\mathrm{DR}}(\mathfrak{X}/S))$$

of sheaves of k-Lie algebras. For a section D of $\mathcal{D}er_k(\mathcal{O}_S)$, $\nabla(\mathrm{D})$ is defined as the composition

$$\underline{\mathcal{H}}^1_{\mathrm{DR}}(\mathfrak{X}/S) \xrightarrow{Д} \Omega^1_{S/k} \otimes_{\mathcal{O}_S} \underline{\mathcal{H}}^1_{\mathrm{DR}}(\mathfrak{X}/S) \xrightarrow{\mathrm{D}\otimes 1} \mathcal{O}_S \otimes_{\mathcal{O}_S} \underline{\mathcal{H}}^1_{\mathrm{DR}}(\mathfrak{X}/S) \cong \underline{\mathcal{H}}^1_{\mathrm{DR}}(\mathfrak{X}/S)$$

(restricted to the open set where D is defined). One has that

$$\nabla(\mathrm{D})\,(f\omega) = \mathrm{D}(f)\,\omega + f \cdot \nabla(\mathrm{D})\,(\omega),$$

f, ω sections of $\mathcal{O}_S$ and $\underline{\mathcal{H}}^1_{\mathrm{DR}}(\mathfrak{X}/S)$, respectively.

The structure $(\underline{\mathcal{H}}^1_{\mathrm{DR}}, Д)$ (or $(\underline{\mathcal{H}}^1_{\mathrm{DR}}, \nabla)$) is called the *Gauss-Manin connection* of the family $\mathfrak{X}/S$.

If S is affine, then $\mathbf{H}^1_{\mathrm{DR}}(\mathfrak{X}/S)$ carries the structure of an $\mathrm{Der}_k\, k[S]$-module. In particular for $S = \operatorname{Spec} \mathbb{C}[t_1, \ldots, t_r, \Pi_t^{-1}]$ we see that $\mathbf{H}^1_{\mathrm{DR}}(\mathfrak{X}/S)$ is a $\mathbb{C}[S, \mathrm{D}] = \mathbb{C}[S]\,[\mathrm{D}_1, \ldots, \mathrm{D}_r]$-module. In § 1 we considered the étale cycloelliptic curve families $\mathfrak{Y}/S$. We defined and investigated a special $\mathbb{C}[S, \mathrm{D}]$-structure of $H^1_{\mathrm{DR}}(\mathfrak{Y}/S) = \mathbf{H}^1_{\mathrm{DR}}(\mathfrak{Y}/S)$. This is exactly the Gauss-Manin structure of $\mathbf{H}^1_{\mathrm{DR}}(\mathfrak{Y}/S)$. It is easy to see that this construction of § 1 works generally for étale families. In order to be precise we give the following

2.4.1. DEFINITION. Let k be a field, $\pi: \mathfrak{U} \to S$ a smooth k-morphism, $\mathfrak{U}$, S smooth affine k-varieties. We call $\mathfrak{U}/S$ an *étale family*, if there exists a decomposition of π

$$\pi: \mathfrak{U} \xrightarrow[\pi_1]{} (\mathbb{A}^m_S)' \hookrightarrow \mathbb{A}^m_S \xrightarrow[p]{} S,$$

where π_1 is étale and p is the natural projection coming from the embedding $k[S] \hookrightarrow k[S]\,[X_1, \ldots, X_m]$. A covering $\{\mathfrak{U}_\alpha\}$ of a smooth family $\mathfrak{X}/S$ is called a *relative étale covering*, if $X = \bigcup_\alpha \mathfrak{U}_\alpha$, $\mathfrak{U}_\alpha$ open in $\mathfrak{X}$, and $\mathfrak{U}_\alpha/S$ is an étale family for each α.

In our cases there exist such relative étale coverings (see [24], IV. 2.2.12) because our families are smooth.

2.4.2. EXAMPLE. Let $\overline{\mathfrak{Y}}/S$ be a generalized Picard curve family. Then one has a relative étale covering $\{\mathfrak{U}_\alpha\}$ consisting of three étale families $\mathfrak{U}_0$, $\mathfrak{U}_1$, $\mathfrak{U}_2$ over S:

$$\mathfrak{U}_0 = \mathfrak{Y} = \operatorname{Spec} \mathbb{C}[t_1, \ldots, t_{n-1}, \Pi_t^{-1}, x, y, \Pi_x^{-1}] = \overline{\mathfrak{Y}} \setminus \mathfrak{S},$$

$$\mathfrak{U}_1 = \operatorname{Spec} \mathbb{C}[S]\,[x, y, (\partial F/\partial x)^{-1}], \qquad F = Y^n - \prod_{i=-1}^{n-1} (X - t_i),$$

$$\mathfrak{U}_2 = \operatorname{Spec} \mathbb{C}[S]\,[u, v, (\partial G/\partial v)^{-1}], \qquad G = U \cdot \prod_{-1 \leq i \neq 0} (U - t_i V) - V$$

(set $U = \frac{W}{Y}$, $V = \frac{X}{Y}$ in the homogeneous equation for $\bar{\mathfrak{Y}}$). The relative étale decompositions are

$$\mathfrak{Y} = \mathfrak{U}_0 \to \mathbf{A}_S^1[x] \to S, \qquad \mathfrak{U}_1 \to \mathbf{A}_S^1[y] \to S, \qquad \mathfrak{U}_2 \to \mathbf{A}_S^1[u] \to S,$$

where $\mathbf{A}_S^1[\xi] = \operatorname{Spec} \mathbf{C}[S]\,[\xi]$. The proof is a simple exercise, which is left to the reader.

In general one has for an étale family $\mathfrak{U}/S$ a diagram of type 1.3.a with $\mathfrak{U}$ instead of $\mathfrak{Y}$ and

$$\Omega_{k[\mathfrak{U}]/k[S]} = \bigoplus_{i=1}^{m} k[\mathfrak{U}]\, \mathrm{d}x_i, \qquad \mathbf{A}_S^m = \operatorname{Spec} k[S]\,[x_1, \ldots, x_m],$$

where m is the relative dimension of the family. The action of $\mathrm{D} \in \operatorname{Der}_k k[S]$ is defined on $k[\mathfrak{U}]$ and $\Omega^1_{\mathfrak{U}/S}$ by the trivial extensions $\mathrm{D}(x_i) = 0$ and $\mathrm{D}(\mathrm{d}x_i) = 0$, respectively. Taking the quotation of $\Omega^1_{\mathfrak{U}/S}$ by $\mathrm{d}_{\mathfrak{U}/S}(\mathcal{O}_S)$ one gets the action of D on $\underline{\mathcal{H}}^1_{\mathrm{DR}}(\mathfrak{U}/S) = \mathcal{H}^1_{\mathrm{DR}}(\mathfrak{U}/S)$. By going through the Čech-bicomplex of $\Omega^{\cdot}_{\mathfrak{X}/S}$ and a relative étale covering $\{\mathfrak{U}_\alpha/S\}$ of $\mathfrak{X}/S$ one can "patch together" the actions of D on $k[\mathfrak{U}_\alpha]$ and $\Omega^1_{\mathfrak{U}_\alpha/S}$ to an action on $\underline{\mathcal{H}}^1_{\mathrm{DR}}(\mathfrak{X}/S)$ (see Katz/Oda [37]). The construction is compatible with base changes, in particular with the base change to the general point $\operatorname{Spec} \mathbf{C}(S) \to \operatorname{Spec} \mathbf{C}[S]$ ($k = \mathbf{C}$ for simplicity). That means that $\mathbf{H}^1_{\mathrm{DR}}(\mathfrak{X}_{\mathbf{C}(S)}/\mathbf{C}(S))$ is endowed with the structure of a Gauss-Manin connection, which restricts to the $\mathbf{C}[S]\,[\mathrm{D}]$-structure of the $\mathbf{C}[S]$-lattice $\mathbf{H}^1_{\mathrm{DR}}(\mathfrak{X}/S)$ of the $\mathbf{C}(S)$-vector space $\mathbf{H}^1_{\mathrm{DR}}(\mathfrak{X}_{\mathbf{C}(S)}/\mathbf{C}(S))$. The first construction of a Gauss-Manin connection was in fact given by Manin for curves over function fields (e.g. $\mathbf{C}(S)$) with differentiations (see [43]).

The construction works also for relative log-complexes (see Katz [36]). So we also have available *Gauss-Manin connections of type* $\underline{\mathcal{H}}^1_{\mathrm{DR}}(\mathfrak{X}/S(\log \mathfrak{S}))$. The structure extends by means of tensor multiplication with $\mathcal{O}_S^{\mathrm{an}}$ to the analytic Gauss-Manin connection

$$\nabla^{\mathrm{an}} \colon \mathcal{D}er_{\mathbf{C}}(\mathcal{O}_S^{\mathrm{an}}) \to \mathcal{E}nd_{\mathbf{C}}(\mathcal{H}^{1,\mathrm{an}}_{\mathrm{DR}}(\mathfrak{X}/S)).$$

So we dispose for analytic open subsets U of $S(\mathbf{C})$ of (analytic) sheaves of $\mathcal{D}_U$-modules

$$\underline{\mathcal{H}}^{1,\mathrm{an}}_{\mathrm{DR},U}(\mathfrak{X}/S(\log \mathfrak{S})) = \underline{\mathcal{H}}^{1,\mathrm{an}}_{\mathrm{DR}}(\mathfrak{X}/S(\log \mathfrak{S}))|_U,$$

$$\underline{\mathcal{H}}^{1,\mathrm{an}}_{\mathrm{DR},U}(\mathfrak{X}/S) \qquad (\text{for } \mathfrak{S} = \emptyset).$$

These are connections in the sense of 2.2.21 by Proposition 2.3.4.

Now we return to our smooth cycloelliptic curve families $\bar{\mathfrak{Y}}/S = \mathfrak{X}/S$, say of type $(n, \mathfrak{b})$, defined in 2.1.7. We can summerize our results in the following global structure theorem. This theorem has an algebraic and an analytic version. We give a simultaneous formulation of these two interpretations. So $\mathcal{D}$ can be understood as the *sheaf* $\mathcal{D}_S$ *of algebraic differential operators* or as the *sheaf of analytic differential operators* $\mathcal{D}_U$, U an open analytic subset of $S(\mathbf{C})$, say a polycylindre. $\underline{\mathscr{H}}^{1\cdot}_{\mathrm{DR}}$ denotes the algebraic hypercohomology group $\underline{\mathscr{H}}^{1}_{\mathrm{DR}}$ or the analytic one $\mathscr{H}^{1,\mathrm{an}}_{\mathrm{DR},U}$, respectively. If $\mathscr{F}$ is a sheaf of submodules of $\underline{\mathscr{H}}^{1}_{\mathrm{DR}}(\bar{\mathfrak{Y}}/S(\log \mathfrak{S}))$ $= \mathscr{H}^{1}_{\mathrm{DR}}(\mathfrak{Y}/S)$, then $\mathscr{F}_{\cdot l}$ and $\mathscr{F}_{\mathrm{prim.}}$ denote the intersection of $\mathscr{F}$ with $\mathscr{H}^{1}_{\mathrm{DR}}(\mathfrak{Y}/S)_l$ and $\mathscr{H}^{1}_{\mathrm{DR}}(\mathfrak{Y}/S)_{\mathrm{prim.}}$, respectively (see Definitions 1.3.18 and 1.4.35). Their analytifications are denoted by $\mathscr{F}^{\mathrm{an}}_{l,U}$ or $\mathscr{F}^{\mathrm{an}}_{\mathrm{prim.},U}$, respectively. $\mathscr{F}^{\cdot}_{l}$ and $\mathscr{F}^{\cdot}_{\mathrm{prim.}}$ admit both, the algebraic and the analytic interpretations.

2.4.3. PROJECTIVE STRUCTURE THEOREM. *Let $\bar{\mathfrak{Y}}/S$ be the cycloelliptic smooth projective plane curve family of type $(n, \mathfrak{b})$. Then:*

(i) *For $k, l \in \mathbb{Z}, l \neq 0$, $\mathscr{D} \cdot \overline{x^k\,\mathrm{d}x/y^l}$ is a principal Euler-Picard sheaf of submodules of $\underline{\mathscr{H}}^{1\cdot}_{\mathrm{DR}}(\bar{\mathfrak{Y}}/S(\log \mathfrak{S})) = \underline{\mathscr{H}}^{1\cdot}_{\mathrm{DR}}(\mathfrak{Y}/S) = \mathscr{H}^{1\cdot}_{\mathrm{DR}}(\mathfrak{Y}/S)$ of type $(k, l, n, \mathfrak{b})$.*

(ii)
$$\mathscr{D}^{r^2} \xrightarrow{\varepsilon} \mathscr{D} \to \mathscr{D} \cdot \overline{x^k\,\mathrm{d}x/y^l} \to 0$$
is a complex of sheaves of $\mathscr{D}$-modules, where $\varepsilon = \{\varepsilon_{ij}^{k,l,n,\mathfrak{b}}\}_{i,j=1}^{r}$ is the Euler-Picard system of differential operators defined in 1.1.7.

(iii) *$\underline{\mathscr{H}}^{1\cdot}_{\mathrm{DR}}(\bar{\mathfrak{Y}}/S(\log \mathfrak{S}))$ is a sheaf of Euler-Picard modules.*

(iv) *If $\overline{\mathrm{d}x/y^l}$ is primitive,* i.e. *$n \nmid b_i$ for $i = -1, 0, \ldots, r$ and $n \nmid lb = l \sum_{i=-1}^{r} b_i$, then $\underline{\mathscr{H}}^{1\cdot}_{\mathrm{DR}}(\bar{\mathfrak{Y}}/S(\log \mathfrak{S}))_l$ is a principal primitive subconnection of $\underline{\mathscr{H}}^{1\cdot}_{\mathrm{DR}}(\bar{\mathfrak{Y}}/S(\log \mathfrak{S}))$.*

(v) *$\underline{\mathscr{H}}^{1\cdot}_{\mathrm{DR}}(\bar{\mathfrak{Y}}/S(\log \mathfrak{S}))_{\mathrm{prim.}}$ is a primitive Euler-Picard subconnection of $\mathscr{H}^{1\cdot}_{\mathrm{DR}}(\bar{\mathfrak{Y}}/S(\log \mathfrak{S}))$.*

(i′) *If $\overline{x^k\,\mathrm{d}x/y^l}$ as in* (i) *is of second kind, then $\mathscr{D} \cdot \overline{x^k\,\mathrm{d}x/y^l}$ is a principal Euler-Picard sheaf of submodules of $\underline{\mathscr{H}}^{1\cdot}_{\mathrm{DR}}(\bar{\mathfrak{Y}}/S)$ of type $(k, l, n, \mathfrak{b})$.*

(iv′) *If $\overline{\mathrm{d}x/y^l}$ is primitive and of second kind, then*
$$\underline{\mathscr{H}}^{1\cdot}_{\mathrm{DR}}(\bar{\mathfrak{Y}}/S)_l = \mathscr{D} \cdot \overline{\mathrm{d}x/y^l} = \underline{\mathscr{H}}^{1\cdot}_{\mathrm{DR}}(\mathfrak{Y}/S)_l$$
is a primitive principal subconnection of $\underline{\mathscr{H}}^{1\cdot}_{\mathrm{DR}}(\bar{\mathfrak{Y}}/S)$. In particular the sequence of sheaves of $\mathscr{D}$-modules
$$\mathscr{D}^{r^2} \xrightarrow{\varepsilon} \mathscr{D} \to \mathscr{D} \cdot \overline{\mathrm{d}x/y^l} = \underline{\mathscr{H}}^{1\cdot}_{\mathrm{DR}}(\bar{\mathfrak{Y}}/S)_l \to 0$$
is exact with ε defined by $\{\varepsilon_{ij}^{0,l,n,\mathfrak{b}}\}_{i,j=1}^{r}$.

(vi) *Under the assumptions of* (iv′) *$\underline{\mathscr{H}}^{1\cdot}_{\mathrm{DR}}(\bar{\mathfrak{Y}}/S)_l = \mathscr{D} \cdot \overline{\mathrm{d}x/y^l}$ is a $\mathscr{D}$-direct summand of $\underline{\mathscr{H}}^{1\cdot}_{\mathrm{DR}}(\bar{\mathfrak{Y}}/S)$.*

(vii) *For a generalized Picard curve family*
$$\bar{\mathfrak{Y}}/S\colon Y^n = \prod_{i=-1}^{n-1} (X - t_i)$$
$\underline{\mathscr{H}}^{1\cdot}_{\mathrm{DR}}(\bar{\mathfrak{Y}}/S)$ is a primitive Euler-Picard connection with the $\mathscr{D}$-direct decomposition
$$\underline{\mathscr{H}}^{1\cdot}_{\mathrm{DR}}(\bar{\mathfrak{Y}}/S) = \bigoplus_{0<l<n} \mathscr{H}^{1\cdot}_{\mathrm{DR}}(\mathfrak{Y}/S)_l = \bigoplus_{0<l<n} \mathscr{D} \cdot \overline{\mathrm{d}x/y^l}$$
in primitive principal subconnections.

Proof. (i) The identity $\underline{\mathscr{H}}^1_{\mathrm{DR}}(\hat{\mathfrak{Y}}/S(\log \mathfrak{S})) = \mathscr{H}^1_{\mathrm{DR}}(\mathfrak{Y}/S)$ comes from Diagram (2.3.b). Proposition 1.4.1 yields the type $(k, l, n, \mathfrak{b})$ of $\mathscr{D} \cdot \overline{x^k\, dy/y^l}$.

(ii) is another formulation of (i): recall the definitions in 1.1, especially (1.1.2) (algebraic language), and definitions 2.2.21 (analytic language).

(iii) Here one just has to use Definition 1.1.7 (algebraic language), the definitions in 2.2.21 (analytic language), Corollary 1.4.19 and Proposition 1.4.8.

(iv) For the definition of $\underline{\mathscr{H}}^1_{\mathrm{DR}}(\hat{\mathfrak{Y}}/S(\log \mathfrak{S}))_l = \mathscr{H}^1_{\mathrm{DR}}(\mathfrak{Y}/S)_l$ we refer the reader back to (1.3.18). Now (iv) follows from Theorem 1.4.23.

(v) is essentially the result already described in Example 2.2.23.

(i′) By Proposition 2.3.11(i) we have

$$\underline{\mathscr{H}}^1_{\mathrm{DR}}(\hat{\mathfrak{Y}}/S) = \underline{\mathscr{H}}^1_{\mathrm{DR}}(\mathfrak{Y}/S)_{2.\mathrm{kind}}.$$

Since $\underline{\mathscr{H}}^1_{\mathrm{DR}}(\hat{\mathfrak{Y}}/S)$ is a sheaf of $\mathscr{D}$-modules the sheaf $\mathscr{D} \cdot \overline{x^k\, dx/y^l}$ is a subsheaf of $\underline{\mathscr{H}}^1_{\mathrm{DR}}(\hat{\mathfrak{Y}}/S)$. Now (i′) follows from (i).

(iv′) In order to verify the identities in (iv′) we remark that:

(a) $\underline{\mathscr{H}}^1_{\mathrm{DR}}(\hat{\mathfrak{Y}}/S)_l \subseteqq \underline{\mathscr{H}}^1_{\mathrm{DR}}(\mathfrak{Y}/S)_l$

by Lemma 2.3.8;

(b) $\mathscr{D} \cdot \overline{dx/y^l} \subseteqq \underline{\mathscr{H}}^1_{\mathrm{DR}}(\mathfrak{Y}/S)_{2.\mathrm{kind}} = \underline{\mathscr{H}}^1_{\mathrm{DR}}(\hat{\mathfrak{Y}}/S)$;

(c) $\underline{\mathscr{H}}^1_{\mathrm{DR}}(\mathfrak{Y}/S)_l = \mathscr{D} \cdot \overline{dx/y^l}$

by Theorem 1.4.23(i). It follows that

$$\mathscr{D} \cdot \overline{dx/y^l} \subseteqq \underline{\mathscr{H}}^1_{\mathrm{DR}}(\hat{\mathfrak{Y}}/S)_l \subseteqq \mathscr{H}^1_{\mathrm{DR}}(\mathfrak{Y}/S)_l = \mathscr{D} \cdot \overline{dx/y^l}.$$

So the inclusions have to be identities. The primitivity and the connection property come from (iv). For the exactness of the sequence we refer to the definitions in 1.1.4 (algebraic language) and 2.2.21 (analytic language).

(vi) $\underline{\mathscr{H}}^1_{\mathrm{DR}}(\hat{\mathfrak{Y}}/S)_l = \mathscr{H}^1_{\mathrm{DR}}(\mathfrak{Y}/S)_l$ is a $\mathscr{D}$-direct summand of $\mathscr{H}^1_{\mathrm{DR}}(\mathfrak{Y}/S)$ by (1.4.34) and (1.4.35). So we have a $\mathscr{D}$-direct decomposition

$$\underline{\mathscr{H}}^1_{\mathrm{DR}}(\mathfrak{Y}/S) = \mathscr{H}^1_{\mathrm{DR}}(\hat{\mathfrak{Y}}/S)_l \oplus \mathscr{M}.$$

Now we intersect with $\underline{\mathscr{H}}^1_{\mathrm{DR}}(\hat{\mathfrak{Y}}/S)$ and obtain the $\mathscr{D}$-direct decomposition

$$\underline{\mathscr{H}}^1_{\mathrm{DR}}(\hat{\mathfrak{Y}}/S) = \underline{\mathscr{H}}^1_{\mathrm{DR}}(\hat{\mathfrak{Y}}/S)_l \oplus \left(\mathscr{M} \cap \underline{\mathscr{H}}^1_{\mathrm{DR}}(\hat{\mathfrak{Y}}/S)\right).$$

(vii) From Example 2.3.12, 1.5.1, (1.4.35) and (iv′) it follows that

$$\begin{aligned} \underline{\mathscr{H}}^1_{\mathrm{DR}}(\hat{\mathfrak{Y}}/S) &= \bigoplus_{0<l<n} \Big(\bigoplus_{0 \leqq k<n} \mathscr{O}_S \cdot \overline{x^k\, dx/y^l} \Big) = \mathscr{H}^1_{\mathrm{DR}}(\mathfrak{Y}/S)_{\mathrm{prim.}} \\ &= \bigoplus_{0<l<n} \mathscr{H}^1_{\mathrm{DR}}(\mathfrak{Y}/S)_l = \bigoplus_{0<l<n} \mathscr{D} \cdot \overline{dx/y^l}. \end{aligned}$$

The proof of Theorem 2.4.3 is finished. As we will see in the next section the $\mathscr{D}$-direct decompositions have important consequences for finding all solutions of several Euler-Picard systems of differential equations.

2.5. *Fundamental Solutions of Some Euler-Picard Systems*

Now we can state the main result of Chapter II. We consider smooth projective cycloelliptic curve families

$$\mathfrak{Y}/S\colon Y^n = \prod_{i=-1}^{r} (x - t_i)^{b_i}.$$

We restrict our attention to analytic subfamilies $\mathfrak{Y}_U$ over small polycylindres $U \subset S(\mathbb{C})$, which we also denote by $\mathfrak{Y}/U$. If $\overline{\omega}$ is a section of $\mathscr{H}^{1,\mathrm{an}}_{\mathrm{DR}}(\mathfrak{Y}/U)$ over U and α is a section over U of the (locally constant) homology sheaf $H_1(\mathfrak{Y}/S, \mathbb{Z})$ of $\mathfrak{Y}/S$, then we can form the integral function

$$\int_\alpha \overline{\omega}\colon U \to \mathbb{C}, \qquad s \mapsto \int_{\alpha_s} \overline{\omega}_s .$$

$\overline{\omega}_s$ is a class in $H^1(\mathfrak{Y}_s, \mathbb{C})$ which can be represented by a differential form ω_s of the second kind on $\mathfrak{Y}_s$ (see Proposition 2.3.11). Since $\alpha_s \in H_1(\mathfrak{Y}_s, \mathbb{Z})$ the integral $\int_{\alpha_s} \omega_s = \int_{\alpha_s} \overline{\omega}_s$ is a holomorphic function on U (see [53], I, § 15). If δ is a section of the sheaf of differential operators $\mathscr{D}_U$ over U, then

2.5.1. $$\delta \int_\alpha \overline{\omega} = \int_\alpha \delta\overline{\omega}.$$

We are interested in sections of type $\overline{\omega} = \overline{x^k \mathrm{d}x/y^l}$. If $x^k\, \mathrm{d}x/y^l$ is of the second kind on each $\mathfrak{Y}_s$, $s \in U$, then we write $\int_\alpha x^k\, \mathrm{d}x/y^l$ instead of $\int_\alpha \overline{x^k\, \mathrm{d}x/y^l}$.

2.5.2. THEOREM. *Let $\mathfrak{Y}/S$ be a smooth projective cycloelliptic curve family as above and U a small open polycylindre in $S(\mathbb{C})$. Then the following statements are true:*

(i) *If $x^k\, \mathrm{d}x/y^l$ is of the second kind over U, $\alpha \in \Gamma(U, H_1(\mathfrak{Y}/S, \mathbb{Z}))$, then the holomorphic function $\int_\alpha x^k\, \mathrm{d}x/y^l$ is a solution of the Euler-Picard differential equation system of type $(k, l, n, \mathfrak{b})$.*

(ii) *$x^k\, \mathrm{d}x/y^l$ is of the second kind over U if $\mathfrak{Y}/S$ is a primitive general cycloelliptic curve family and $k \geq 0$, $l < n$.*

(iii) *If $\mathfrak{Y}/S$ is a primitive general cycloelliptic curve family and $0 < l < n$, then there exist relative cycles $\alpha_0, \alpha_1, \ldots, \alpha_r \in \Gamma(U, H_1(\mathfrak{Y}/S))$ such that*

$$\int_{\alpha_0} \mathrm{d}x/y^l, \quad \int_{\alpha_1} \mathrm{d}x/y^l, \ldots, \quad \int_{\alpha_r} \mathrm{d}x/y^l$$

is a fundamental system of solutions of type $(0, l, n, (1, 1, \ldots, 1))$.

(iv) *Under the assumptions of (iii) the corresponding Euler-Picard system of differential operators $\{\varepsilon_{ij}\}_{i,j=1}^{r}$ is a fundamental system of partial differential operators for the class of holomorphic functions $\int_\alpha \mathrm{d}x/y^l$, $\alpha \in \Gamma(U, H_1(\mathfrak{Y}/S))$.*

Proof. (i) follows from (2.5.1) and from Theorem 2.4.3(i) or (ii).

(ii) The proof of 2.3.13 works for general smooth cycloelliptic curve families. Therefore $x^k\,dx/y^l$ is regular on $\ddot{\mathfrak{Y}}_s$. We show that $\overline{\mathfrak{Y}}_s \setminus \ddot{\mathfrak{Y}}_s$ consists of one point only. Then the residues of $x^k dx/y^l$ in each point of $\overline{\mathfrak{Y}}_s$ have to be zero by the residue theorem. We only have to check the cases

A) $n < b = r + 2$ and C) $n > b = r + 2$

because our general families are primitive, i.e. $(b, n) = 1$.

A) The equation of the plane projective family is

$$Y^n W^{b-n} = \prod_{i=-1}^{r} (X - t_i W),$$

and $\overline{\mathfrak{Y}}_s \setminus \ddot{\mathfrak{Y}}_s$ consists of the point $(w : x : y) = (0 : 0 : 1)$. We set $u = w/y$, $v = x/y$ and investigate the resolution of singularities of the affine part of the curve $\overline{\mathfrak{Y}}_s$ described by

2.5.3. $$u^{b-n} = \prod_{i=-1}^{r} (v - s_i u).$$

The only singularity is the point $(u, v) = (0, 0)$. The resolution of the singularity $(0, 0)$ consists of points $P_1, \ldots, P_g$ of the spectrum of the normalization of $\mathbb{C}[u, v]$, which is equal to the normalization of $\mathbb{C}[u]$ in $\mathbb{C}(u, v)$. From equation (2.5.3) it follows that $[\mathbb{C}(u, v) : \mathbb{C}(u)] \leqq b$. Let ν be the valuation of $\mathbb{C}(u)$ corresponding to u $\big(\nu(u) = 1\big)$. The extension to $\mathbb{C}(u, v)$ is also denoted by ν. We know that

2.5.4 $$b \geqq [\mathbb{C}(u, v) : \mathbb{C}(u)] = \sum_{i=1}^{g} e_i,$$

where e_i denotes the branch order of P_i over $O \in \operatorname{Spec} \mathbb{C}[u]$. If we can find an element ξ of $\mathbb{C}(u, v)$ such that $\nu(\xi) = 1/b$, then the branch order e_i of one of our points must be $\geqq b$. Then it follows from (2.5.4) that $g = 1$, which implies that $\overline{\mathfrak{Y}}_s \setminus \ddot{\mathfrak{Y}}_s$ consists only of one point.

From (2.5.3) it follows that

2.5.5. $$b - n = (b - n)\,\nu(u) = \sum_{i=-1}^{r} \nu(v - s_i u).$$

It is immediately clear that $\nu(v) < \nu(u) = 1$ because otherwise we would obtain the contradiction

$$b - n = \sum_i \nu(v - s_i u) \geqq \sum_i \min\,\{\nu(v), \nu(s_i u)\} \geqq \sum_i 1 = r + 2 = b.$$

From $\nu(v) < \nu(u)$ it follows that $\nu(v - s_i u) = \nu(v)$, hence

$$b - n = b \cdot \nu(v), \qquad \nu(v) = (b - n)/b$$

by (2.5.5). But $(b - n, b) = (n, b) = 1$. Therefore there are integers l, m such that

$$l(b - n) + mb = 1, \qquad l(b - n)/b + m = 1/b.$$

If we set $\xi = v^l u^m$, then $\nu(\xi) = 1/b$.

C) The equation is

$$Y^n = W^{n-b} \prod_{i=-1}^{r} (X - s_i W).$$

The point $\overline{\mathfrak{Y}}_s \setminus \mathfrak{Y}_s$ has coordinates $(w : x : y) = (0 : 1 : 0)$. Setting $v = y/x$, $u = w/x$ we examine the resolution of singularities of the point $(u, v) = (0, 0)$ on the affine curve

2.5.6. $$v^n = u^{n-b} \prod_{i=-1}^{r} (1 - s_i u).$$

The degree of $\mathbb{C}(u, v)$ over $\mathbb{C}(u)$ is n. We look for an element $\xi \in \mathbb{C}(u, v)$ with $\nu(\xi) = 1/n$, ν as in the part A). From (2.5.6) follows

$$n\nu(v) = (n - b)\, \nu(u) = n - b, \qquad \nu(v) = (n - b)/n.$$

Now we conclude as in case A).

(iii) $\mathcal{D} \cdot \overline{\mathrm{d}x/y^l}$ is a primitive Euler-Picard connection by (ii) and Theorem 2.4.3(iv'). The morphism ε in the exact sequence in (iv') is a primitive simple quadratic system of differential operators as already remarked in 2.2.21. The corresponding system of differential equations has a fundamental system of solutions $(F_0, F_1, \ldots, F_r)$ by Proposition 2.2.18. Now we prove that each solution of ε in $\mathcal{O}_U$ is a $\mathbb{C}$-linear combination of functions of integral type $\int_\alpha \mathrm{d}x/y^l$. Then we can replace F_i by $\int_{\alpha_i} \mathrm{d}x/y^l$, $i = 0, 1, \ldots, r$. We make use of the "thickened" de Rham duality for our family $\mathfrak{Y}/U$, which is described in [53], I, Prop. of § 15. From there we need

2.5.7. $\mathcal{H}om_{\mathcal{D}_U}\big(\mathcal{H}^{1,\mathrm{an}}_{\mathrm{DR}}(\mathfrak{Y}/U),\, \mathcal{O}_U^{\mathrm{an}}\big)$ is generated over $\mathbb{C}$ by sections of type

$$\int_\alpha . : \Gamma\big(U, \mathcal{H}^{1,\mathrm{an}}_{\mathrm{DR}}(\mathfrak{Y}/U)\big) \to \Gamma(U, \mathcal{O}_U^{\mathrm{an}}), \qquad \varpi \mapsto \int_\alpha \varpi.$$

So each section of the sheaf in 2.5.7 can be written as $\sigma = \sum_{i \in I} c_i \int_{\alpha_i} .$, I finite, $\alpha_i \in \Gamma\big(U, H_1(\mathfrak{Y}/U)\big)$. Now F_i can be understood as a section σ_i of $\mathcal{H}om_{\mathcal{D}_U}(\mathcal{D}_U \cdot \overline{\mathrm{d}x/y^l}, \mathcal{O}_U)$ $(\overline{\mathrm{d}x/y^l} \mapsto F_i)$ because

$$\mathcal{D}_U^{r^2} \xrightarrow{\varepsilon} \mathcal{D}_U \to \mathcal{O}_U, \qquad 1 \mapsto F_i$$

is a complex and the sequence

2.5.8. $$\mathcal{D}_U^{r^2} \xrightarrow{\varepsilon} \mathcal{D}_U \to \mathcal{D}_U \cdot \overline{\mathrm{d}x/y^l} \to 0$$

is an exact sequence of $\mathscr{D}_U$-module sheaves (see Theorem 2.4.3(iv′)). By Theorem 2.4.3(vi) there exists a $\mathscr{D}_U$-direct decomposition

$$\underline{\mathscr{H}}^{1,\mathrm{an}}_{\mathrm{DR}}(\mathfrak{Y}/U) = \mathscr{D}_U \cdot \overline{\mathrm{d}x/y^l} \oplus \mathscr{M},$$

which induces a direct decomposition

$$\mathscr{H}om_{\mathscr{D}_U}\big(\underline{\mathscr{H}}^{1,\mathrm{an}}_{\mathrm{DR}}(\mathfrak{Y}/U), \mathscr{O}_U\big) = \mathscr{H}om_{\mathscr{D}_U}(\mathscr{D}_U \cdot \overline{\mathrm{d}x/y^l}, \mathscr{O}_U) \oplus \mathscr{H}om_{\mathscr{D}_U}(\mathscr{M}, \mathscr{O}_U).$$

Therefore σ_i is the restriction of a section

$$\sigma = \sum_{j\in J} c_j \int_{\alpha_j} \cdot\cdot$$

Applying σ_i to $\overline{\mathrm{d}x/y^l}$ we find

$$F_i = \sigma_i(\overline{\mathrm{d}x/y^l}) = \sum_{j\in J} c_j \int_{\alpha_j} \mathrm{d}x/y^l, \qquad c_j \in \mathbb{C}.$$

(iv) Now we can set without loss of generality

$$F_0 = \int_{\alpha_0} \mathrm{d}x/y^l, \quad F_1 = \int_{\alpha_1} \mathrm{d}x/y^l, \ldots, \quad F_r = \int_{\alpha_r} \mathrm{d}x/y^l.$$

By the definition above of the F_i's these functions form a general solution of ε in $\mathscr{O}_U^{r+1}$; that means that the sequence

$$\begin{array}{ccccccc} & & & & \mathscr{O}_U^{r+1} & & \\ & & & & \cup & & \\ \mathscr{D}_U^{r^2} & \xrightarrow{\varepsilon} & \mathscr{D}_U & \to & \mathscr{D}_U \cdot \left(\int_{\alpha_0} \mathrm{d}x/y^l, \ldots, \int_{\alpha_r} \mathrm{d}x/y^l\right) & \to & 0 \\ & & & & \| \qquad\qquad \| & & \\ & & 1 & \mapsto & (F_0, \quad \ldots, \quad F_r) & & \end{array}$$

is exact. Further ε is a primitive simple system of differential operators because (2.5.8) is exact and $\mathscr{D}_U \cdot \overline{\mathrm{d}x/y^l}$ is primitive (Theorem 2.4.3(iv′)). Altogether we see that $\{\varepsilon_{ij}\}$ is a fundamental system of differential operators for the functions of type $\int_\alpha \mathrm{d}x/y^l$ in the sense of the remark after (2.2.20).

REMARK. It should be checked more generally on which smooth projective cycloelliptic curve families $\mathfrak{Y}/S$ the forms $P(x)\mathrm{d}x/y^l$, $P(x)$ a polynomial in x, are of the second kind. Then we would have available solutions of integral type $\int_\alpha P(x)\,\mathrm{d}x/y^l$ of Euler systems $\{\varepsilon_{ij}\}_{i\neq j}$ of type $(l, n, \mathfrak{b})$.

We refer now the reader back to the end of the introduction. The clear results 1., 2., 3., 4. stated in 0.3 without any formalism are now proved by Theorem 2.5.2. Observe that 4. is an amazing version of Hilbert's Nullstellensatz for systems of partial differential equations. The Euler-Picard operators ε_{ij} play the same role as a basis of a prime ideal corresponding to an (irreducible) algebraic variety.

Bibliography

References

[1] Appell, P., Sur les fonctions hypergéométriques de deux variables et sur des équations linéaires aux dérivées partielles, C. R. Acad. Sci. Paris **90**, 296, 731, 977 (1880).

[2] Appell, P., Fonctions hypergéométriques et hypersphériques, polynômes d'Hermite, Gauthier-Villars, Paris 1926.

[3] Artin, E., Theory of braids, Ann. of Math. **48**, 101–126 (1947).

[4] Baily, W. L., Jr., Borel, A., Compactification of arithmetic quotients of bounded symmetric domains, Ann. of Math. **84**, 442–528 (1966).

[5] Borel, A., Introduction aux groupes arithmétiques, Hermann, Paris 1969.

[6] Borel, A., Some metric properties of arithmetic quotients of symmetric spaces and an extension theorem, J. Diff. Geom. **6**, 543–560 (1972).

[7] Bourbaki, N., Algèbre commutative, Hermann, Paris 1965.

[8] Chevalley, C., Introduction to the Theory of Algebraic Functions of one Variable, Math. Surveys VI, Amer. Math. Soc., New York 1951.

[9] Darboux, G., Leçons sur la théorie générale des surfaces et les applications géométriques du calcul infinitésimal, t. 2, Gauthier-Villars, Paris 1889.

[10] Dedekind, R., Schreiben an Herrn Borchardt über die Theorie der elliptischen Modulfunktionen, J. reine angew. Math. **83**, 265–292 (1877).

[11] Deligne, P., Théorème de Lefschetz et critères de dégénérescence de suites spectrales, Publ. Math. I.H.E.S. **35** (1969).

[12] Deligne, P., Equations différentiels à points singuliers réguliers, Lecture Notes Math. **163** (1970), Springer.

[13] Deligne, P., Théorie de Hodge II, Publ. Math. I.H.E.S. **40** (1971).

[14] Deligne, P., Mostow, G. D., Monodromy of Hypergeometric Functions and Non-Lattice Integral Monodromy, I.H.E.S./M/82/46 (1983).

[15] Euler, L., Specimen transformationis singularis serierum, Nova Acta Acad. Petrop. **12**, 58–70 (1801).

[16] Euler, L., Institutiones calculi integralis, t. III, Petrop. 1769.

[17] Feustel, J. M., Kompaktifizierung und Singularitäten des Faktorraumes einer arithmetischen Gruppe, die in der zweidimensionalen Einheitskugel wirkt, Diplomarbeit, Humboldt-Univ. Berlin 1976.

[18] Feustel, J. M., Über die Spitzen von Modulflächen zur zweidimensionalen komplexen Einheitskugel, Prepr. Ser., Akad. Wiss. DDR, ZIMM 1977.

[19] Feustel, J. M., Klassifikation der elliptischen Fixpunkte bezüglich der Wirkung der Picardschen Modulgruppe auf die komplexe Einheitskugel, Prepr. Ser. 30/81, Akad. Wiss. DDR, I. Math. 1981.

[20] Fuchs, L., Gesammelte mathematische Werke, Bd. 1, 3, Berlin 1904–1909.

[21] Gauss, C. F., Disquisitiones generales circa seriem infinitam ..., Werke III, 125–162, Göttingen 1866.
[22] Gauss, C. F., Werke X/1, 243–245, Göttingen 1917.
[23] Griffiths, P., Harris, J., Principles of Algebraic Geometry, John Wiley, New York 1978.
[24] Grothendieck, A., Dieudonné, J., Eléments de géométrie algébrique, IV, Publ. Math. I.H.E.S. **20**, **24**, **28**, **32** (1964–1967).
[25] Hall, M., Jr., The Theory of Groups, Macmillan Comp., New York 1959.
[26] Hammond, F., The Hilbert modular surface of a real quadratic field, Math. Ann. **200**, 25–45 (1973).
[27] Hartshorne, F., Algebraic Geometry, Graduate Texts in Math., Springer-Verlag, Berlin–Heidelberg–New York 1977.
[28] Hemperly, J. C., The parabolic contribution to the number of linearly independent automorphic forms on a certain bounded domain, Amer. J. Math. **94**, 1078–1100 (1972).
[29] Holzapfel, R.-P., Arithmetische Kugelquotientenflächen I/II, Seminarber. Humboldt-Univ. Berlin **14** (1979).
[30] Holzapfel, R.-P., Arithmetische Kugelquotientenflächen V/VI, Seminarber. Humboldt-Univ. Berlin **21** (1980).
[31] Holzapfel, R.-P., A class of minimal surfaces in the unknown region of surface geography, Math. Nachr. **98**, 211–232 (1980).
[32] Holzapfel, R.-P., Invariants of arithmetic ball quotient surfaces, Math. Nachr. **103**, 117–153 (1981).
[33] Holzapfel, R.-P., Arithmetic curves of ball quotient surfaces, Ann. Glob. Analysis and Geometry **1**, No. 2, 21–90 (1983).
[34] Hurwitz, A., Über Riemannsche Flächen mit gegebenen Verzweigungspunkten, Math. Ann. **39**, 1–61 (1892).
[35] Igusa, J.-I., On Siegel modular forms of genus two, Amer. J. Math. **84**, 175–200 (1962).
[36] Katz, N. M., Algebraic solutions of differential equations (p-curvature and the Hodge filtration), Inv. math. **18**, 1–118 (1972).
[37] Katz, N. M., Oda, T., On the differentiation of de Rham cohomology classes with respect to parameters, J. Math. Kyoto Univ. **8-2**, 199–213 (1968).
[38] Klein, F., Gesammelte mathematische Abhandlungen, vol. 3, Springer, Berlin 1923
[39] Klein, F., Vorlesungen über die hypergeometrische Funktion, Springer, Berlin 1933.
[40] Kuribayashi, A., Komiya, K., On Weierstrass points and automorphisms of curves of genus three, SLN **687**, 253–299 (1978).
[41] Lehner, J., Discontinuous Groups and Automorphic Functions, Math. Surveys **18**, Amer. Math. Soc., Providence, Rhode Island 1964.
[42] Livné, R. A., On Certain Covers of the Universal Elliptic Curve, Thesis, Harvard Univ., Cambr., Mass. 1981.
[43] Manin, Yu. I., Algebraic curves over fields with differentiation, Isv. Akad. Nauk SSSR, Ser. Mat. **22**, 737–756 (1958) (russ.).
[44] Maruyama, M., On Classification of Ruled Surfaces, Lect. in Math. **3**, Kyoto Univ., Kinokumiya, Tokyo 1970.
[45] Mostow, G. D., Existence of Non-Arithmetic Monodromy Groups, I.H.E.S./M/81/35 (1981).

[46] Mostow, G. D., Complex Reflection Groups and Non-Arithmetic Monodromy, Math. Arbeitstagung, Univ. Bonn 1981.
[47] Mumford, D., The topology of normal singularities of an algebraic surface and a criterion for simplicity, Publ. Math. I.H.E.S. **9**, 229–246 (1961).
[48] Mumford, D., Lectures on Curves on an Algebraic Surface, Princeton Univ. Press 1966.
[49] Mumford, D., Pathologies III, Amer. J. Math. **79**, 94–104 (1967).
[50] Mumford, D., Hirzebruch's proportionality theorem in the non-compact case, Inv. math. **42**, 239–272 (1977).
[51] Mumford, D., Algebraic Geometry, I. Complex Projective Varieties, Springer-Verlag, Berlin–Heidelberg–New York 1976.
[52] Mumford, D., Introduction to Algebraic Geometry, Mim. notes, Harvard Univ. 1967.
[53] Pham, F., Singularités de systèmes différentielles de Gauss-Manin, Progress in Math., vol. 2, Birkhäuser-Verlag, Basel–Stuttgart 1980.
[54] Picard, E., Sur une extension aux fonctions de deux variables du problème de Riemann relatif aux fonctions hypergéométriques, Ann. Sc. Ecole Norm. Sup. **10**, 305 to 322 (1881).
[55] Picard, E., Sur des fonctions de deux variables indépendentes analogues aux fonctions modulaires, Acta math. **2**, 114–135 (1883).
[56] Picard, E., Sur les formes quadratiques ternaires indéfinies à indéterminées conjuguées et sur les fonctions hyperfuchsiennes correspondantes, Acta math. **5**, 121–182 (1884).
[57] Picard, E., Sur les fonctions hyperfuchsiennes provenant des séries hypergéométriques de deux variables, Ann. Ecole Norm. Sup., 3e ser. **62**, 357–384 (1885).
[58] Picard, E., Théorie de courbes et des surfaces algébriques, Gauthier-Villars, Paris 1931.
[59] Pund, O., Algebra, Sammlung Schubert VI, Göschen, Leipzig 1899.
[60] Riemann, B., Ueber die Fortpflanzung ebener Luftwellen von endlicher Schwingungsweite, Abh. d. Kgl. Ges. d. Wiss. Göttingen **8**, 3–25 (1860).
[61] Schwarz, H. A., Über diejenigen Fälle, in welchen die Gaussische hypergeometrische Reihe eine algebraische Funktion ihres vierten Elementes darstellt, Crelles J. **75**, 292–335 (1873).
[62] Serre, J.-P., Cours d'arithmétique, Presses Univ. de France, Paris 1970.
[63] Shimura, G., On purely transcendental fields of automorphic functions of several variables, Osaka J. Math. **1**, No. 1, 1–14 (1964).
[64] Shimura, G., Introduction to the Arithmetic Theory of Automorphic Functions, Iwanami Shoten Publ., Princeton Univ. Press 1971.
[65] Švarčman, O. V., On the Factor Space of an Arithmetic Discrete Group Acting on the Complex Ball (russian), Thesis, MGU, Moscou 1974.
[66] Zink, Th., Über die Anzahl der Spitzen einiger arithmetischer Untergruppen unitärer Gruppen, Math. Nachr. **89**, 315–320 (1979).

Additional Literature

Alezais, R., Sur une classe de fonctions hyperfuchsiennes, Ann. Ecole Norm. Sup. **19**, 261–323 (1902).
Kaneko, J., Monodromy group of Appell's system (F_4), Tokyo J. Math. **4**, No. 1, 35–54 (1981).
Katz, N. M., An overview of Deligne's work on Hilbert's twenty-first problem, in: "Mathematical Developments Arising from Hilbert Problems" (F. E. Browder, ed.), Proc. Symp. Pure Math., **28**, Amer. Math. Soc., 537–557 (1976).

Lauricella, M., Sulla funzioni ipergeometriche a piu variabili, Rend. Palermo **7**, 111–158 (1893).

Le Vavasseur, R., Sur le système d'équation aux dérivées partielles simultanées auxquelles satisfait la séries hypergéométriques à deux variables, J. Fac. Toulouse **7**, 1–205 (1896).

Manin, J. I., Rational points on algebraic curves over function fields, Izv. Acad. Nauk SSSR, Ser. Math., **27**, 1395–1440 (1963) (russ.).

Mostow, G. D., On a remarkable class of polyhedra in complex hyperbolic space, Pacific J. Math. **86**, No. 1, 171–276 (1980).

Mostow, G. D., Existence of a nonarithmetic lattice in $\mathbb{SU}(2, 1)$, Proc. Nat. Acad. Sci. USA **78**, 3029–3033 (1978).

Mostow, G. D., Generalized Picard lattice arising from half-integral conditions, Preprint 1983.

Oda, T., Introduction to Algebraic Analysis on Complex Manifolds, in: Advanced Studies in Pure Mathematics, 1, North-Holland Publ. Comp., Amsterdam–New York–Oxford 1983.

Picard, E., Sur les fonctions hyperfuchsiennes provenant des séries hypergéométriques de deux variables, Bull. Soc. Math. France **15**, 148–152 (1887).

Pochhammer, L., Über hypergeometrische Funktionen höherer Ordnung, J. Math. **71**, 316–362 (1870).

Shiga, H., One attempt to the $K3$ modular function I, Ann. Sc. Norm. Sup., Cl. di Sci., Ser. IV, **6**, 609–635 (1979).

Shiga, H., One attempt to the $K3$ modular function II, Ann. Sc. Norm. Sup., Cl. di Sci., Ser. IV, **8**, 157–182 (1981).

Shiga, H., One attempt to the $K3$ modular function, III, A relation between the versal deformation of an exceptional singularity and a family of elliptic surfaces, Preprint, Sc. Norm. Sup. Pisa, Chiba Univ. 1983.

Terada, T., Problème de Riemann et fonctions automorphes provenant des fonctions hypergéométriques de plusiers variables, J. Math. Kyoto Univ. **13**, 557–578 (1973).

Terada, T., Quelques propriétés géométriques de domaine de F_1 et le group de tresses colorées, Publ. RIMS **17**, 95–111 (1981).

Terada, T., Fonctions hypergéométriques F_1 et fonctions automorphes I, J. Math. Soc. Japan **35**, 451–475 (1983).

Yoshida, M., On the confluence of regular and irregular singularities, Japan J. Math. **6**, No. 1, 165–172 (1980).

Yoshida, M., Euler integral transformations of hypergeometric functions of two variables, Hiroshima Math. J. **10**, 329–335 (1980).

Yoshida, M., Construction of a moduli space of Gauss hypergeometric differential equations, Funkcialaj Ekvacioj **24**, 1–10 (1981).

Yoshida, M., Volume formula for certain discrete reflection groups in $\mathbb{PU}(2, 1)$. Mem. Fac. Sci., Kyushu Univ., Ser. A, **36**, No. 1, 1–11 (1982).

Yoshida, M., Discrete reflection groups in the parabolic subgroup of $\mathbb{SU}(n, 1)$ and generalized Cartan matrices of Euclidean type, J. Fac. Sci. Univ. Tokyo, Sec. IA, **30**, No. 1, 25–52 (1983).

Yoshida, M., Orbifold – uniformizing differential equations, Preprint 1983.

Yoshida, M., Yamazaki, T., On Hirzebruch's examples of surfaces with $c_1^2 = 3c_2$, Preprint 1983.

Added in proof:

[i] Feustel, J.-M., Darstellung von Picardschen Modulformen durch Thetakonstanten, to appear.

[ii] Hirzebruch, F., Chern numbers of algebraic surfaces — An example, Math. Ann. **266**, 351—356 (1984).

[iii] Holzapfel, R.-P., Arithmetische Kugelquotientenflächen III/IV, Seminarber. Humboldt-Univ. Berlin **20** (1979).

[iv] Holzapfel, R.-P., Chern numbers of algebraic surfaces — Hirzebruch's examples are Picard modular surfaces, Math. Nachr. **126**, 31—49 (1986).

[v] Holzapfel, R.-P., On the Nebentypus of Picard modular forms, to appear.

[vi] Mumford, D., Tata Lectures of Theta I, II, Progress in Mathematics **28**, **43**, Birkhäuser, Boston 1983.

[vii] Shiga, H., On the Representation of the Picard's Modular Function by Theta Constants, Preprint, Chiba Univ. 1984.

Index